Fabrication, Characteristics and Applications of Multifunctional Polymer Nanocomposites

Fabrication, Characteristics and Applications of Multifunctional Polymer Nanocomposites

Guest Editor

Jianhua Zhang

MDPI

Basel • Beijing • Wuhan • Barcelona • Belgrade • Novi Sad • Cluj • Manchester

Guest Editor
Jianhua Zhang
Department of Polymer
Science and Engineering
Tianjin University
Tianjin
China

Editorial Office
MDPI AG
Grosspeteranlage 5
4052 Basel, Switzerland

This is a reprint of the Special Issue, published open access by the journal *Materials* (ISSN 1996-1944), freely accessible at: https://www.mdpi.com/journal/materials/special_issues/Y7CS9Z5K43.

For citation purposes, cite each article independently as indicated on the article page online and as indicated below:

Lastname, A.A.; Lastname, B.B. Article Title. *Journal Name* **Year**, *Volume Number*, Page Range.

ISBN 978-3-7258-8129-1 (Hbk)
ISBN 978-3-7258-8130-7 (PDF)
https://doi.org/10.3390/books978-3-7258-8130-7

Contents

About the Editor

Jianhua Zhang

Jianhua Zhang is an Associate Professor at the Department of Polymer Science and Engineering, School of Chemical Engineering and Technology, Tianjin University, China. He received his B.S. in Chemical Engineering, M.S. and Ph.D. degrees in Materials Science from Tianjin University in 2004, 2007 and 2010, respectively. He started his independent research career as a Lecturer at Tianjin University in 2010. He stayed in the group of Prof. Liangfang Zhang at the University of California, San Diego from 2014 to 2015, and then he was promoted to Associate Professor in 2015. His research interests focus on the design, functionalization and fabrication of functional polymers for biomedical and environmental applications as well as oilfield exploitation. He has published over 150 SCI papers in prestigious journals. His work has garnered more than 5000 citations (H-index 43), including 2 ESI highly cited papers and 10 cover papers. He holds 10 authorized invention patents. He served as a Reviewer for many leading journals (over 800 reviews) and received the Global Peer Review Awards from Publons in 2018 and 2019. He was an Editorial Board Member for *Materials* and a Peer Reviewer for the National Natural Science Foundation of China.

materials

Editorial

Multifunctional Polymer Nanocomposites

Yinan Zeng [1] and Jianhua Zhang [1,2,*]

[1] Department of Polymer Science and Engineering, School of Chemical Engineering and Technology, Tianjin University, Tianjin 300350, China; zyn757810425@tju.edu.cn
[2] Tianjin Key Laboratory of Membrane Science and Desalination Technology, Tianjin University, Tianjin 300350, China
* Correspondence: jhuazhang@tju.edu.cn

Polymer nanocomposite materials are a type of composite material that physically or covalently incorporates nanosized particles or nanostructures into a matrix of cross-linked polymer networks. The polymer nanocomposites integrate the inherent flexibility, processability, and chemical stability of polymer matrices with the unique physicochemical, mechanical, and functional properties of nanoscale fillers (e.g., carbon nanotubes, graphene, metal oxides, layered silicates, and quantum dots) [1–9]. The incorporation of nanomaterials into polymer matrices can render new functionalities to composites and produce superior physicochemical properties absent in individual components, offering an efficient route to enhance the physicochemical properties and thus expand the application scopes of both polymer materials and nanomaterials [10–13]. Owing to their versatile and customizable characteristics, multifunctional polymer nanocomposites have found widespread and rapidly expanding applications across numerous high-tech fields, including aerospace, electronics and optoelectronics, automotive engineering, biomedicine, environmental engineering, and smart textiles [14–22]. This Special Issue gathers and highlights recent research achievements in the design, fabrication, characterizations and applications of multifunctional polymer nanocomposites. This Special Issue includes eight papers in total, of which five are research articles and three are review papers, which are expected to be able to serve as a bridge for academic exchange and technological innovation in the field of polymer nanocomposites. The contributions of this collection are summarized below.

The key characteristics of multifunctional polymer nanocomposites stem from the synergistic coupling of polymer matrices and nanofillers, manifesting in a remarkable combination of enhanced mechanical, thermal, electrical, optical, magnetic, and barrier properties—even at ultra-low nanofiller loadings (typically <5 wt%). Thermally, these nanocomposites exhibit elevated glass transition temperatures, thermal decomposition temperatures, and thermal conductivity, addressing the intrinsic thermal instability of many pristine polymers. The enhanced thermostability of polymer nanocomposites can ensure the efficient exploration of the oil and gas resources under harsh conditions featuring high temperature, high pressure, and high salinity in deep formations. The paper of Zhang et al. [23] comprehensively summarized the design, preparation and applications of polymer nanocomposites as additives in drilling operation. Particular focus was placed on the strategies and structure–property–performance relationships of polymer nanocomposites to enhance the performance of drilling fluid under high-temperature, high-pressure, and high-salinity conditions in the deep formation. Future perspectives for achieving excellent comprehensive performance are also discussed to motivate future contributions and explore new possibilities. Polypyrrole (PPy) as a kind of biocompatible

 https://doi.org/10.3390/ma19061165

conductive polymer has excellent conductivity, excellent redox properties, biocompatibility, easy synthesis, and environmental stability, which has potential applications in fields such as biomedical systems, flexible electronics, and bioelectronics. Zhao et al. [24] summarized the research progress of electrospun PPy nanocomposite fiber materials and their related applications in the field of tissue engineering. They comprehensively analyzed the topological characteristics of three composite fiber architectures—randomly distributed, aligned, and core–shell structures—and elucidated their application mechanisms in nerve regeneration, skin repair, bone mineralization, and myocardial tissue reconstruction (e.g., facilitating oriented cell migration and regulating differentiation through specific signaling pathway activation).

The three-dimensional hydrogels and nanomaterials have shown a very high potential for medical therapeutic and diagnostic applications. However, some inherent limitations of hydrogels and nanomaterials inhibited their widespread applications. The incorporation of nanomaterials in a three-dimensional polymeric hydrogel matrix as an innovative means to obtain nanocomposite hydrogels with improved properties and multiple functionalities has gained significant attention in the biomedical field. The review of Chen et al. [25] presented a comprehensive overview of the integration of functionalized hydrogels and flexible electronics from the viewpoint of materials science, investigating their promising applications as well as the technical challenges involved in treating diabetic wounds. Additionally, it proposes potential future research pathways, aiming to establish a theoretical basis for facilitating technological advancement and clinical application in this domain. Li et al. [26] fabricated a hydrogel by grafting 4-formylphenylboronic acid (FPBA) onto carboxymethyl chitosan (CMCS) and mixed it with oxidized sodium alginate (OAlg) to form a dual-dynamic covalent hydrogel (CPOA), which can effectively load exosomes and form hydrogel nanocomposites (CPOA@Exos). The CPOA@Exos can not only show good injectability and self-healing ability due to its dual-dynamic covalent crosslinking network, but also sustainedly release exosomes at the diabetic wound site while effectively scavenging excess reactive oxygen species (ROS). The CPOA@Exos system as a promising therapeutic strategy for the management of diabetic wounds was demonstrated to be able to modulate inflammatory responses in diabetic wounds, promote macrophage polarization toward the M2 phenotype, and accelerate wound repair. In Kalska-Szostko et al. [27], they first prepared copper and silver nanoparticles through an eco-friendly green chemistry method using beetroot extract. And then they demonstrated that the green-synthesized copper and silver nanoparticles can be effectively embedded into a polypropylene polymer matrix, while maintaining their physicochemical stability and antibacterial performance. Importantly, the nanocomposites incorporating green-synthesized copper nanoparticles displayed remarkably improved bactericidal and fungicidal effects relative to those containing silver nanoparticles or copper nanoparticles fabricated via traditional chemical reduction techniques. These findings verify that the synthetic route exerts a critical influence on determining the biological activity of the final polymer nanocomposites.

The multifunctional polymer nanocomposites have also been developed for drug and gene delivery systems, tissue engineering scaffolds, biosensors, and medical imaging contrast agents. The paper of Wang et al. [28] synthesized a series of core–shell-structured calcium peroxide/poly(ethylene glycol)@silica (CPO@SiO_2) nanoparticles with tunable oxygen-generating properties. The nano-CPO core acts as the oxygen source to improve hypoxia, while the poly(ethylene glycol) and SiO_2 shell layer serves as the physical barrier to improve biocompatibility and control the oxygen-generating rate. As a result, the CPO@ SiO_2 hybrid nanoparticles can support cell survival under hypoxia, effectively decrease oxidative stress damage, and reduce the levels of expression of hypoxia-induced superoxide

 https://doi.org/10.3390/ma19061165

dismutase and malondialdehyde for repairing damaged tissue. Yang et al. [29] synthesized N-doped graphene quantum dots (N-GQDs) via a one-step hydrothermal route using cellulose as the starting material and ethylenediamine as the nitrogen dopant. The entire procedure eliminated the use of strong acids or bases and prevented contamination from inorganic salt byproducts and heavy metal ions. The as-synthesized N-GQDs exhibited excellent photostability and salt tolerance. And then they prepared a new fluorescent sensing platform with high selectivity and sensitivity for the detection of 6-mercaptopurine (6-MP) via the fluorescence inner filter effect of N-GQDs. Furthermore, this probe can be effectively applied to the determination of 6-MP in human urine samples, verifying that the N-GQD-based fluorescent sensor possesses outstanding practical application potential.

The introduction of inorganic nanomaterials as fillers into polymer matrixes can endow the resulting polymer nanocomposites with unique mechanical properties, optical properties, magnetic properties, and/or electrical properties. The paper of Iritani et al. [30] developed highly functional zinc oxide nanoparticles (ZnONPs) by surface modification with 6-amino-1-hexanol bearing both amino and hydroxyl functional groups. The composite films of polylactic acid (PLA) reinforced with the functionalized ZnONPs were prepared and then evaluated. The composite containing functionalized ZnONPs showed greater tensile strength in comparison to films loaded with unmodified ZnONPs. Moreover, the elevated water contact angle values suggested that surface modification improved the hydrophobic characteristics of the composite films. This work validated the feasibility of using surface-modified ZnONPs as reinforcing additives, underscoring an effective strategy for the design of advanced eco-friendly materials.

In summary, multifunctional polymer nanocomposites have emerged as a cornerstone of advanced material science, driven by their unique fabrication flexibility, synergistic multi-property characteristics, and broad application potential. With the continuous advancement of nanotechnology, filler modification strategies, and composite processing techniques, the design and synthesis of polymer nanocomposites with more precise and diverse functionalities will continue to evolve. Future research directions will focus on developing sustainable and bio-based polymer matrices and nanofillers, scaling up fabrication processes with high efficiency and low cost, and exploring novel multifunctional combinations to address emerging challenges in energy, environment, healthcare, and advanced manufacturing. As such, these materials will play an increasingly pivotal role in driving technological innovation and sustainable development across global industries.

Author Contributions: Y.Z.: writing—review and editing, manuscript revision; J.Z.: writing—original draft preparation, writing—review and editing, supervision, critical comments, suggestions, manuscript revision, and funding acquisition. All authors have read and agreed to the published version of the manuscript.

Data Availability Statement: Not applicable.

Acknowledgments: We gratefully acknowledge the financial support provided by the Natural Science Foundation of Tianjin City (23JCYBJC01860).

Conflicts of Interest: The authors declare no conflicts of interest.

References

1. Giraud, L.; Tourrette, A.; Flahaut, E. Carbon Nanomaterials-Based Polymer-Matrix Nanocomposites for Antimicrobial Applications: A Review. *Carbon* **2021**, *182*, 463–483. [CrossRef]
2. Sharma, S.K.; Pujari, P.K. Role of Free Volume Characteristics of Polymer Matrix in Bulk Physical Properties of Polymer Nanocomposites: A Review of Positron Annihilation Lifetime Studies. *Prog. Polym. Sci.* **2017**, *75*, 31–47. [CrossRef]

 https://doi.org/10.3390/ma19061165

3. Gupta, S.; Henson, A. Engineered Nanocomposites through Embedding of Smaller "Organic Inorganic" Nanoparticles in Thermoplastic Poly(2-Vinylpyridine) Polymer Matrix. *Compos. Part B Eng.* **2024**, *272*, 111207. [CrossRef]
4. Anwer, A.H.; Ahtesham, A.; Shoeb, M.; Mashkoor, F.; Ansari, M.Z.; Zhu, S.; Jeong, C. State-of-the-Art Advances in Nanocomposite and Bio-Nanocomposite Polymeric Materials: A Comprehensive Review. *Adv. Colloid Interface Sci.* **2023**, *318*, 102955. [CrossRef]
5. Lipovka, A.; Petrov, I.; Fatkullin, M.; Murastov, G.; Ivanov, A.; Villa, N.E.; Shchadenko, S.; Averkiev, A.; Chernova, A.; Gubarev, F.; et al. Photoinduced Flexible Graphene/Polymer Nanocomposites: Design, Formation Mechanism, and Properties Engineering. *Carbon* **2022**, *194*, 154–161. [CrossRef]
6. Baishya, H.; Dutta, J.; Kumar, S. Nanofillers Reinforcing Biopolymer Composites for Sustainable Food Packaging Applications: A State-of-the-Art Review. *Adv. Funct. Mater.* **2025**, *35*, 2503819. [CrossRef]
7. Wen, Y.H.; Yuan, J.M.; Ma, X.M.; Wang, S.R.; Liu, Y.C. Polymeric Nanocomposite Membranes for Water Treatment: A Review. *Environ. Chem. Lett.* **2019**, *17*, 1539–1551. [CrossRef]
8. Liang, S.F.; Yuan, C.R.; Nie, C.; Liu, Y.Z.; Zhang, D.C.; Xu, W.C.; Liu, C.W.; Xu, G.F.; Wu, S. Photocontrolled Reversible Solid-Fluid Transitions of Azopolymer Nanocomposites for Intelligent Nanomaterials. *Adv. Mater.* **2024**, *36*, 2408159. [CrossRef]
9. Xu, Q.Y.; Qu, H.B.; Wu, Z.H.; Zhao, Z.; Ge, S.S.; Luo, J.T.; Zuo, B. Relaxation-Enhanced Polymer Nanocomposites Induced by Bound Polymer Loops on the Particle Surfaces. *Nat. Commun.* **2025**, *16*, 9870. [CrossRef]
10. Mishra, K.; Devi, N.; Siwal, S.S.; Zhang, Q.B.; Alsanie, W.F.; Scarpa, F.; Thakur, V.K. Ionic Liquid-Based Polymer Nanocomposites for Sensors, Energy, Biomedicine, and Environmental Applications: Roadmap to the Future. *Adv. Sci.* **2022**, *9*, 2202187. [CrossRef]
11. Dananjaya, V.; Marimuthu, S.; Yang, R.; Grace, A.N.; Abeykoon, C. Synthesis, Properties, Applications, 3d Printing and Machine Learning of Graphene Quantum Dots in Polymer Nanocomposites. *Prog. Mater. Sci.* **2024**, *144*, 101282. [CrossRef]
12. Xu, H.Z.; Hao, Z.W.; Wang, C.W.; Deng, J.Y.; Wang, T.J.; Zhang, J.H. In Situ Emulsion Polymerization to Multifunctional Polymer Nanocomposites: A Review. *Macromol. Chem. Phys.* **2023**, *224*, 2300185. [CrossRef]
13. Zhao, F.L.; Yao, D.; Guo, R.W.; Deng, L.D.; Dong, A.J.; Zhang, J.H. Composites of Polymer Hydrogels and Nanoparticulate Systems for Biomedical and Pharmaceutical Applications. *Nanomaterials* **2015**, *5*, 2054–2130. [CrossRef] [PubMed]
14. Shen, X.; Zheng, Q.B.; Kim, J.K. Rational Design of Two-Dimensional Nanofillers for Polymer Nanocomposites toward Multifunctional Applications. *Prog. Mater. Sci.* **2021**, *115*, 100708. [CrossRef]
15. Jeong, M.J.; Kim, S.H.; Kim, D.; Jung, S.W.; Kim, T.; Oh, Y.; Zhang, Y.L.; Seong, D.G. Cnt Film-Integrated Polyimide Composites: Scalable Joule Heated Infusion Process and Multifunctional Aerospace Applications. *Compos. Part B Eng.* **2026**, *315*, 113471. [CrossRef]
16. Salehpour, N.; Mahdavi, P.; Nojavan, S. Advancements in Nanomaterials for Water and Wastewater Treatment and Monitoring. *Trends Environ. Anal. Chem.* **2026**, *49*, e00299. [CrossRef]
17. Rajni; Taruna; Udayasri, A.; Raghav, N.; Bendi, A.; Tomar, R. Revolutionizing Wastewater Treatment: Polymeric Metal Oxide Nanocomposites for Effective Dye and Heavy Metal Removal. *Chem. Eng. J.* **2025**, *511*, 161694. [CrossRef]
18. Cosola, A.; Roppolo, I.; Frascella, F.; Napione, L.; Barrera, G.; Tiberto, P.; Turbant, F.; Arluison, V.; Caldelari, I.; Mercier, N.; et al. 4D Printing of Multifunctional Devices Induced by Synergistic Role of Magnetite and Silver Nanoparticles in Polymeric Nanocomposites. *Adv. Funct. Mater.* **2024**, *34*, 2406226. [CrossRef]
19. Zhao, W.; Li, C.L.; Chang, J.; Zhou, H.M.; Wang, D.S.; Sun, J.J.; Liu, T.Q.; Peng, H.; Wang, Q.F.; Li, Y.A.; et al. Advances and Prospects of Raft Polymerization-Derived Nanomaterials in Mri-Assisted Biomedical Applications. *Prog. Polym. Sci.* **2023**, *146*, 101739. [CrossRef]
20. Roy, S.; Ghosh, B.D.; Mishra, S.; Goh, K.L.; Kim, J. Customized Extrusion Nozzle Assisted Robust Nylon 6/Mwcnt Nanocomposite Based Triboelectric Nanogenerators for Advanced Smart Wearables. *Chem. Eng. J.* **2024**, *493*, 152598. [CrossRef]
21. Ye, Z.P.; Li, S.Y.; Zhao, S.Y.; Deng, L.D.; Zhang, J.H.; Dong, A.J. Textile Coatings Configured by Double-Nanoparticles to Optimally Couple Superhydrophobic and Antibacterial Properties. *Chem. Eng. J.* **2021**, *420*, 127680. [CrossRef]
22. Zhao, F.L.; Wu, D.; Yao, D.; Guo, R.W.; Wang, W.W.; Dong, A.J.; Kong, D.L.; Zhang, J.H. An Injectable Particle-Hydrogel Hybrid System for Glucose-Regulatory Insulin Delivery. *Acta Biomater.* **2017**, *64*, 334–345. [CrossRef] [PubMed]
23. Wakeel, S.; Aslam, A.; Zhang, J.H. Advances in Polymer Nanocomposites for Drilling Fluids: A Review. *Materials* **2025**, *18*, 4809. [CrossRef] [PubMed]
24. Hao, L.; Yu, D.M.; Hou, X.Y.; Zhao, Y.X. Advances in Polypyrrole Nanofiber Composites: Design, Synthesis, and Performance in Tissue Engineering. *Materials* **2025**, *18*, 2965. [CrossRef] [PubMed]
25. Liu, Z.J.; Zhang, H.P.; Li, Y.Q.; Xu, S.X.; Fu, N.; Wang, F.; Chen, W.S. Hydrogel-Flexible Electronics Integrated Platforms for Diabetic Wound Management. *Materials* **2026**, *19*, 509. [CrossRef] [PubMed]
26. Lin, D.; Li, Z.H.; Hao, J.Y.; Xu, X.B.; Li, X.Q.; Feng, Y.; Lu, X.C.; Yao, F.L.; Zhang, H.; Li, J.J. A Dual-Dynamic Crosslinked Polysaccharide-Based Hydrogel Loaded with Exosomes for Promoting Diabetic Wound Healing. *Materials* **2026**, *19*, 445. [CrossRef]

 https://doi.org/10.3390/ma19061165

27. Wasilewska, A.; Bielicka, M.; Klekotka, U.; Markiewicz, G.; Jalbrzykowski, M.; Lewandowska, W.; Swiecicka, I.; Kalska-Szostko, B. Antimicrobial Properties of Polymer-Based Nanocomposites Modified by Nanoparticles Produced by Green Chemistry. *Materials* **2026**, *19*, 251. [CrossRef]
28. Xie, X.L.; Sun, X.; Lin, W.M.; Yang, X.F.; Wang, R.C. Preparation and Properties of Calcium Peroxide/Poly(Ethylene Glycol)@Silica Nanoparticles with Controlled Oxygen-Generating Behaviors. *Materials* **2025**, *18*, 2568. [CrossRef]
29. Xu, Q.; Dong, J.Y.; Yan, G.Q.; Yi, R.N.; Yang, X.J. Synthesis of N-Doped Graphene Quantum Dots from Cellulose and Construction of a Fluorescent Probe for 6-Mercaptopurin Quantitative Detection. *Materials* **2024**, *17*, 5852. [CrossRef]
30. Nakagawa, H.; Iritani, K. Preparation of Particle-Reinforced Resin Using Highly Functional Zno Particle Filler Driven by Supramolecular Interactions. *Materials* **2025**, *18*, 2986. [CrossRef]

materials

MDPI

Review

Advances in Polymer Nanocomposites for Drilling Fluids: A Review

Shahbaz Wakeel [1], Ammara Aslam [1] and Jianhua Zhang [1,2,*]

1 Key Laboratory of Systems Bioengineering of the Ministry of Education, Department of Polymer Science and Engineering, School of Chemical Engineering and Technology, Tianjin University, Tianjin 300350, China; shahbazwakeel410@tju.edu.cn (S.W.); ammaraaslam18@gmail.com (A.A.)

2 Tianjin Key Laboratory of Membrane Science and Desalination Technology, Tianjin University, Tianjin 300350, China

* Correspondence: jhuazhang@tju.edu.cn

Abstract: Hydrocarbon exploration and extraction increasingly rely on drilling fluids that guarantee operating safety and efficiency, particularly in ultra-deep, high-temperature, and unconventional reservoirs. Traditional drilling fluids, especially for water-based muds (WBMs), have several problems, including excessive fluid loss, severe swelling in shale and instability in high-pressure/high-temperature (HPHT) conditions. Polymer nanocomposites (PNCs) are new types of drilling fluid additives that combine the vast surface area and reactivity of nanoparticles (NPs) with the structural flexibility and stability of polymers. This combination enhances rheology, reduces filtrate loss, and, most importantly, creates hydrophobic and pore-blocking barriers that prevent shale from swelling. This review highlights important improvements in drilling fluids with PNCs regarding exceptional rheological properties, low fluid loss, and improved suppression of the shale swelling. The particular focus was placed on the specific mechanisms and role that PNCs play in enhancing shale stability, as well as their responsibilities in improving rheology, heat resistance, and salt tolerance. Current advancements, persistent hurdles, and prospective prospects are rigorously evaluated to emphasize the scientific and industrial trajectories for the development of next-generation, high-performance drilling fluids. Moreover, the current challenges and future opportunities of PNCs in drilling fluids are discussed to motivate future contributions and explore new possibilities.

Keywords: drilling fluids; polymeric filtrate reducer; shale swelling; polymer nanocomposites; temperature resistance; salt tolerance

1. Introduction

Drilling fluids, often referred to as drilling muds, play a crucial role in oil and gas exploration by controlling pressure in the hole, transporting cuttings, lubricating the drill bit, and maintaining the wellbore stability. They are primarily divided into oil-based muds (OBMs) and water-based muds (WBMs). OBMs are relatively stable in terms of heat and chemicals, but they are detrimental to the environment and the economy because they are poisonous and expensive to clean up [1]. WBMs have become preferred because of their lower environmental impact and lower disposal costs. Still, they have operational issues, such as losing too much fluid, expanding shale, and being unstable in high-pressure/high-temperature (HPHT) conditions. Figure 1 illustrates the important issues of typical drilling fluids, including fluid loss, instability at high temperatures, and shale swelling. To resolve the above problems and achieve better performance of drilling fluids during drilling operations in complex geological conditions of deep formations, some additives have been developed and widely employed to change the physicochemical properties and improve the applicable properties of drilling fluids, especially for WBMs. However, the complex geological conditions in deep formations with high temperature, high pressure, and high salinity result in high requirements for the performance of drilling fluids. Therefore, it is imperative to explore a new methodology and techniques to accelerate the search for more effective WBMs and related additives to address these issues of WBMs and ensure the efficient drilling and exploration of oil and gas resources mainly distributed in deep layers [2].

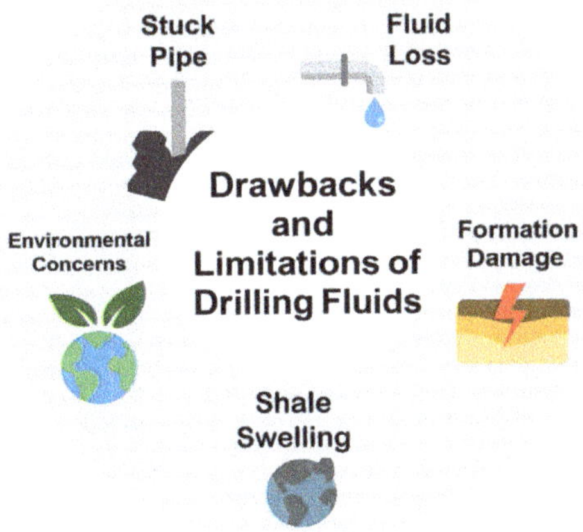

Figure 1. Drawbacks and limitations of drilling fluids.

Various additives are demonstrated to decide the physicochemical properties and functional characteristics of WBMs. The additives have become indispensable, especially for drilling activities in deep formations, unconventional reservoirs, and harsh circumstances. The functional additives serve critical functions such as minimizing fluid loss, maintaining wellbore stability, ensuring drilling safety and mitigating environmental impacts. Polymeric WBMs improve drilling fluid rheology, fluid loss, and wellbore stability. Shale swelling, low ROP, and formation damage plague traditional WBMs. They benefit from natural or synthetic additions. Despite their benefits, polymeric additives must be examined for cost, environmental impact, and drilling site characteristics. Table 1 not only compares the pros and cons of natural and manufactured polymers, but also describes the primary functions, and operational limitations of polymeric additives utilized in drilling fluid formulations. Biopolymers such as XG, GG, and starch exhibit favorable rheological modification capabilities, ecological compatibility, and effective fluid loss mitigation; however, they are often constrained by thermal degradation, microbial susceptibility, and limited solubility under high-pressure, high-temperature (HPHT) conditions. Most natural polymers fail at 150 °C or higher hot drilling temperatures due to thermal degradation. Many modification strategies have been researched to increase natural polymer performance in high-temperature situations [3,4]. Ionic crosslinking, esterification, etherification, salinization, hydroxymethylation, and free radical polymerization were studied. Modified natural and synthetic polymers, including HPAM, KPAM hydrogel, PHPA, PAC, HEC, and CMC, demonstrate enhanced thermal stability, shale inhibition, and improved rheological control. Still, their performance can be compromised by ion sensitivity, elevated viscosity, and environmental concerns related to monomer toxicity. Advanced polysaccharides such as chitosan, scleroglucan, welan gum, and schizophyllan offer promising viscoelastic behavior, salt tolerance, and biodegradability. Yet, they remain prone to oxidative degradation and reduced structural integrity under biologically active or thermally extreme environments. The optimal selection of these additives must therefore consider the thermodynamic and geochemical constraints of the target formation to ensure functional efficiency, environmental safety, and economic viability [5]. Synthetic additives like polyacrylate (PAA) and PAM were introduced to improve the performance of fluids. Adding these chemicals enhanced the filtering capabilities and viscosity of water-based drilling fluids (WBDFs), rendering them more adept at enduring high-temperature and deep-well drilling conditions [6]. However, these improvements were not enough to solve WBDFs problems with performance instability, especially during changing drilling conditions with different temperatures, pressures, and salt levels [7].

Parallel to the aforementioned developments in polymer additives, nanomaterials as one of the most promising materials have attracted enormous interest and achieved substantial progress for a diverse array of applications in various fields in recent decades, due to their nano-sized characteristics, tunable morphology, versatile modification possibilities, unique physicochemical properties and well-defined multifunctionalities. Various nanoparticles (NPs) have also exhibited great potential in drilling fluids [8,9]. For example, inorganic nanoparticles such as SiO_2, mesoporous silica nanoparticles (MSNs), and clay, help stabilize shale by making it less permeable and stronger as a filter cake. Their tiny size allows them to effectively plug micro- and nanopores, creating a denser and less permeable barrier that stops filtrate invasion and improves wellbore sealing. This action makes shale minerals less hydrated, thereby increasing the wellbore's stability. Alongside shale stabilization, integrating NPs with lost circulation materials has shown efficacy in reducing filtrate loss in drilling fluids [10,11]. Including NPs in inversion emulsion muds, in conjunction with low-cost additives, greatly reduced filtrate loss and yielded a thinner filter cake than muds containing just NPs or low-cost additives alone. NPs reduce filter cake permeability by occupying interstices between nano- and micron-sized particles. Integrating NPs into the drilling fluid produces a more consistent and denser filter cake, reducing fluid loss and enhancing the overall efficacy of the drilling process. Figure 2 demonstrates that the incorporation of NPs decreases filtrate loss volume and improves the sealing of the filter cake in comparison to conventional drilling fluids devoid of NPs. The image illustrates the disparities in fluid loss behavior between muds containing NPs and those without, showcasing the efficacy of NPs in reducing filtrate loss during drilling operations [12,13]. Despite these attractive features, NPs are associated with difficulties related to stability and aggregation. Due to their increased energy density and small sizes, NPs are prone to aggregation in the mud system. Such an inclination to aggregate is due to strong interactions, such as metallic or covalent contacts, and weaker van der Waals interactions that cause groupings of particles [14].

Table 1. Key polymeric additives relevant to water-based drilling fluids (WBDFs) and their functional performance.

Polymer	Primary Functions in WBDFs	Advantages	Limitations/Challenges
Xanthan Gum (XG)	Viscosifier, suspension aid, improves gel strength	Exceptional rheology control at low shear rates; eco-friendly and biodegradable	Thermal degradation above 150 °C; viscosity loss in high-salinity environments
Polyanionic Cellulose (PAC)	Fluid-loss reducer, filtrate-control agent	Maintains stable filter cake; good salt tolerance; improves wellbore stability	Sensitive to Ca^{2+}/Mg^{2+} contamination; limited performance above 160 °C
Carboxymethyl Cellulose (CMC)	Filtrate control and rheology modifier	Biodegradable; forms thin, low-permeability filter cakes	Degrades above 150 °C; reduced efficiency in highly saline systems
Partially Hydrolyzed Polyacrylamide (PHPA)	Shale encapsulation, viscosity enhancement	Provides very good cutting suspension and shale inhibition	Decomposes under HPHT; limited solids tolerance
Hydrolyzed Polyacrylamide (HPAM)	Rheology enhancer and clay stabilizer	High viscosity at low dosage; improves drilling rate	Mechanically unstable under shear; acrylamide monomer toxicity concerns
Polyanionic Starch (PAS)/Modified Starch	Filtrate-loss control and wellbore stabilization	Cost-effective; biodegradable; good salt tolerance	Loses structure above 180 °C without crosslinking or nanofiller support
Guar Gum (GG)	Viscosifier and lubricity enhancer	Provides high viscosity at low concentrations; bio-based and renewable	Poor thermal stability (>150 °C); bacterial degradation risk
Scleroglucan/Welan Gum	Rheology modifier for HPHT wells	High shear and thermal resistance; stable gel formation	Costly; susceptible to oxidation in long-term storage
Chitosan Derivatives	Rheology and fluid-loss modifier	Biodegradable; enhances thixotropy and film-forming capability	Thermal degradation above 160 °C; limited salt tolerance
Potassium Polyacrylamide (KPAM)	Shale-inhibiting polymer, viscosity enhancer	Improves wellbore stability; effective in low-solids muds	Overuse increases yield stress and pumping energy
Acrylamide-co-AMPS Copolymer	HPHT filtrate-control additive	Good thermal (≥200 °C) and salt resistance; stable in brine	Relatively expensive; requires dispersion control
Polysaccharide–Nanocomposite Blends (e.g., XG–SiO_2)	Combined rheology and filtrate control	Enhanced thermal stability (>200 °C) and improved shale inhibition	Preparation complexity; dispersion stability must be maintained

Currently, the development of drilling fluid technology has increasingly centered on polymer nanocomposites (PNCs), which amalgamate the elevated surface reactivity of nanoparticles (NPs) with the stability and structural adaptability of polymers. This synergy enables PNCs to outperform single polymers or nanoparticles by reducing the amount of filtrate lost, enhancing the material's flow, and resisting damage from heat and salt [15]. The polymer matrix is important because it prevents nanoparticles from clumping together, ensuring they are evenly distributed and remain stable over time. Furthermore, the NPs improve the performance of the polymer because of the molecular chain entanglement upon adsorption onto the polymer matrix surface [16]. There are multiple mechanisms through which interaction between NPs and polymers may occur, including electrostatic attraction, hydrophobic interactions, steric repulsion, hydrogen bonding, and electrostatic repulsion [17,18]. These interactions result in new nanocomposite materials that are extraordinary and often exhibit important improvements compared to their building blocks. These enhanced characteristics include advanced performance, resistance to hostile temperatures, and the ability to withstand increased salinity levels [19]. Across the studies summarized in Tables 1 and 2, PNC additives generally increase YP relative to base WBM and lower HPHT filtrate loss under saline/thermal stress; the magnitude depends on NP type/size and surface treatment, with silica- and graphene-based systems most consistent under high salinity [20]. For many of the issues that traditional drilling fluids cause, PNCs offer a consistent fix. PNCs can increase wellbore stability and dramatically slow fluid loss [21]. For example, shale swelling remains one of the most important challenges in drilling operations. It often leads to wellbore instability, pipe sticking, and structural collapse. Under HPHT and saline conditions, conventional inhibitors such as potassium chloride or simple polymers provide only partial protection. PNCs mitigate these effects through two complementary mechanisms: (i) forming nanopore-blocking barriers that minimize filtrate invasion, and (ii) generating hydrophobic layers that reduce water uptake and the resulting volumetric expansion of shale minerals. PNCs can also provide improved thermal stability, guaranteeing that the drilling fluid is proper even under demanding circumstances [22,23]. Better lubricating qualities of PNCs also help lower the coefficient of friction (CoF) between the drill bit and the wellbore, lowering equipment wear and increasing efficiency. Reduced fluid loss and enhanced sealing capacity produced by the synergy between the NP and the polymer matrix also help avoid fluid intrusion into adjacent formations and strengthen wellbore integrity [24,25]. In sum, polymer nanocomposites (PNCs) offer a reliable and technically sound solution to many of the challenges associated with traditional drilling fluids, including excessive filtrate loss, low temperature resistance, and shale swelling. Through the synergistic interaction between nanoparticles and polymer matrices, PNCs enhance rheological stability, reduce filtrate invasion, and strengthen wellbore sealing.

Table 2. A comparison of polymer–nanoparticle integration methods, detailing their compatibility, processing characteristics, and resultant fluid performance.

Method	Polymer Type	Nanoparticle Compatibility	Dispersion Quality	Scalability	Impact on Stability & Shale Inhibition
Solution Mixing	Water-soluble polymers (PAM, PAC, CMC)	SiO_2, TiO_2, GO	Good at lab scale; solvent removal may cause agglomeration	Moderate	Ensures high uniformity; improved filtrate control, but limited long-term NP stability
Melt Mixing	Thermoplastics and modified gums	SiO_2, ZnO, CNTs	Good physical dispersion; thermal shear may cause NP clustering	High	Produces durable, temperature-resistant composites; effective for rheology but moderate shale inhibition
In situ Polymerization	Acrylamide, AMPS, or acrylic copolymers	TiO_2, ZnO, GO, α-LAP	Higher chemical bonding and NP immobilization	Moderate to high	Best dispersion and stability; strong barrier formation and shale inhibition under HPHT

This review aims to summarize and rigorously assess the most recent advancements in the formulation, processes, and practical applications of PNCs-based drilling fluids. It specifically looks at (i) modification strategies and how they affect rheological behavior, fluid-loss control, and thermal stability; (ii) the role of inorganic and metallic nanoparticles in improving wellbore integrity and filter cake structure; and (iii) the synergistic mechanisms by which PNCs improve shale inhibition and drilling efficiency. We believe that this review article will offer some ideas on the design and use of PNCs for researchers working in this area.

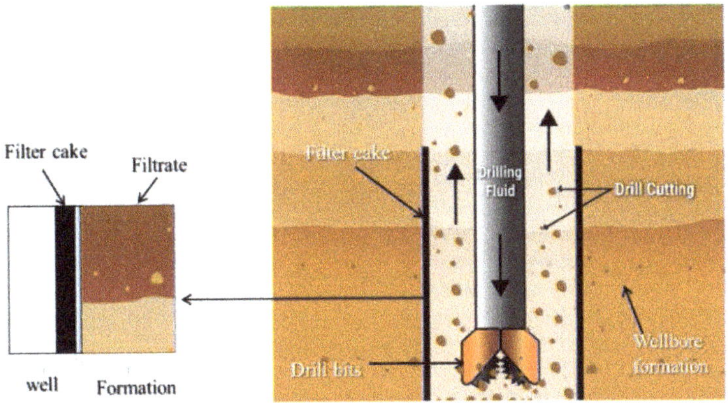

Figure 2. Schematic representation of filter-cake microstructure in conventional and nanoparticle-enhanced water-based drilling fluids (WBDFs) [26].

2. Overview of PNCs

Drilling fluids serve several essential functions, including transporting cuttings, controlling pressure, stabilizing the wellbore, and preventing fluid from escaping. All of these things are more challenging to accomplish under HPHT and saline conditions. PNCs have become advanced, multifunctional additives that can help address these problems. Their unique polymer–nanoparticle synergy enables more effortless movement of cuttings, maintains viscosity and gel strength, makes the material resistant to heat, and creates nano barriers that prevent filtrate invasion and shale swelling. This section discusses the formation, structure, and operation of PNCs in drilling fluids.

2.1. Fundamentals and Preparation of PNCs

PNCs are hybrid materials made of nanoparticles (NPs < 100 nm) that are spread out in a polymer matrix. This makes them better at filtering, rheology, and thermal properties. The polymer matrix ensures that the nanoparticles are evenly distributed and stable. The nanoparticles add more surface activity, barrier strength, and chemical resistance.

NPs are incorporated into a polymer matrix using a variety of processes during the manufacturing of PNCs to guarantee a uniform distribution and optimal performance of the nanocomposite material. The most often used techniques for PNC preparation with their impact on shale stability and inhibition are represented in Table 2.

The Solution Mixing technique dissolves the polymer in a solvent and distributes the NPs among the polymer solution. By following dispersion, the solvent evaporates, forming a solid nanocomposite layer or substance. Although this method guarantees good dispersion of the NPs inside the polymer matrix, it may make it challenging to maintain their stability in the solution.

The melt mixing method combines the polymer with NPs in the melt state, in which the polymer is heated above its melting point. Mixing the NPs with the molten polymer guarantees their uniform distribution over the matrix. Thermoplastic polymers are processed mostly using this technique since it is reasonably affordable and appropriate for mass manufacturing. Still, it can be challenging to regulate the dispersion of NPs in the molten form.

In situ polymerization involves the direct polymerization of monomers when NPs are present. Clusters of polymer chains form around dispersed nanoparticles, including SiO_2, TiO_2, CNTs, GO, ZnO, and other metal nanoparticles (such as gold or silver). Among the several advantages of this approach are the high dispersion and robust interactions between the nanoparticles and the polymer. In many cases, nanoparticles initiate or accelerate the polymerization reaction, which in turn strengthens the contact. Nanocomposites made of thermoset-based polymers are commonly created using this method. As a result, the composite's mechanical, chemical, thermal, and adhesive properties are all enhanced. Every preparation technique has benefits and drawbacks; the type of polymer, NPs, and desired qualities of the resultant nanocomposite material will determine the method of choice. These preparation techniques aim to ensure homogeneous dispersion of the NPs inside the polymer matrix, thereby ensuring that both components' synergistic benefits are completely realized and producing a material displaying exceptional performance under demanding conditions [27]. PNCs are a type of high-tech material that can contain both organic and inorganic nanoparticles. The primary factor determining the type of PNC is the type of nanoparticles employed.

2.1.1. Strategic Integration of Polymers and Nanoparticles for Advanced Drilling Fluids

The type of polymers and NPs best suited for drilling depends on the environment in which the drilling will take place. Polyanionic cellulose (PAC) and carboxymethyl cellulose (CMC) are environmentally friendly and help manage viscosity. Polyacrylamide (PAM) and AMPS copolymers enhance the material's resistance to salt and temperature fluctuations. Inorganic nanoparticles (SiO_2, clays, MSNs) make the filter cake denser and seal the pores more effectively. Metal and metal oxide nanoparticles (TiO_2, ZnO, Fe_3O_4) enhance the material's resistance to heat and improve its ability to control fluid loss by crosslinking and strengthening the structure. This combination of organic polymers and inorganic/metallic nanofillers enables the creation of next-generation drilling fluids that perform well under varying field conditions and maintain their rheological and filtration properties.

2.1.2. PNCs Made of Polymers and Inorganic NPS

Inorganic NP_S are widely used in PNCs to enhance the tensile strength, chemical resistance, and thermal stability of the composites. These NPS, including SiO_2, quantum dots (QDs), and clays, are particularly beneficial under demanding conditions, improving the viscosity, wellbore stability, and filtration control of composite materials. Inorganic NPS play a crucial role in making PNCs suitable for a range of industrial applications, from structural composites to wellbore drilling fluids.

Silica NPS are among the most commonly used inorganic NPS due to their large surface area, good thermal stability, and biocompatibility. These properties allow SiO_2 to enhance the mechanical, viscosity, and chemical resistance of composites. SiO_2 NPS are widely used in structural composites, medicinal applications, WBMs, and lubricants, contributing to their overall performance. Additionally, mesoporous silica nanoparticles (MSNs) are ideal for drug delivery and catalysis applications due to their high surface area and adjustable pore diameters. These features make MSNs useful in adsorption systems, catalysis, and controlled drug delivery [8,27].

2.1.3. PNCs Made of Polymers and Metal NPS or Metal Oxide NPS

In PNCs, metal and metal oxide NPS are used to enhance mechanical strength, thermal stability, and electrical conductivity. These NPS also possess biocidal and catalytic properties, making them ideal for applications in biomedical devices, electronics, and catalysis. Standard metal and metal oxide NPS in PNCs include Cu, Ni, Co, Pt, Ag, Au, Fe_3O_4, Fe_2O_3, TiO_2, and ZnO. Metallic NPS, such as copper, nickel, and gold, improve the electrical conductivity, catalytic activity, and biocompatibility of PNCs. These properties make them useful in various applications, including electronic devices, drug delivery systems, and biosensors. The high conductivity and catalytic properties of metallic NPS enable advancements in fields such as electronics and healthcare. Platinum and silver are also commonly used for their unique catalytic and biocidal properties, making them essential in biomedical applications and environmental remediation [28–30].

2.1.4. Properties and Advantages of PNCs

Drilling fluids can be greatly improved with the use of PNCs, particularly in extreme drilling conditions. NPSs, such as Gr, SiO2, or clay, are added to PNCs to enhance valuable properties, including resistance to heat and salt. For drilling fluids to remain stable and perform well in extreme situations, such as the high temperatures and salty environments encountered during deep drilling, these qualities are crucial. Because it is now more resistant to heat, the drilling fluid will retain its properties even in extremely harsh environments. Because of its enhanced salt resistance, the fluid can be used effectively in areas with high salt concentrations. Additionally, PNCs can thicken drilling fluids, which is crucial for controlling the viscosity of the fluid [31–33]. The enhanced thickening ability facilitates the suspension of drill cuttings and other materials, leading to streamlined and expedited drilling operations.

Drilling fluids treated with PNCs are not only more resistant to cold and salt, but their properties can be fine-tuned by adjusting the nanoparticle type and concentration. Due to this adaptability, tailored fluids that perform effectively in various drilling conditions can be developed. Drilling fluids that contain PNCs not only perform better, but they also reduce the likelihood of wellbore instability and improve the whole operation's efficiency. PNCs can alter the flow of fluids, strengthen suspensions, and increase the stability of fluids in harsh environments, all of which contribute to improved drilling fluid performance in challenging scenarios.

By including NPS like GO, CNTs, and SiO_2, PNCs can increase the fluid's viscosity and yield point. Cuttings can be suspended and moved more easily as a result. This alteration is necessary to enable wellbore stability and efficient hole cleaning. Additionally, the presence of NPS helps avoid issues such as shale swelling and wellbore instability, which can be costly and time-consuming to rectify. The construction of a protective barrier by the NPS stabilizes the wellbore and reduces fluid leakage into the formation, thereby further enhancing the well's integrity [23,34,35].

2.2. PNCs in Drilling Fluids

PNCs have demonstrated good performance in drilling fluid effectiveness, and adding NPs to a polymer matrix enhances fluid loss, wellbore stability, and filter cake formation in drilling operations. Instead of basic polymers, PNCs combine the benefits of the polymer matrix with NPs to improve mechanical, thermal, and chemical performance [36–38].

In drilling fluids, the primary function of PNCs is to reduce filtrate loss, which is crucial for maintaining wellbore integrity. Traditional drilling fluids often encounter issues such as fluid loss into the formation, which can lead to instability and potential operational problems. By adding NPS to polymer matrices, PNCs fill the gaps between nano- and micron-sized particles, creating a denser and more uniform filter cake. This helps to prevent fluid loss, maintaining the efficiency of the drilling operation. Studies have shown that

PNCs, when combined with low-cost additives, greatly reduce fluid loss and form thinner filter cakes compared to fluids with just NPS or additives alone.

An important challenge for PNCs, however, is the stability and aggregation of the NPS within the fluid. Due to their small size and high surface energy, NPS tend to aggregate, which can compromise their effectiveness. Aggregation causes a reduction in the fluid's viscosity and alters its thermal properties, negatively impacting the overall performance of the drilling fluid. To address this, surface modification techniques have been developed to enhance the dispersion and stability of NPS in the polymer matrix. These modifications include the use of surfactants, polymers, and silane functional groups, which prevent aggregation and ensure the NPS remain uniformly distributed throughout the fluid [39,40]. These enhancements allow PNCs to maintain their performance, even under challenging HPHT conditions.

In the most recent Study, in situ polymerization technologies are being used to create unique PNC additives that can help with drilling situations that are becoming increasingly challenging. Figure 3 illustrates the synthesis of α-laponite (α-LAP) clay-based polymer nanocomposites through in situ polymerization of monomers such as ACMO, AMPS, and DMDAAC. The process involves the ultrasonic dispersion of α-LAP, followed by temperature- and pH-controlled polymerization, resulting in a PBMA-co-LAP nanocomposite with enhanced mechanical strength, thermal stability, and rheological properties.

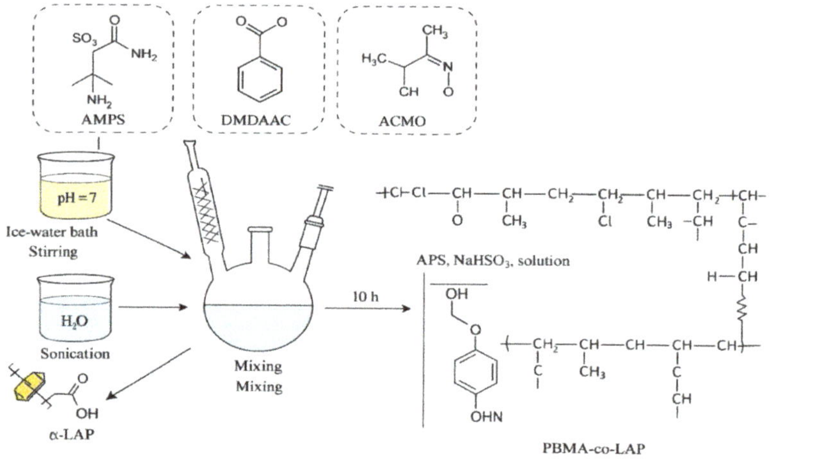

Figure 3. Schematic diagram of process steps for developing PNCs PDADA-LAP for drilling fluid. Reproduced with permission [40]. Copyright 2024, Elsevier.

Modern hydrocarbon exploration aims to stabilize wells, limit fluid loss, and inhibit shale from expanding. Adding these nanocomposites to drilling fluid formulae directly helps with all three of these goals. Drilling in deep, unconventional, and extreme settings becomes possible with the specific combination of organic monomers and inorganic nanoparticles, as seen in the image. Figure 3 shows a diagram of how to make PNCs using α-laponite clay nanoparticles [41]. This approach demonstrates how customized interactions between polymers and nanoparticles can yield composites that function more effectively under HPHT conditions for controlling rheology and filtration. It highlights the technological advantages of nanoparticles and polymer matrices, as well as the need for further research to understand and enhance their interaction in real-life downhole scenarios.

2.2.1. PNCs for Improving Rheological Properties

The effectiveness of drilling fluids in various operations, such as cuttings transport, wellbore stability, filter cake formation, and fluid loss prevention, is primarily determined by their rheological properties. Key properties such as yield point (YP), plastic viscosity (PV), apparent viscosity (AV), and gel strength (GS) directly affect a drilling fluid's ability to suspend and transmit cuttings, generate stable filter cakes, lubricate the drilling bit, and prevent fluid loss into the formation.

In drilling fluids, PNCs are specifically utilized to improve viscosity, yield point, flow behavior, and fluid stability. The high surface area and reactivity of the NPs ensure that the polymer matrix exhibits more interaction, leading to enhanced rheological properties that conventional drilling fluid additives cannot match. The primary function of PNCs in drilling fluids is to improve the rheological properties by introducing NPs such as SiO_2, TiO_2, GO, CNTs, and metal oxides into the polymer matrix. These NPs provide a synergistic effect, enhancing the fluid's viscoelastic properties and enabling it to maintain stability under extreme conditions. By incorporating NPS, PNCs enhance the fluid's ability to adapt to varying shear rates, ensuring optimal fluid flow and performance during drilling operations.

The mechanism by which PNCs improve the rheological properties of drilling fluids is primarily attributed to the interaction between the polymer matrix and the NPs. NPs help the fluid shear-thin, which is important for drilling. Shear-thinning behavior makes a fluid less viscous as the shear rate increases, making it easier to drill through narrow drill pipes (Figure 4). This property helps PNCs maintain fluid flow while ensuring that cuttings are properly suspended and transported to the surface [42,43]. As a result, PNCs help prevent common drilling issues such as pipe sticking, inadequate cuttings transport, and fluid loss to the formation.

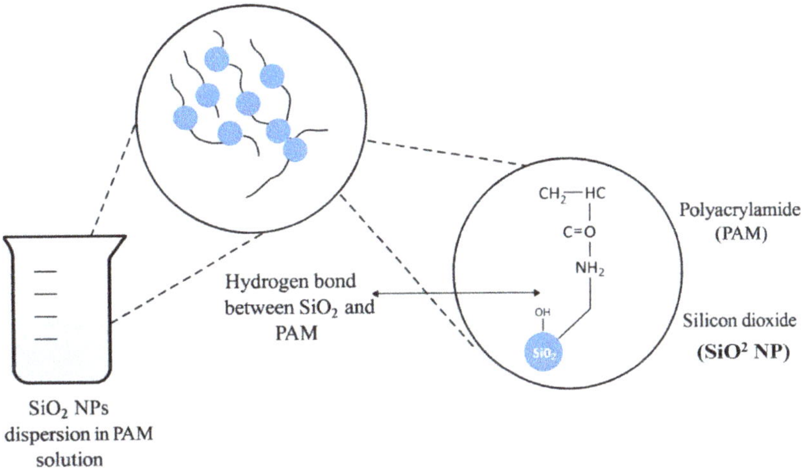

Figure 4. Formation of bonds between PAM and NPs. Reproduced with permission [42]. Copyright 2022, Elsevier.

Researchers [44–46] examined the use of a titania–polyacrylamide nanocomposite (TiO_2-PAM NC) in WBMs and observed improvements in the yield point (YP) and plastic viscosity (PV). The addition of TiO_2-PAM NC also enhanced the shear-thinning rate, meaning the fluid exhibited better adaptability to varying shear rates during the drilling process. As the concentration of TiO_2-PAM NC increased in the base fluid, the apparent viscosity (AV) decreased, leading to improved fluid flow and enhanced cuttings suspension.

Moreover, the addition of TiO_2 NPs helped form a low-permeability filter cake, which reduced fluid loss to the formation, improving overall wellbore stability.

Similarly, researchers [45] developed a polyacrylamide-grafted polyethylene–silica nanocomposite (PAM-g-PEG SiO_2 NC) and demonstrated that the PNCs exhibited better rheological properties than pure polyacrylamide (PAM). The PNCs showed increased PV, YP, AV, and GS, which contributed to improved suspension capabilities and better cutting transport. The SiO_2 NPs enhanced the viscosity of the fluid, allowing it to adapt to varying drilling conditions. Furthermore, the PNCs exhibited improved filtration properties, with the formation of a low-permeability filter cake that reduced fluid loss to the formation [47,48].

One of the most important challenges for conventional drilling fluids is maintaining rheological stability at high temperatures and pressures. For instance, researchers [49,50] tested a polyethylene–nanosilica composite (PEG-SiO_2 NC) modified with sodium dodecyl sulfate (SDS) in WBMs at temperatures of 78 °F and 250 °F. The study found that the SDS-modified PNCs significantly improved AV and rheological stability at both temperatures [50]. The SDS-modified PNCs particles interacted with the bentonite clay in the mud, leading to enhanced viscosity and structure stability. Moreover, the filtration properties of the fluid improved, with reduced fluid loss and the formation of a denser, more permeable filter cake. Figure 5 illustrates that utilizing filtrate reducers composed of nano-laponite-based PNCs improves the performance of drilling fluid in challenging drilling situations. Particles of disc-shaped nano-laponite that are roughly 20 nm wide and 1 nm thick [51]. Polymerization transforms these nanoparticles into a complex nanocomposite, which helps reduce the amount of filtrate. When these nanocomposite materials are mixed with the drilling fluid, they react with the clay particles already present in the wellbore. When PNCs stick to the surfaces of clay particles through both hydration and adsorption, as seen in the lower left, they form a dense and compact mud cake at the interface with the geological deposit (filter paper/stratum). This deep mud cake creates a physical barrier that prevents fluids from entering the formation, responding to the urgent need to avoid fluid leakage and stabilize the wellbore. Figure 5 shows how nano-laponite PNCs adhere to clay surfaces and form compact filter cakes, which directly supports their role in preventing fluid loss.

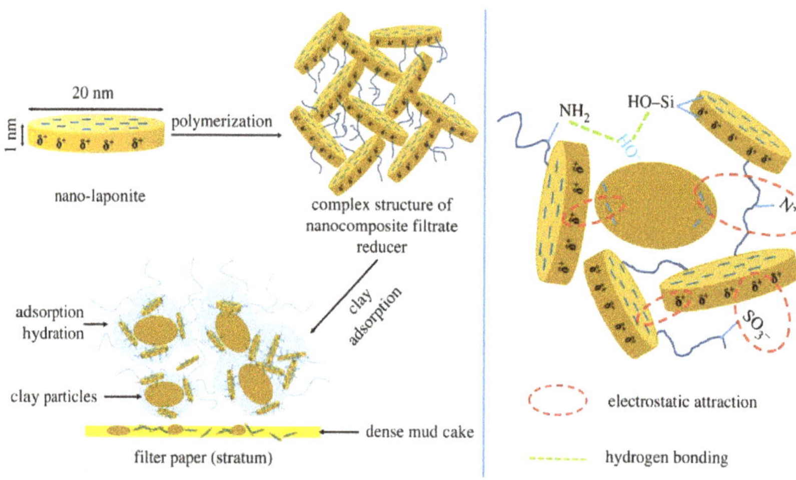

Figure 5. The process by which a polymer nanocomposite may be used to reduce fluid loss [51].

To improve the nanocomposite's performance, the molecules interact, as seen in the right panel. The polymer matrix and laponite discs influence the clay surfaces through multiple mechanisms, including electrostatic attraction (indicated by the red dashed circles) and hydrogen bonding (represented by the green dashed lines) between functional groups such as NH_2, HO-Si, and SO_3^-. The stable nanoparticles and well-encapsulated clay particles result from the strong interactions that form between their surfaces. In addition to strengthening the mud cake structure, these robust physicochemical interactions prevent the shale from becoming too saturated and elongating, which enhances the thermal stability, mechanical strength, and rheological properties of the drilling fluid [52,53].

Similarly, Researchers [54,55] demonstrated that the inclusion of ZnO-PAM NC in WBMs led to an 18.8% improvement in PV and a 16.7% increase in YP. The ZnO-PAM NC enhanced the rheological properties of the fluid and reduced fluid loss to the formation. These improvements were attributed to the synergistic interactions between the ZnO NPs and the polymer matrix. Although PNCs offer important enhancements in rheological properties, there are still challenges that need to be addressed. One of the primary concerns is the aggregation of NPs within the polymer matrix, which can compromise the effectiveness of PNCs. Aggregation leads to a decrease in the surface area and dispersion of the NPs, affecting the overall rheological performance of the fluid. To overcome this issue, surface modification techniques, such as the use of surfactants and polymers, have been developed to improve the dispersion and stability of the NPs.

Research regularly demonstrates that PNCs improve rheology more efficiently than traditional polymers, especially under HPHT settings. Composites made of silica and graphene exhibit better shear-thinning and viscosity control, whereas systems composed of TiO_2 and ZnO yield more stable structures. Nanoparticle aggregation remains a problem, and surface modification is often necessary to maintain optimal performance. Future research should focus on modifying techniques that are scalable and economical. Table 3 shows that nanocomposites made of silica and graphene have better rheological properties than conventional polymer additives.

Comparative studies show that PNCs retain viscosity and yield point better than traditional polymer additives under HPHT settings. PNCs composed of silica and graphene have stronger shear-thinning performance, while composites of TiO_2 and ZnO stabilize structures and prevent fluid leakage. When compared to conventional polymer additives like PAM or PHPA, PNCs consistently demonstrate their ability to withstand higher temperatures and salt. This indicates that they are suitable for deep-sea and unusual drilling situations. However, differences in performance also demonstrate that the choice of nanofiller has an important impact on the balance between rheology, thermal stability, and environmental compatibility.

Table 3. Overview of current studies demonstrating improvements in the rheology of drilling fluids via PNCs.

Polymer Used	Nanofiller Used	Tested Rheology Parameters	Test Conditions	Main Findings	Benefits	Reference
ANDP (AM, AMPS, DMDAAC)	SiO_2, Metal Oxides	PV, AV, YP, GS	150–200 °C, 500 psi	At 200 °C, ANDP improved AV, PV, and YP over the base fluid.	High thermal stability and salt resistance prevent clay aggregation.	[54]
C-g-AA-NH2	KCl	PV, AV, YP	25 °C, 100 psi	2 wt% AA-C-g-AA-NH2 outperformed 10 wt% KCl in rheology.	Effective KCl replacement at a lower dosage.	[55]
SP-GO	GO	PV, AV, YP, GS	78–250 °F	0.4 wt% gave best performance; improved PV and YP.	Lubricates and enhances rheology in salty, high-temp conditions.	[56]
PEG-SiO_2 NC + SDS	SiO_2, SDS	PV, AV, YP, GS	78–250 °F, 500 psi	At 1.5 g, boosted PV by 75%, YP by 100%, and AV by 81.8%.	SDS enhances dispersion and thermal stability.	[57]
PEG–SiO_2	SiO_2	PV, AV, YP, GS	78–250 °F, 500 psi	Slight decrease in AV/PV vs. base fluid, but good thermal performance.	Stable rheology at high temperature and pressure; compatible with mud systems.	[58]
PAAG NC	SiO_2	PV, AV, YP	25–180 °C, 500 psi	1.0 wt% improved rheology by 46–80%.	Good salt resistance and thermal stability up to 240 °C.	[59]
Hydrophobic polymer NC	Bentonite	PV, YP, GS	25 °C, 100 psi	PNCs improved viscosity and structure due to the interaction with bentonite.	Promotes uniform dispersion and consistent fluid behavior.	[60]
PP-SiO_2 vs. PHPA	SiO_2	AV, PV, YP, GS	25–150 °C, 500 psi	PNCs exhibited greater YP and thermal performance compared to PHPA.	Sustains properties under heat without degradation.	[61]
PVP/SWCNTS	CNTs	AV, PV, YP, GS	25 °C, 100 psi	Small drops in AV and PV; shear-thinning behavior.	Good mud stability with little performance change.	[62]
CuO/PAM NC	ZnO	AV, PV, YP, GS	25 °C, 100 psi	Big rheology improvements in fresh and salt water.	Stable and effective across fluid types.	[63]

Table 3. *Cont.*

Polymer Used	Nanofiller Used	Tested Rheology Parameters	Test Conditions	Main Findings	Benefits	Reference
XG–SiO_2	SiO_2	AV, PV, YP, GS	25–120 °C, 500 psi	Outperformed XG and other base muds.	Strong rheological performance even at HPHT.	[64]
PP-SiO_2–NH2	SiO_2	AV, PV, YP, GS	25 °C, 100 psi	Improved PV from 11 to 15 mPa·s and YP from 9 to 14 Pa.	Good thermal resistance and rheology stability.	[65]
PSt MMA nano clay	Clay (Laponite)	AV, PV, YP, GS	25 °C, 100 psi	AV/PV increased by 44–51%.	Stable up to 120 °C; enhances mud rheology.	[66]
Nano-LS-g-PAM-AMPS	SiO_2	AV, PV, YP, GS	200 °C, 600 psi	Strong AV/PV/YP increase with thermal/salt resistance.	Durable at high temperatures and salinity.	[67]
Laponite NC polymer	Laponite	AV, PV, YP, GS	Up to 260 °C	Maintains viscosity after 72 h of aging.	Extreme thermal stability.	[68]
Laponite copolymer	Laponite	AV, PV, YP	210 °C, 500 psi	Stable AV (19 cP), PV (15 cP), YP (8 lbf).	High-temp stability.	[69]
PEG–NS	SiO_2	PV, YP, GS	78 °F	PV improved by 14% and YP by 50%.	Stable up to 200 °C.	[70]
SMA/SiO_2	SiO_2	PV, YP, GS	78 °F	Minor rheology change; filtration control improved.	Reasonable fluid loss control.	[71]
CNT/ZnO/polymers	CNTs, ZnO	AV, PV, YP, GS	Up to 400 °F	Rheology remained stable post-aging.	High-temp durability.	[72]
Core–shell NC	SiO_2	PV, AV, GS	120 °C, 100 psi	Rheology close to commercial CMC.	Reasonable control and competitiveness.	[73]
MES/Polystyrene NC	Polystyrene	PV, AV, GS	150 °C, 500 psi	Exhibits shear thinning and uniform velocity.	Stable thermal and flow behavior.	[74]
TiN NPs in KCl	Titanium Nitride (TiN)	Cof, AV, PV, YP, GS	25 °C	YP improved by 122%, PV by 17%.	Good cutting, transport, and hole cleaning.	[75]
Hydrophobic PEG–SiO_2	SiO_2	PV, YP, GS	250 °F	AV/YP/PV all increased.	Improved NPs integration and fluid loss control.	[76]
PAM–PEG–SiO_2	SiO_2	PV, YP, GS	95 °C, 500 psi	AV and YP improved greatly.	Enhances stability up to 95 °C.	[77]

Table 3. *Cont.*

Polymer Used	Nanofiller Used	Tested Rheology Parameters	Test Conditions	Main Findings	Benefits	Reference
PAM–ZnO NC	ZnO	AV, PV, YP, GS	80–150 °F	18–17% increases in PV and YP.	Works well even at low dosages.	[78]
EDA–G graphene NC	Gr	PV, AV, YP, GS	150 °C, 500 psi	Good rheology; salt- and calcium-resistant.	Effective in extreme salinity.	[79]
SFDL core–shell NC	SiO_2	PV, AV, YP, GS, Cof	Up to 446 °F	Massive gains in AV, PV, and YP; Cof dropped significantly.	Prevents polymer breakdown at very high temperatures.	[80]
TiO_2–PAM NC	Titanium Dioxide TiO_2	PV, AV, YP, GS	25 °C	273% PV increase with 14 g PNCs.	Good cross-linking and high stability.	[81]
PAM/Al_2O_3 NC	Alumina (Al_2O_3)	PV, YP, AV, GS	150 °C, 500 psi	PV exceeded 300 cp at a 4 wt% concentration.	Long-term mud stability and compatibility.	[82]
ZnO–Clay NC	ZnO, Clay	PV, YP, AV, GS	110–370 °F, up to 18,500 psi	PV and YP were greatly enhanced at 370 °F.	Good under extreme HPHT.	[83]

2.2.2. PNCs for Controlling Filtrate Loss

Filtrate loss is one of the most critical challenges in drilling operations, affecting both wellbore stability and reservoir productivity. The process of filtrate loss occurs when the liquid phase of the drilling fluid seeps into the formation through the pore spaces. This results in the formation of a filter cake on the borehole wall, which acts as a seal, preventing further fluid infiltration. While the formation of the filter cake is a necessary function, excessive filtrate loss can result in thicker filter cakes, increasing the potential for operational issues such as pipe sticking, torque, drag, and the creation of tight holes [84].

The filter cake thickness (FCT) is directly related to the volume of filtrate lost during the drilling process. As more fluid is lost to the formation, the filter cake becomes thicker, and operational challenges become more pronounced. Excessive filtrate loss leads to increased fluid consumption, higher operational costs, and the potential for formation damage. Additionally, high filtrate loss leads to the contamination of formation fluids, which is detrimental to the reservoir's productivity. It is widely accepted that filtrate loss increases with rising temperature and decreasing viscosity [85,86]. When the viscosity of the fluid decreases, it becomes more challenging to control filtrate loss, especially under extreme conditions like HPHT. To address the challenges associated with filtrate loss, there has been an increasing interest in the development of PNCs as additives in drilling fluids Table 4.

Table 4. Current analysis of PNCs materials utilized as shale rheological properties, and viscosities in (WBMs) drilling.

Material Used (PNCs)	Polymer Used	Nanofillers Used	Key Findings	Shale Inhibition Mechanism	Reference
PNSC + NH_2	PNSC	Nanohydroxide (NH_2)	The percentage of shale retained after the hydration test was 75.3% with PNSC + NH_2, compared to 33.6% with base mud and 58.4% with KCl.	Uses physical plugging and hydrogen bonding to reduce water contact between the shale and the fluid.	[87]
PEG-NS	Polyethylene glycol (PEG)	SiO_2	PEG-NS exhibited larger plugging capacity and retarded shale hydration, with slower pressure growth than JHC.	Physical plugging and disruption of hydrogen bonds to reduce shale hydration and water absorption.	[88]
AR/SiO_2 NC	Acrylic Resin (AR)	SiO_2	Reduced API FL by 53.6% and API FCT by 23.1%, proving effective in inhibiting shale hydration.	Forms substantial physical barriers by sealing microcracks and nanopores in shale through AR/SiO_2's core–shell structure.	[89]

The large surface area and high surface reactivity of the nanoparticles promote uniform filter cake formation, a critical factor in minimizing filtrate loss. As shown in Figure 6, recent studies have employed advanced molecular design techniques to develop polyzwitterionic PNCs that function effectively as filtration-control additives in water-based drilling fluids under high-salinity and elevated-temperature conditions. The top portion illustrates the process of creating a zwitterionic polymer network (AASV) through the copolymerization of acrylamide (AM), AMPS, sodium styrene sulfonate (SSS), and a phenyl-methylene-imidazole zwitterionic monomer. Researchers synthesized a polymer that shows an important anti-polyelectrolyte effect, unlike regular polyelectrolytes. This polymer stabilizes the colloidal matrix, making it easier for water molecules to bind to it due to its higher ionic strength. Adding the phenyl-methylene-imidazole structure greatly stiffens the polymer backbone and enhances its ability to bind salt, resulting in increased stability at high temperatures [90]. This demonstrates, for instance, that prolonged exposure to high salinity and temperatures considerably reduces HTHP and API filtering losses. These molecular design strategies represent an important shift in the formulation of drilling fluids. They lay the groundwork for the scientifically sound development of next-generation additives tailored to the challenging conditions encountered in modern hydrocarbon exploration and extraction. Figure 6 is aschematic illustration depicting the synthesis and mechanism of zwitterionic copolymer (AASV) in enhancing the filtration control performance of drilling fluid under high-salinity conditions. Reproduced with permission [91]. Figure 6 provides a molecular-level view of how zwitterionic PNCs are designed. It demonstrates how ionic interactions enhance filtration control when salinity levels are high. This supports the idea that chemical design methods are just as important as physical plugging when it comes to preventing fluid loss.

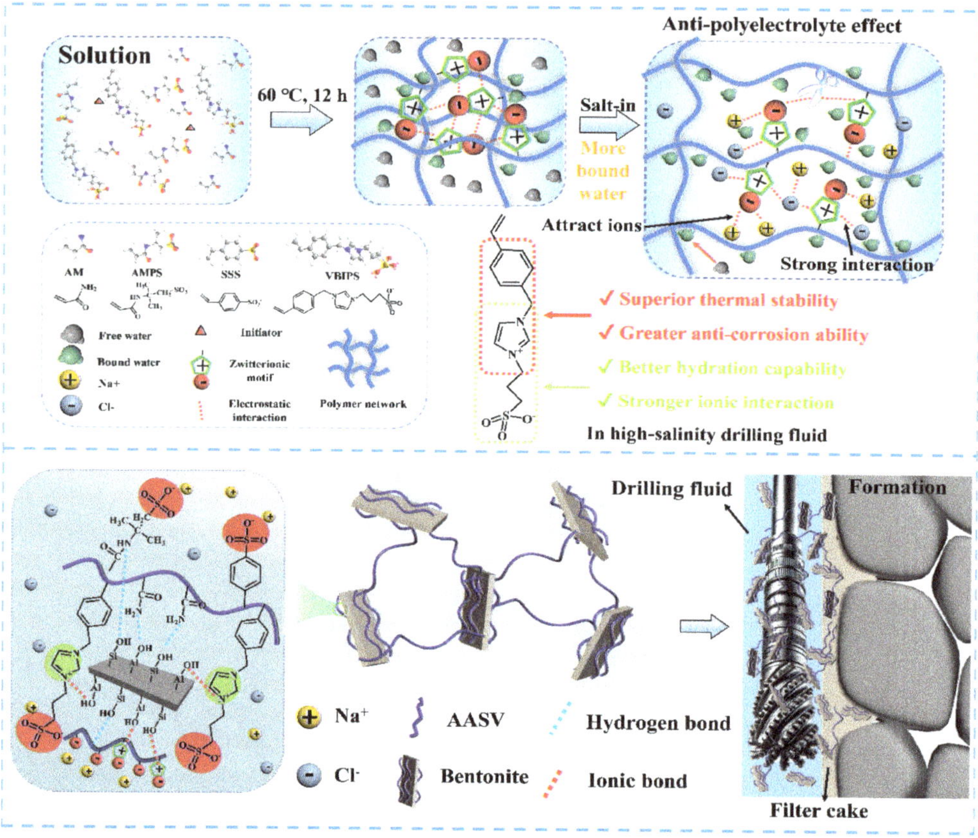

Figure 6. A schematic illustration depicting the zwitterionic copolymer (AASV) synthesis and its mechanism in enhancing the filtration control performance of drilling fluid under high salinity conditions. Reproduced with permission [92]. Copyright 2025, Elsevier.

In addition to enhancing filter cake formation, PNCs also improve the rheological properties of the drilling fluid. The polymer matrix in PNCs provides flexibility and adhesion to the drilling fluid, ensuring that the filter cake adheres firmly to the formation surface. The NPs further improve the adhesion properties of the fluid, enhancing the stability of the filter cake and preventing it from being displaced under the pressure exerted by the drilling fluid. This improvement in filter cake adhesion helps prevent the filter cake from breaking apart, which can lead to filtrate infiltration into the formation. By enhancing both the viscosity of the drilling fluid and the adhesion of the filter cake, PNCs help ensure that the filter cake remains stable and effective throughout the drilling process.

A recent study has highlighted the potential of PNCs in controlling filtrate loss and improving filter cake formation. Researcher [92] developed a PNCs additive for bentonite-based WBMs and sodium carbonate-based WBMs. The PNCs were synthesized using acrylamide, AMPS, dimethyl diallyl ammonium chloride (DMDAAC), and modified nano-laponite. The results of this study showed that the PNCs greatly reduced filtrate loss after thermal aging at 150 °C, with the fluid loss of the PNC-modified fluid being 6.4 mL, compared to 10.4 mL for the base bentonite-based fluid. This reduction in filtrate loss was attributed to the colloidal protection provided by the PNCs, which stabilized the clay particle aggregates and formed a denser filter cake that effectively sealed the formation and prevented further fluid penetration. Table 5 shows that nanoclay-based composites reduce the most filtrate, indicating that they could be beneficial in HPHT drilling.

Table 5. Recent studies demonstrate that adding PNCs to WBMs reduces fluid loss.

Materials Used	Test Conditions	Main Results	Reference
Used a polymer mix (AM, AMPS, DMDAAC) with nano-laponite	Tested in both freshwater and saltwater mud at 150–200 °C	2% PNCs reduced fluid loss in saltwater mud from 280 mL to 72 mL after heating for 16 h.	[92]
Combined salt-based polymer with GO	Concentrations of 0.1–0.5% at 78 and 250 °F	The blend improved fluid filtration performance at both temperatures.	[93]
Mixed polypropylene beads with silica	Used 8 g PP and 1 g silica at 78 and 250 °F	Reduced fluid loss by 24% and slightly decreased filter cake thickness.	[94]
Created a nanocomposite using AM, AMPS, NVP, and GO	Evaluated at 25, 150, and 180 °C	1% dosage maintained strong filtration control even with high salt content (up to 25%).	[23]
Blended PVP with CNTs	Tested 1% and 5% at 25 °C	Achieved 18–23% reduction in fluid loss.	[95]
Used polypropylene and PHPA with modified silica	Tested at 25 and 120 °C	Improved high-temperature stability and reduced fluid loss by about 18%.	[51]
Combined PAM-AMPS with nano-lignite	LPLT: 78 °F/HPHT: 356 °F	Aging produced good fluid loss results (7.1 mL LPLT, 30 mL HPHT)	[96]
Used PSt MMA and nano-clay	Tested at 78 and 250 °F	Reduced fluid loss by up to 65% under high-temperature conditions.	[97]
Developed a copolymer with laponite	HPHT at 356 °F and 500 psi	Halved the fluid loss from 34 mL to 16 mL.	[8]
Combined PAM with copper oxide	Used 1–10 g doses at 25 and 120 °C	Lowered fluid loss and produced a thinner filter cake.	[98]

Table 5. *Cont.*

Materials Used	Test Conditions	Main Results	Reference
Created a nanocomposite with PEG and nano-silica	1% tested at 78°f	Cut fluid loss by over 15%.	[34]
Mixed methyl ester sulfonate with nano-polystyrene	Tested at both low (78 °F) and high (250 °F) temperatures	Reduced fluid loss by 22% at LPLT and 61% at HPHT due to a synergistic effect.	[34]
Blended SMA with silica	2% tested at 78 °F	Reduced API fluid loss by 22%.	[99]
Combined CNT, ZnO, and synthetic polymers	Tested up to 400 °F	Maintained stable mud and created thin, less permeable filter cakes.	[100]
Used carboxymethyl cellulose with polystyrene	LPLT test at 25 °C	It resulted in the lowest fluid loss volume in this category.	[101]
Combined PEG and silica	Tested at 78 and 250 °F	2% dosage performed well even after thermal aging.	[102]
Used PAM with ZnO	Tested at 78 °F and 150 °F	Reduced fluid loss by 12.7% (LPLT) and 23% (HPHT).	[21]
Created PAM-silica nanocomposite	0.1–1.0% tested at up to 203 °F	0.7% dosage provided stable and effective filtration.	[103]
Used AM, AMPS, MA, and St with silica	Tested from 78 °F to 446 °F	Withstood extreme conditions and effectively controlled fluid loss.	[104]
Blended PAM with titanium dioxide	Doses from 1–14 g at 25 °C	Reduced fluid loss from 53 mL to 19 mL at the highest dose.	[105]

When comparing different types of PNCs, nanoclay- and GO-based ones lose the most fluid at high temperatures. Silica- and polymer-blend composites, on the other hand, show more moderate gains but make the filter cake more stable. The results show that single nanofillers can help limit fluid loss, while hybrid systems (such as PAM–GO and PAM–ZnO) work more effectively to improve both rheology and filtrate control in a balanced manner. Therefore, the integration method, whether it involves hybridization or surface modification, appears to be the most crucial factor in achieving optimal performance from drilling fluid. By creating thin, low-permeability filter cakes, PNCs effectively prevent filtrate loss. Nanoclay- and GO-based systems work best when the temperature is high. Hybrid composites, on the other hand, increase both rheology and filtration in a balanced way. One important problem is that there hasn't been enough field-scale testing, which highlights the need for pilot studies in genuine drilling operations.

In another study, researcher [106] synthesized a GO-PAM NC and tested it in WBMs. The results of this study demonstrated that incorporating GO-PAM NC into the drilling fluid improved the rheological properties of the fluid and reduced filtrate loss compared to using GO nanosheets. The study found that the PAM matrix enhanced particle adhesion, while the GO NPs sealed the pore spaces in the filter cake, preventing mud cake erosion and reducing filtrate loss. The incorporation of GO-PAM NC was shown to minimize mud cake erosion by over 60% under dynamic shear conditions, demonstrating the PNC's effectiveness in improving filter cake stability at high shear rates. Figure 7 shows a diagram of hyperthermal drilling fluids that contain nano-scale polymer-modified drilling additives (Nano PM-DAD). It also demonstrates their stability at high temperatures, their interaction with each other, and their interaction with the wellbore. The diagram illustrates the effects on several parts of the left side when they are exposed to heat at 200 °C for an extended period. Bentonite particles tend to clump together and lose their colloidal stability and performance if they are not appropriately handled. PM-DAD particles, on the other hand, are prone to breaking down when exposed to such high temperatures. The smaller particle size and higher temperature resistance of Nano PM-DAD, on the other hand, allow bentonite platelets to intercalate. Using this intercalation process to make composites makes them less likely to break down or adhere to each other when exposed to heat. This makes them ideal for use in high-temperature environments [95].

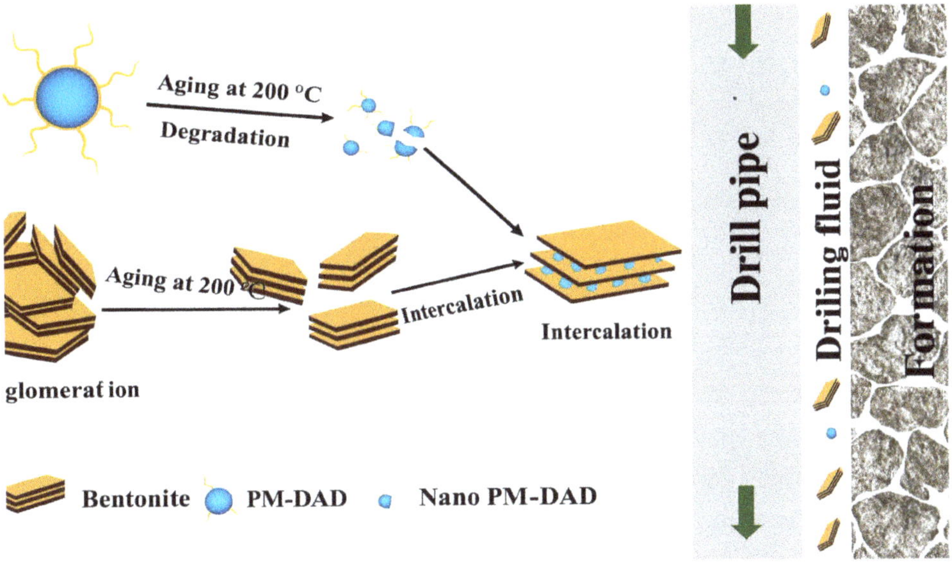

Figure 7. Schematic diagram of ZSHNM filtration control mechanism. Reproduced with permission [107] Copyright 2024, Elsevier.

Further research by [108] tested the filtrate loss control capabilities of a Sodium Dodecyl Sulfate (SDS)-modified polyethylene–nanosilica composite (PEG-SiO_2 NC) in WBMs. The study found that the addition of the PNCs resulted in greatly reductions in both filtrate loss and filter cake thickness (FCT). The optimal concentration of the PNCs was found to be 1.5 g, which formed a low-permeability filter cake that effectively sealed the formation and prevented further penetration of the filtrate. The results of this study confirm the ability of PNCs to provide pore-sealing integrity and improve fluid loss control under both API and HPHT conditions, highlighting the importance of PNCs in high-pressure and high-temperature drilling environments (Figure 7).

Another benefit of PNCs is their chemical resistance, which enables them to perform well in environments with high salinity, contamination, or unconventional drilling conditions (Table 6). The high surface area of the NPs enhances the dispersion and stability of the PNCs, ensuring that they remain active and effective throughout the drilling process. This ability to perform well in challenging environments makes PNCs a highly versatile additive for improving fluid loss control in various drilling applications.

Table 6. Current analysis of PNCs materials utilized as fluid loss agents and viscosities in WBM drilling.

Material Used (PNCs)	Polymer Used	Nanofillers Used	Key Findings	Shale Inhibition Mechanism	Reference
PAM-SiO_2 NC	PAM	SiO_2	Achieved the highest shale stability (86.6%) over 48 h, outperforming commercial inhibitors (74.7%) and KCl (49.2%).	Disrupts hydrogen bonds in water through PAM-SiO_2 NC's OH groups and physically blocks microcracks and nanopores in shale.	[109]
C-g-AA-NH_2	Acrylic acid (AA)	Aminomethyl, Amyl	Achieved a high anti-swelling ratio of 95.2% with AA-AAm-C-Amyl, surpassing KCl (55.4%).	Combines carbon core nanopore plugging and strong polymer adsorption to inhibit swelling.	[110]
AP-ZnO NC	Acrylic Polymer (AP)	ZnO	Shale recovery improved to 97% with AP-ZnO NC compared to 81% with BF; the swelling rate reduced from 14.29% to 6.69%.	ZnO reduces mud invasion by plugging pores and cracks, while AP encapsulates the shale surface, preventing water interaction.	[111]

2.2.3. PNCs for Inhibiting Shale Swelling

Shale instability due to swelling remains an important challenge in drilling operations. Percentage of shale retained after hydration test: Sensitive clay minerals in shale formations, such as smectite, montmorillonite, and kaolinite, exhibit high hydration potential when exposed to water-based drilling fluids, resulting in rapid volumetric expansion and loss of structural integrity. These minerals have a high affinity for water. When they encounter water-based drilling fluids, they absorb water, causing the shale to swell and potentially leading to operational problems such as pipe sticking, formation damage, and wellbore instability. The problem is particularly severe in tight drilling windows in depleted reservoirs or deep-sea drilling operations, where even minor formation issues can have substantial consequences. The primary mechanism through which PNCs inhibit shale swelling is by increasing the hydrophobicity of the shale and physically blocking pore throats in the shale's structure. PNCs consist of a polymer matrix and NPs, such as SiO_2, GO, CNTs, and CNCs [23]. These NPs are highly reactive and possess an important surface area, allowing them to interact directly with the shale's surface, reduce water absorption, and prevent the shale from swelling. When PNCs are added to the drilling fluid, they adhere to the surface of the shale, forming a protective layer that prevents water from penetrating the shale and causing it to swell.

Shale Swelling Mechanisms and Mineralogical Basis

The hydration and swelling of clay minerals, such as smectite and montmorillonite, primarily cause the instability of shale. These minerals have a layered structure with voids between the layers that can hold water. These clays have a high cation exchange capacity (CEC), which means that interlayer cations (Na^+, K^+, Ca^{2+}) can swap places with ions in the drilling fluid. This causes the volume to grow. When polar water molecules get between the layers, they break down the electrostatic interaction between the sheets and push them apart. An osmotic imbalance between the clay pore fluid and the external drilling mud also causes water to flow in, exacerbating the swelling. The effects of surface charge are also important, clay platelets with a negative charge attract hydration shells and more cations, which makes dispersion and structural weakening worse. WBMs do not restrict shale formations from being water-sensitive, while PNCs physically block nanopores and produce hydrophobic barriers to prevent hydration, osmotic, and ion exchange. This makes them a stronger and more chemically compatible solution for formations that are prone to shale.

Different types of nanoparticles interact with shale minerals in various ways. Silica nanoparticles (SiO_2) mainly stop shale from getting wet by neutralizing the charges on the surface. Their negatively charged surfaces can interact with the diffuse double layer that surrounds clay platelets. This compresses the layer and lowers the repulsive forces that cause interlayer swelling. GO renders clay surfaces hydrophobic due to its amphiphilic nature. This makes it harder for water to stick to the surface by forming a barrier layer that prevents hydration shells from penetrating the interlayers. Carbon nanotubes (CNTs) have a high aspect ratio and a tubular shape, which enables them to bridge. This makes it harder for water to get in. These mechanisms work together to improve the polymer matrix: the polymer adheres to the nanoparticle surface, facilitating its adhesion, while the nanoparticles activate the prevention of hydration, ion exchange, and fluid invasion. This synergy is why PNCs work better than single-component additions in strata that are prone to shale.

By placing all the shale-related failures (instability, pipe sticking, collapse) here, rather than in the introduction, the discussion clarifies that the hydration process directly causes these operational problems. The PNCs that are offered (Figure 8) show how filter cakes form in shale strata that have been treated with PNCs. This diagram is closely related to the conversation about how to prevent shale from expanding, as it illustrates how interactions between nanoparticles and polymers block pore mouths and maintain formation stability.

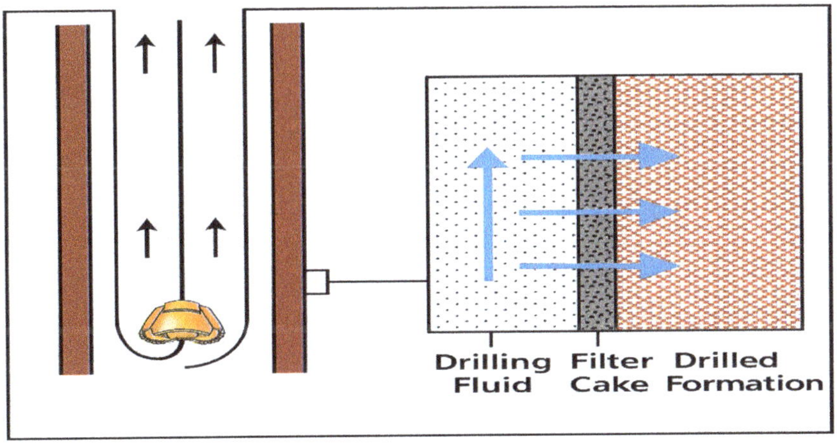

Figure 8. Filter cake formation in drilled formations by drilling fluid [23].

Among the various nanoparticles utilized for stabilizing bentonite in high-temperature drilling fluids, nano-silica (SiO_2 nanoparticles) is particularly notable. When incorporated as a polymer-modified additive, nano-silica not only resists thermal degradation but also efficiently intercalates between bentonite layers, maintaining colloidal stability and improving filtration control in challenging downhole environments. Additionally, the polymer matrix in PNCs serves to stabilize the NPs, ensuring they remain evenly dispersed in the drilling fluid. This ensures that the PNCs can continue to adhere to the shale surface and provide effective shale stabilization over time. By stabilizing the NPs, the polymer matrix also helps prevent nanoparticle aggregation, which could reduce the effectiveness of the PNCs in stabilizing the shale.

Recent studies have demonstrated the efficacy of PNCs in inhibiting shale swelling and stabilizing shale formations during drilling operations. Resercher [112] developed a PNCs additive for bentonite-based WBMs and sodium carbonate-based WBMs. The PNCs were synthesized using acrylamide, AMPS, dimethyl diallyl ammonium chloride (DMDAAC), and modified nano-laponite. The study revealed that the addition of PNCs improved the fluid retention and reduced the filtrate loss. The PNCs enhanced the structural stability of the shale by reducing shale swelling and mud cake erosion. When tested at 150 °C, the PNCs reduced the filtrate loss from 10.4 mL to 6.4 mL, demonstrating the effectiveness of PNCs in stabilizing shale and reducing filtrate loss in high-temperature environments.

Another promising study by [113] focused on a Sodium Dodecyl Sulfate (SDS)-modified polyethylene–nanosilica composite (PEG-SiO_2 NC). The results showed that these PNCs reduced filtrate loss and filter cake thickness (FCT), while improving the percentage of shale retained after the hydration test. The PNCs formed a low-permeability filter cake that effectively sealed the formation and prevented further penetration of the filtrate. The study indicated that the PNCs enhanced the hydration resistance of the shale and improved the overall shale stabilization in WBMs. This study further corroborates the role of PNCs in shale stabilization and swelling inhibition during drilling operations [114].

Hydrophobicity is a critical factor in shale swelling inhibition. Hydrophobic NPs such as SWCNTs can reduce shale swelling by repelling water and preventing water absorption into the shale matrix. In addition to hydrophobicity, PNCs also act through physical blocking mechanisms. NPs in the PNCs can physically block the pores and microcracks in the shale, preventing water from entering the shale structure. SWCNTs/PVP (single-walled carbon nanotubes and Polyvinylpyrrolidone) nanocomposites, for instance, exhibit a higher ability to stabilize shale by adsorbing onto the shale surface and blocking microcracks. FTIR analysis of SWCNTs/PVP-treated shale showed too much reduction in the -OH stretch peak, indicating decreased water adsorption and increased hydrophobicity (Figure 9). These findings suggest that PNCs with hydrophobic properties can effectively reduce shale swelling and enhance shale stability in water-based drilling fluids. Table 7 shows that graphene-based nanocomposites always work better than silica-based systems at stopping shale from swelling because they are hydrophobic.

Mechanism of Shale Swelling

Shale formations, which contain an important proportion of clay minerals, are prone to swelling when exposed to water-based drilling fluids (WBDFs). The primary mechanism driving this swelling is the interaction between water and the clay minerals, especially montmorillonite, which is highly hydrophilic. These clay minerals have a layered structure that contains cations (such as sodium, calcium, or potassium) in the interlayer spaces. When exposed to water, these cations are hydrated, causing an increase in the distance between clay layers, a process known as crystalline swelling.

There are two main types of swelling mechanisms in clay minerals: crystalline swelling and osmotic swelling. In crystalline swelling, water molecules form layers around the cations in the interlayer spaces, increasing basal spacing. This expansion is typically limited to a few layers of water due to electrostatic repulsion between the layers. Osmotic swelling occurs when there is a difference in ionic concentration between the interlayer cations and the surrounding water. This leads to the diffusion of water molecules into the interlayer space to balance the ion concentration, greatly increasing the swelling pressure between the layers.

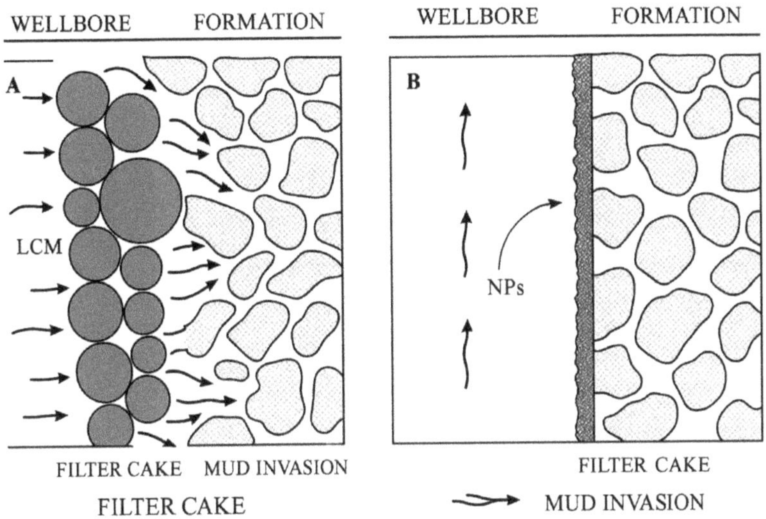

Figure 9. A schematic illustration of mud losses during drilling in the scenarios of (**A**) conventional Lost Circulation Material (LCM) and (**B**) Non-Newtonian Polymer (NP). Reproduced with permission [7] (Copyright 2015, Springer Nature).

The swelling of shale can have severe consequences for wellbore stability during drilling operations. As the shale absorbs water, the volume of the clay minerals increases, causing mechanical failure in the wellbore wall. This can lead to borehole collapse, sloughing, or excessive enlargement of the hole, all of which complicate drilling operations. Furthermore, the dispersion of fines from the swollen shale can contaminate the drilling fluid, degrade its performance and increase the risk of lost circulation or stuck pipe incidents.

To mitigate the adverse effects of shale swelling, chemical inhibitors are incorporated into drilling fluids. These inhibitors work by modifying the interaction between water and clay minerals, either by encapsulating clay particles to prevent water absorption or by neutralizing the surface charge of the minerals to reduce their tendency to swell. Advanced inhibitors, such as ionic liquids (ILs) and deep eutectic solvents, offer enhanced inhibition capabilities. These compounds function through hydrogen bonding, electrostatic attraction, and hydrophobic shielding, providing a more effective barrier against water penetration and reducing swelling pressures.

Cationic organic polymers (COPs) notably quaternary polyamines and quaternary polyacrylamides adsorb multipoint onto negatively charged clays, immobilizing PNCs and forming a barrier to water ingress. Hyperbranched, amine-terminated designs enhance binding via protonated primary amines, suppressing Na-bentonite hydration and dispersion. Their cationic nitrogen anchor to clay while hydrophobic backbones create a network that limits fines migration and reduces water uptake, outperforming single-cation inhibitors in WBDFs.

Graphene- and CNT-based PNCs are very good at stopping shale from expanding because they repel water and plug pores. Silica-based PNCs are cheap, but they don't work as well. Hybrid systems that mix polymers with oxide or carbon fillers are more stable. Even though the lab results were good, there is still some doubt about how stable the system will be in the long run when conditions change downhole.

PNCs offer a promising solution for inhibiting shale swelling and stabilizing shale formations during drilling operations. By incorporating NPs such as SiO_2, GO, CNTs, and CNCs into a polymer matrix, PNCs improve the hydrophobicity of the shale and reduce water absorption, thereby preventing shale swelling. The ability of PNCs to block pores, seal microcracks, and increase shale stability makes them a valuable additive in water-based drilling fluids. As drilling conditions become increasingly challenging, the use of PNCs will continue to play a crucial role in improving shale stabilization, reducing filtrate loss, and enhancing the overall performance of drilling fluids. Future research and development of PNC formulations will continue to provide more effective and cost-efficient solutions for shale swelling inhibition and wellbore stabilization.

Comparative analyses indicate that graphene- and CNT-based nanocomposites exhibit the most effective long-term suppression of shale swelling, attributed to their hydrophobic characteristics and pore-blocking capabilities. Silica-based composites, on the other hand, provide moderate inhibition but are more chemically stable and less expensive. Hybrid PNCs, such as ZnO or GO–PAM systems, achieve both objectives: they enhance shale stability while maintaining control over filtration. This comparative research supports the perspective that hybrid nanocomposites represent the most advantageous pathway for practical field applications.

2.2.4. PNCs for Stabilizing the Wellbore

Wellbore instability is a common and costly issue encountered during drilling operations, especially in challenging environments such as deep-sea drilling or depleted reservoirs. The primary cause of wellbore instability is the reaction between water-sensitive clay minerals in shale formations and the drilling fluids, which leads to problems like sloughing, pipe sticking, formation damage, and wellbore collapse. These issues can result in the loss of circulation, poor wellbore cleaning, and, in extreme cases, formation failure. Therefore, stabilizing the wellbore is crucial to ensure the integrity of the drilling operation. To address these challenges, PNCs have emerged as effective additives in drilling fluids, offering solutions for wellbore stabilization. Figure 9 illustrates how traditional loss circulation materials (LCMs) and nanoparticle-based filter cakes effectively seal and filter drilling fluids. At the contact between the wellbore and the formation, conventional LCMs typically form a rough filter cake with large, irregular particles. This filter cake allows a good amount of muck to enter the formation's porous structure due to the relatively large particles, which also result in poor sealing. Hydrocarbon exploration is inherently risky, especially in complex conditions were insufficient plugging can facilitate fluid escape and reduce wellbore stability. Figure 9 shows the utilization of NPs in the drilling fluid system, including nano-silica (SiO_2 NPs), nano-titania (TiO_2 NPs), and GO nanosheets. The high surface area and diminutive size of these NPs allow them to penetrate even the most microscopic pore spaces found in the formation. The final product is a filter cake with reduced mud invasion, thanks to its thinner, more compact, and less porous structure. This nanoscale sealing improves wellbore stability, decreases fluid loss, and is most effective in environments with high salt, heat, and pressure. This illustration demonstrates that utilizing drilling fluids reinforced with nanoparticles can enhance well control and operational safety in contemporary hydrocarbon exploration. Figure 9 shows the difference between traditional loss circulation materials and nanoparticle-based PNCs. It demonstrates why nanoscale sealing results in cakes that are thinner and less permeable, thereby making the wellbore more stable.

Table 7. Current analysis of PNCs materials utilized as shale swelling inhibitors and viscosifiers in WBM drilling, including thermal stability thresholds and quantitative swelling inhibition metrics.

Material Used (PNCs)	Polymer Used	Nanofillers Used	Thermal Stability Threshold (°C)	Swelling Inhibition/Shale Recovery Metrics	Key Findings	Shale Inhibition Mechanism	Reference
PEI–Gr NC	Polyethyleneimine (PEI)	Graphene (Gr)	180 °C	Shale recovery 65%; swelling reduction ≈ 45% vs. base mud	Maintained stability under HPHT; enhanced hydrophobic sealing	PEI interacts with –OH groups on the shale surface; Gr creates a barrier layer blocking microcracks	[115]
Glu–Gr NC	Glutamic acid (Glu)	Graphene (Gr)	200 °C	Swelling reduction ≈ 52%; shale recovery 70%	Long-term shale stability after 48 h aging at 150 °C	Hydrogen bonding binds Gr to shale; intercalation prevents water penetration into nanopores	[116]
PNS Latex	Polymer (PNS)	Latex NPs	160 °C	API FL reduced from 24 mL to 10.8 mL; shale recovery ≈ 60%	Greater thermal resistance and filtrate control under HPHT conditions	Adsorbs on shale microstructures, plugging cracks and pores	[117]
PAM–SiO_2 NC	Polyacrylamide (PAM)	SiO_2	210 °C	Swelling inhibition ≈ 70%; recovery 86.6% over 48 h	Outperformed commercial inhibitors (KCl 49.2%, base mud 74.7%)	Physical blocking of nanopores and hydrogen bond disruption by SiO_2–OH groups	[118]
C-g-AA–NH_2 NC	Acrylic acid (AA)	Aminomethyl, Amyl	190 °C	Anti-swelling ratio 95.2%; recovery > 80%	Effective plugging and surface adsorption at moderate HPHT	Carbon-core nanopore plugging and strong polymer–shale adsorption	[119]
AP–ZnO NC	Acrylic Polymer (AP)	ZnO	220 °C	Swelling rate reduced from 14.3% to 6.7%; recovery 97%	Good thermal endurance and inhibition under saline conditions	ZnO plugs pores and cracks; AP encapsulates the shale surface and reduces water contact	[120]

Table 7. *Cont.*

Material Used (PNCs)	Polymer Used	Nanofillers Used	Thermal Stability Threshold (°C)	Swelling Inhibition/Shale Recovery Metrics	Key Findings	Shale Inhibition Mechanism	Reference
GO–PAM NC	Polyacrylamide (PAM)	GO	230 °C	Swelling reduction ≈ 60%; shale recovery 90%	Retained integrity after 150 °C aging; thin filter cake formation	GO sheets provide a hydrophobic barrier and improve shale cohesion	[121]
SWCNTs/PVP NC	Polyvinylpyrrolidone (PVP)	SWCNTs	240 °C	Swelling reduction ≈ 68%; FTIR shows reduced –OH absorption	Outstanding shear and thermal resilience under HPHT conditions	Hydrophobic shield and micro-pore plugging mechanism	[122]

Several studies have demonstrated the efficacy of PNCs in stabilizing wellbores during drilling operations. Dong et al. [123] developed a PNC additive for bentonite-based WBMs and sodium carbonate-based WBMs. The PNCs were synthesized using acrylamide, AMPS, dimethyl diallyl ammonium chloride (DMDAAC), and modified nano-laponite. The study demonstrated that the PNCs enhanced wellbore stability by improving the structural integrity of the shale and forming a dense filter cake that prevented the infiltration of drilling fluid into the formation. After thermal aging at 150 °C, the PNCs reduced the filtrate loss from 10.4 mL to 6.4 mL, demonstrating their effectiveness in stabilizing the wellbore under high-temperature conditions. Figure 10 highlights wellbore stabilization strategies with PNCs, showing fracture sealing at the penetration point.

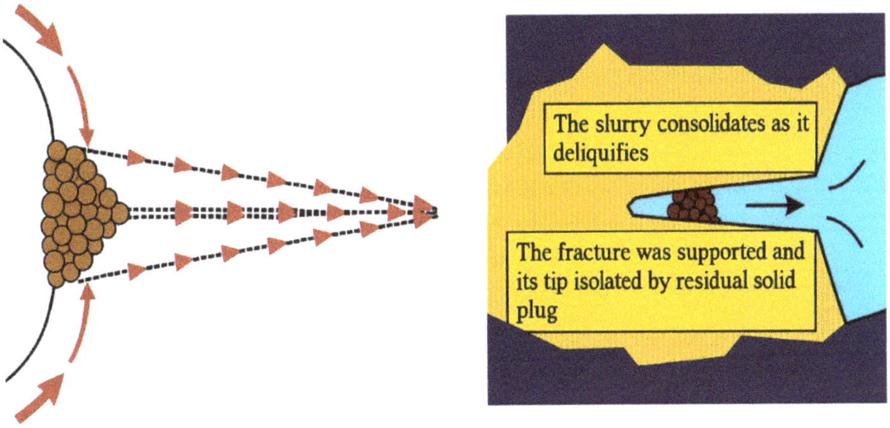

Figure 10. Wellbore stabilizers may separate the wellbore from the fracture at the penetration. Reproduced with permission [124] or permit the fracture to enter the wellbore.

In a similar study, researcher [125] synthesized a GO-PAM NC and tested its wellbore stabilization properties. The results showed that incorporating GO-PAM NC into the drilling fluid significantly improved the rheological properties of the fluid and enhanced wellbore stability. The GO-NPs helped seal the pores in the shale, preventing water penetration and swelling. At the same time, the polyacrylamide (PAM) matrix increased the adhesion between the NPs and the shale surface. This study indicated that PNCs effectively stabilize the wellbore by creating a mechanical barrier that prevents the drilling fluid from infiltrating the shale and protects the formation from damage.

The hydrophobicity of PNCs plays a key role in stabilizing the wellbore, particularly in formations with water-sensitive shale. Hydrophobic NPs such as SWCNTs can reduce shale swelling by preventing water from entering the shale. For instance, the SWCNTs/PVP (Polyvinylpyrrolidone) nanocomposite has demonstrated decent performance in wellbore stabilization. Fourier-Transform Infrared Spectroscopy (FTIR) and Field-Emission Scanning Electron Microscopy (FESEM) analysis showed that the SWCNTs/PVP nanocomposite reduced the -OH stretch peak in the FTIR spectra, indicating a decrease in water adsorption and an increase in the hydrophobicity of the shale surface (Figure 10). This increase in hydrophobicity helps to reduce shale swelling and maintain wellbore stability by preventing the shale from absorbing water, which would otherwise lead to wellbore instability [126,127].

In addition to hydrophobicity, PNCs also prevent wellbore instability through physical blocking mechanisms. NPs in the PNCs can block the pores and microcracks in the shale, physically preventing water from penetrating the shale and causing it to swell. For example, PNSC + NH_2 (Polymer Nanocomposite with Amine Functional Groups) has been shown to effectively plug pores in the shale and prevent water absorption. The hydrophobic shield created by PNCs + NH_2 reduces shale swelling and enhances wellbore stability by limiting water interaction with the shale.

PNCs offer several advantages over traditional wellbore stabilizers in drilling fluids. The dense filter cake formed by PNCs prevents fluid loss and protects the formation from damage. By sealing microcracks and pore throats in the shale, PNCs reduce the risk of shale swelling, formation damage, and wellbore collapse. The NPs in PNCs enhance the mechanical strength of the drilling fluid, making it more effective in stabilizing the wellbore under extreme conditions, such as HPHT environments. PNCs also improve the rheological properties of the fluid, ensuring that it maintains its flow characteristics and particle suspension capabilities even under challenging conditions.

Additionally, PNCs can be customized to fit specific wellbore conditions, including shale type, temperature, and salinity. The polymer matrix in PNCs allows for modification of the fluid's viscosity, yield point, and filtration properties to suit the specific needs of the drilling operation. This versatility makes PNCs an ideal solution for wellbore stabilization in various drilling environments. Researchers increasingly utilize PNCs in drilling fluids to stabilize shale during wellbore development. These nanocomposites block pore throats, generate internal filter cakes, and reduce fluid incursion, lowering wellbore shale permeability. PNCs increase shale's hydrophobicity, limiting cation exchange with WBMs, which causes expansion and instability. PNCs can intercalate in the shale's clay matrix, reducing its expansion and increasing structural stability [46]. A nanocomposite of SWCNTs and PVP was produced and named SWCNTs/PVP NC. They examined the nanocomposite's swelling inhibition mechanism and interaction with the shale core using FESEM and FTIR. In FTIR, SWCNTs/PVP-treated shale had a greatly lower -OH stretch peak region than untreated shale (Figure 11). The brown peak indicates decreased water adsorption and higher hydrophobicity. However, the untreated shale pellet had a larger -OH stretch peak, indicating hydrophilia. These findings demonstrate the potential of SWCNTs/PVP to stabilize shale and recommend its application in semi-hydrophobic drilling fluids. Figure 11 confirms through FTIR analysis that SWCNTs/PVP nanocomposites reduce water adsorption on shale, supporting hydrophobic stabilization mechanisms.

PNCs offer a highly effective solution for wellbore stabilization during drilling operations. By combining polymers with NPs, PNCs provide enhanced mechanical strength, thermal stability, and chemical resistance. The hydrophobic properties and pore-blocking mechanisms of PNCs help prevent shale swelling and maintain wellbore integrity. In contrast, their filter cake-forming ability prevents fluid loss and protects the formation from damage.

Another issue is the shear resistance of polymer nanocomposites, as drilling fluids undergo high shear forces near the drill bit. If these forces shatter nanocomposites, their efficiency may decrease. Researchers are developing shear-resistant nanocomposites that can withstand tremendous shear stress without degrading. To ensure drilling success, nanocomposites can be made more shear-resistant.

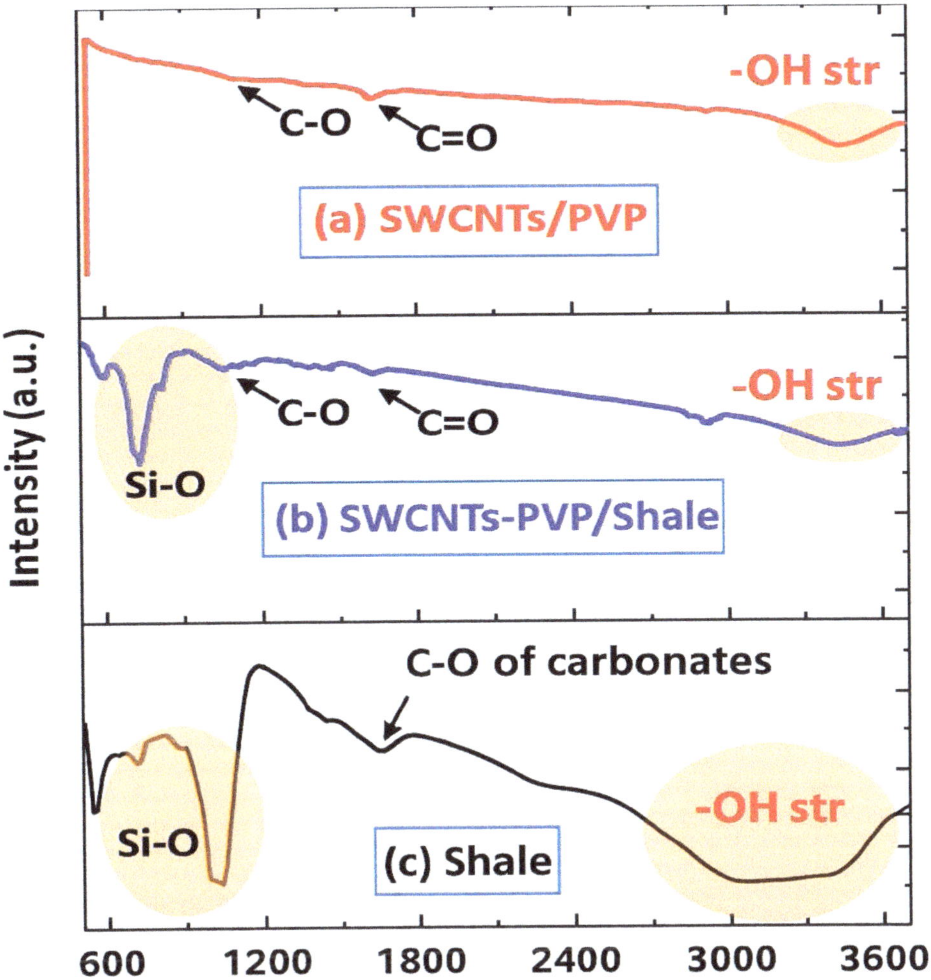

Figure 11. FTIR spectra of (**a**) SWCNTs/PVP nanocomposite, (**b**) SWCNTs/PVP nanocomposite changed with shale, and (**c**) shale pellet that has not been treated. Reproduced with permission [128]. Copyright 2020, Elsevier.

Despite these challenges, PNCs can improve wellbore stabilization. As drilling processes become more complex, nanocomposites will be essential for efficient and reliable drilling. Future studies should focus on addressing cost, environmental impact, and shear resistance to maximize the potential of these materials. Nanocomposite technology enhances performance, affordability, and environmental sustainability, making wellbore stabilization a promising prospect. The structural examination of shale samples treated with SWCNTs/PVP nanocomposites and their effects on the surface characteristics of the shale. Figure 12 illustrates that the imaging approach used for powder sample preparation on a copper stub, devoid of sonication, elucidates the occurrence of SWCNT agglomerates [106,129,130]. The behavior of carbon nanotubes, including van der Waals forces and dipole–dipole interactions, may explain these agglomerates at high magnification. Figure 12 shows SWCNTs as a coherent network on the surface. After disintegrating in water, untreated shale surfaces are rough and porous, with minute particles visible at low resolution. The shale's holes indicate its vulnerability to water absorption. The SWCNTs/PVP nanocomposite's protective layer makes the shale's surface smooth and less porous, minimizing water exposure. This change decreases swelling and dispersion, stabilizing the WBMs shale. The experimental data in Figure 12 show that SWCNTs/PVP NC efficiently obstruct nano-micro-pores, improving shale water resistance and structural integrity during drilling. Figure 12 illustrates, via FESEM, how SWCNT/PVP nanocomposites form a protective layer over shale, thereby reducing porosity and enhancing structural stability.

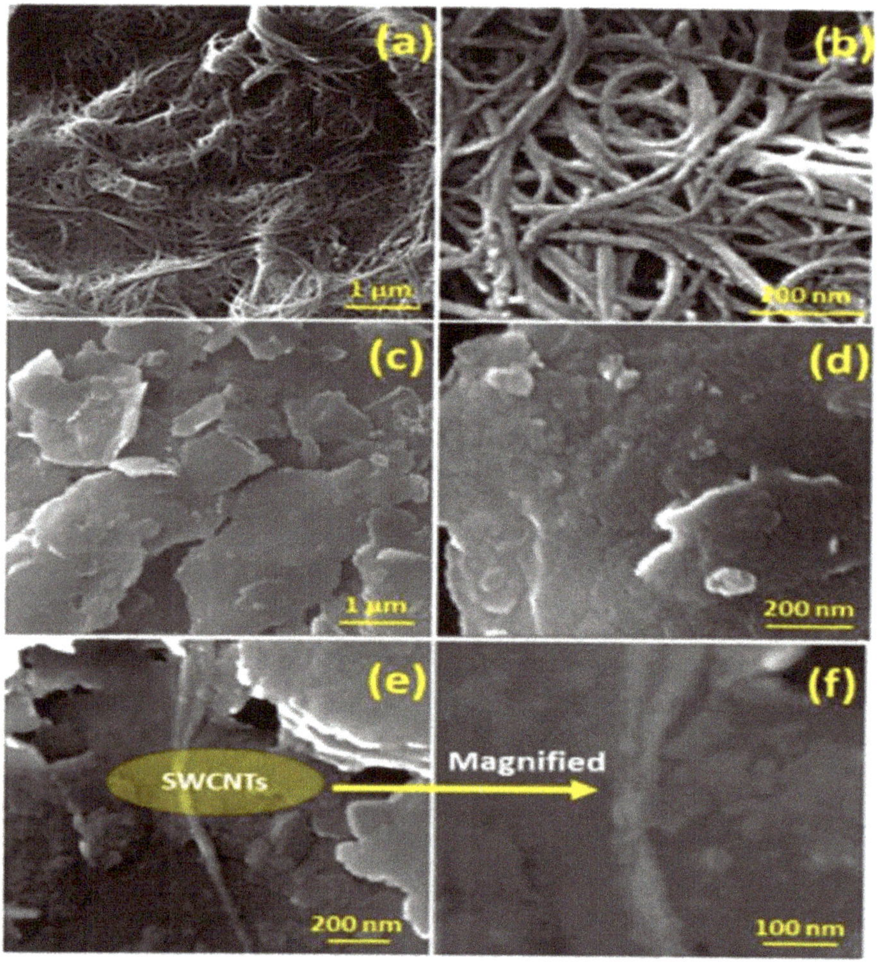

Figure 12. Field emission scanning electron micrographs of (**a**,**b**) pure SWCNTs, (**c**,**d**) untreated shale sample, and (**e**,**f**) SWCNTs/PVP with treated shale. Reproduced with permission [128]. Copyright 2020, Elsevier.

In a similar work, researcher [128] examined the use of modified polyvinyl alcohol (PVA) nanocomposites to enhance shale stability in WBMs. The findings indicated that incorporating a 1.5 w/v% PVA-based nanocomposite markedly improved the inhibitory characteristics of the base fluid. The swelling index of shale decreased by almost 85% after 24 h of contact with the nanocomposite-amended drilling fluid, in contrast to the base fluid, which exhibited an elevated swelling rate. The PVA-based nanocomposite was observed to create a protective layer on the shale surface, restricting the infiltration of water molecules [129]. The barrier formation primarily occurred due to hydrogen bonding between the hydroxyl (-OH) groups on the shale and the PVA chains, which reduced water absorption on the shale particles. This interaction reduced the swelling and diffusion of clay particles in the wellbore.

Besides, at a high temperature (250 °F), the stability of the PVA nanocomposite was maintained, proving its efficacy in adverse conditions. The addition of the nanocomposite did not alter the rheological properties of the drilling fluid, as indicated by the marginal increase in plastic viscosity (PV) and yield point (YP). The results are potent evidence for the use of PVA nanocomposites as a viable shale improvement in WBMs. The research demonstrated that the nanocomposite's protective layer can reduce fluid loss and improve the overall efficacy of the drilling mud under challenging conditions.

Both hydrophobic PNCs (like CNT- and graphene-based ones) and oxide-based systems (like silica and titania) can help stabilize wellbores. Oxides make things more stable thermally and structurally, while hydrophobic nanocomposites prevent swelling. The choice depends on the type of formation and the drilling conditions. Cost, environmental problems, and shear resistance are still issues that need to be solved before widespread use can happen.

The effectiveness of AA-AAm-C-Amyl composites, which consist of acrylic polymer and amyl ester-activated carbon, as shale inhibitors. Study shows that the wellbore was successfully stabilized by both the somewhat hydrophobic AA-AAm-C-Amyl composite (with shale recovery capacities of 95.2% and 97%) and the very hydrophobic AA-AAm-OD-C-Amyl composite (with a 93.7% anti-swelling ratio). The composites were evaluated against KCl, a widely used shale swelling inhibitor, and showed bigger efficacy in mitigating shale hydration. The hydrophobic characteristics of the AA-AAm-OD-C-Amyl composite enhanced its effectiveness in reducing shale swelling. However, the C-g-AA-NH_2 showed inhibitory effects on Na-Bt clay, resembling the smectite constituents of shale formations. These results indicate the capability of these composites to improve drilling fluid efficacy.

Researchers created a PEG-NS composite with dimensions ranging from 110 to 434 nm, which is resistant to high temperatures [130]. They showed that it worked well as a stabilizing agent for shale in traditional WBMs. The composite was tested using rolling dispersion and pore pressure transmission studies. The results showed that PEG-NS limits the transmission of pore pressure and reduces the permeability of shale. Compared to polymeric alcohol (JHC) and KCl, the PEG-NS-based WBMs have effectively alleviated shale swelling at similar concentrations, thus serving as an effective stabilizer and improving the functions of conventional drilling fluids in shale formations.

Researchers [33] investigated how nano-polymer drilling fluids mitigate shale-related problems, focusing on the impact of additives on contact angles and shale expansion. The researchers assessed four different base fluid formulations. BF, BN_4, BP_4, and BPN are all formed by combinations of NPs or polymers. The mixture consisting of ZnO-NPs and associative polymer (AP) (BPN_4) had a maximum contact angle of 64.80°. The increase in contact angle represents the enhanced hydrophobicity of the shale surface, as the nano-polymer composite alters its surface properties. The decrease in shale swelling, from 14.29% using conventional WBMs to 6.69% with the PNCs recipe (BPN_4), was reported in the research. The findings confirm the ability to incorporate ZnO nanoparticles and ammonium phosphate into drilling fluids to improve shale inhibition and minimize formation damage, which is essential for delivering effective drilling operations.

PNCs in drilling fluids demonstrate an important potential in reducing wellbore instability, particularly in shale accumulations [128]. These fluids are formulated to create a low-permeability filter cake that prevents direct contact between the drilling fluid and the shale, thereby avoiding hydration, swelling, and destabilization risks. Moreover, hydrophobic properties are typically enhanced in PNC-based drilling formulations, thereby reducing the shale's ability to absorb water from the fluid. The increased hydrophobicity reduces the cation exchange ability of the shale, thus further hindering harmful interactions. As a result, the stability of the wellbore is enhanced, and PNCs become highly effective in controlling the damage caused by fluids in reactive shale formations, thereby improving overall drilling results and efficiency.

PNCs in drilling fluids have the potential to alleviate problems like shale formation stabilization, enhanced rheology, and improved management of filtrate loss. However, some challenges still hinder their full implementation in drilling processes. The uneven distribution of NPs in the polymer matrix causes the fluids to exhibit varying properties and perform poorly. The uneven spread of NPs often weakens the stability of the drilling fluid, affecting its flow and filtration properties. The performance of PNCs depends on NP dispersion. High-pressure homogenization, magnetic stirrers, ball mills, and ultrasonic baths have been used to disperse NPs. Overall, nanocarbon-based composites like graphene oxide and carbon nanotubes create stronger hydrophobic barriers; however, inorganic fillers like silica or titania provide better thermal and structural stability. Therefore, the PNC system you choose should be tailored to the specific circumstances in the wellbore. For example, use hydrophobic nanocomposites for formations that are rich in shale and sensitive to water, and oxide-based nanocomposites for HPHT settings. This framework for comparison facilitates the selection of PNCs for drilling operations in specific fields.

3. Conclusions and Future Prospects

Polymer nanocomposites (PNCs) represent an important advancement in drilling fluid technology. They exhibit better rheology, reduced filtrate loss, enhanced shale inhibition, and improved thermal and salt tolerance compared to typical polymer-based additives. Their ability to stabilize wellbores and enhance environmental performance shows that they could be used in the next generation of high-performance water-based drilling fluids (WBDFs).

However, to make this happen, both design and field applications must incorporate sustainability, safety, and predictive performance assessment. Life Cycle Assessment (LCA) and circular economy principles are essential tools for determining the environmental impact of PNCs, from sourcing raw materials to disposal or reuse. There aren't many LCA studies on drilling fluids currently available, and those that do exist often employ different methods. This makes it hard to compare sustainability. It is suggested that subsequent studies utilize standardized frameworks, such as ISO 14040/14044 and the EU Product Environmental Footprint (PEF) criteria, to facilitate uniform environmental effect assessments across investigations [131].

To ensure that PNC development aligns with the goals of a circular economy, we need to transition toward biodegradable and waste-derived polymers. Lignin-based polymers, which come from pulping or biomass waste, and cellulose nanocrystals (CNCs), which come from agricultural or forestry waste, are two promising candidates. Lignin-based polymers excel at dispersing and altering rheology, while CNCs are strong, functional on the surface, and biodegradable. These bio-based fillers can replace some synthetic monomers, reduce carbon intensity, and make the product easier to break down at the end of its life, all without compromising its thermal or filtering performance. Starch, chitosan, and modified guar derivatives have also been shown to be very beneficial for the environment and may be used on a large scale in WBDFs.

At the same time, the fact that synthetic nanocomposites are harmful and persist in the environment for an extended period remains an important concern. If polyacrylamide-based systems aren't completely polymerized, residual acrylamide monomers can be detrimental to the environment. Nanoparticle leaching (such as SiO_2, ZnO, or graphene oxide, or GO) can also lead to bioaccumulation or groundwater contamination. Therefore, operational protocols should include strict control of synthesis conditions, comprehensive verification of polymerization, and proper treatment of drilling fluids after use. Additionally, eco-toxicological tests and assessments of the potential for nanomaterials to leach out should be standard reporting criteria for PNC-based fluid formulations.

The next phase of research should focus on developing multiscale modeling and simulation tools that can accurately replicate PNC behavior in dynamic downhole conditions, thereby enhancing the predictability of field performance. Coupled computational fluid dynamics (CFD) and molecular dynamics (MD) simulations can describe how nanoparticles interact with clay minerals, temperature fluctuations, and salinity fluctuations, providing quantitative insights into rheology and filtration mechanisms before field deployment. Predictive models based on machine learning could also accelerate the development of PNCs by linking data on composition from the lab to performance results in the field.

In short, to improve PNCs for drilling fluids, we need a framework that combines high-performance engineering, long-term sustainability, and predictive modeling. Adding biodegradable, waste-derived polymers, such as lignin, cellulose nanocrystals, and chitosan, to nanocomposite matrices can help protect the environment and promote the goals of a circular economy. Addressing toxicity risks through standardized LCA evaluation and nanoparticle leaching control, alongside the development of data-driven predictive tools, will be pivotal for the sustainable deployment of PNC-based drilling fluids in future hydrocarbon exploration and geothermal applications.

Author Contributions: Conceptualization, S.W. and A.A.; writing—original draft preparation, S.W. and A.A.; writing—review and editing, J.Z.; supervision, J.Z.; project administration, J.Z.; funding acquisition, J.Z. All authors have read and agreed to the published version of the manuscript.

Funding: This research was supported by the Natural Science Foundation of Tianjin City (23JCY-BJC01860).

Data Availability Statement: No new data were created or analyzed in this study. Data sharing is not applicable to this article.

Conflicts of Interest: The authors declare no conflicts of interest.

Abbreviations

AV	Apparent Viscosity
CMC	Carboxymethyl Cellulose
CNT	Carbon Nanotube
FCT	Filter Cake Thickness
GG	Guar Gum
GS	Gel Strength
HPAM	Hydrolyzed Polyacrylamide
HPHT	High-Pressure High-Temperature
HEC	Hydroxyethyl Cellulose
LCM	Lost Circulation Material
NP(s)	Nanoparticle(s)

OBM	Oil-Based Mud
PAC	Polyanionic Cellulose
PAM	Polyacrylamide
PHPA	Partially Hydrolyzed Polyacrylamide
PNC(s)	Polymer Nanocomposite(s)
PV	Plastic Viscosity
ROP	Rate of Penetration
SiO_2	Silica
WBM	Water-Based Mud
XG	Xanthan Gum
WBMs	Water-Based Muds
OBMs	Oil-Based Muds
PNCs	Polymer Nanocomposites
HPHT	High-Pressure & High-Temperature
KPAM	Potassium Polyacrylamide
YP	Yield Point
CoF	Coefficient of Friction
GO	Graphene Oxide
CNTs	Carbon Nanotubes
Gr	Graphene
QDs	Quantum Dots
MSNs	Mesoporous Silica Nanoparticles
FESEM	Field-Emission Scanning Electron Microscopy
FTIR	Fourier-Transform Infrared Spectroscopy

References

1. Chen, H.; Xu, C. The impact of cryptocurrencies on China's carbon price variation during COVID-19: A quantile perspective. *Technol. Forecast. Soc. Change* **2022**, *183*, 121933. [CrossRef]
2. Saba, C.S.; Ngepah, N. The impact of artificial intelligence (AI) on employment and economic growth in BRICS: Does the moderating role of governance Matter? *Res. Glob.* **2024**, *8*, 100213. [CrossRef]
3. Yan, L.; Wang, C.; Xu, B.; Sun, J.; Yue, W.; Yang, Z. Preparation of a novel amphiphilic comb-like terpolymer as viscosifying additive in low-solid drilling fluid. *Mater. Lett.* **2013**, *105*, 232–235. [CrossRef]
4. Li, X.; Ju, Y.; Jia, Y.; Liu, F.; Liu, G.; Wang, S.; Wang, H.; Mao, S.; Yang, J.; Du, G. Design of novel temperature-resistant and salt-tolerant acrylamide-based copolymers by aqueous dispersion polymerization. *Des. Monomers Polym.* **2022**, *25*, 220–230. [CrossRef]
5. He, Y.; Gou, S.; Zhou, Y.; Zhou, L.; Tang, L.; Liu, L.; Fang, S. Thermoresponsive behaviors of novel polyoxyethylene-functionalized acrylamide copolymers: Water solubility, rheological properties and surface activity. *J. Mol. Liq.* **2020**, *319*, 114337. [CrossRef]
6. Saleh, T.A.; Rana, A.; Arfaj, M.K.; Ibrahim, M.A. Hydrophobic polymer-modified nanosilica as effective shale inhibitor for water-based drilling mud. *J. Pet. Sci. Eng.* **2022**, *209*, 109868. [CrossRef]
7. Borisov, A.S.; Husein, M.; Hareland, G. A field application of nanoparticle-based invert emulsion drilling fluids. *J. Nanoparticle Res.* **2015**, *17*, 340. [CrossRef]
8. Boyou, N.V.; Ismail, I.; Sulaiman, W.R.W.; Haddad, A.S.; Husein, N.; Hui, H.T.; Nadaraja, K. Experimental investigation of hole cleaning in directional drilling by using nano-enhanced water-based drilling fluids. *J. Pet. Sci. Eng.* **2019**, *176*, 220–231. [CrossRef]
9. Boudouh, D.; Ikram, R.; Jan, B.M.; Metselaar, H.S.C.; Hamana, D.; Kenanakis, G. Synthesis, characterization and filtration properties of ecofriendly Fe3O4 nanoparticles derived from olive leaves extract. *Materials* **2021**, *14*, 4306. [CrossRef]
10. Hajiabadi, S.H.; Aghaei, H.; Kalateh-Aghamohammadi, M.; Shorgasthi, M. An overview on the significance of carbon-based nanomaterials in upstream oil and gas industry. *J. Pet. Sci. Eng.* **2020**, *186*, 106783. [CrossRef]
11. Gbadamosi, A.; Yusuff, A.; Agi, A.; Muruga, P.; Junin, R.; Jeffrey, O. Mechanistic study of nanoparticles-assisted xanthan gum polymer flooding for enhanced oil recovery: A comparative study. *J. Pet. Explor. Prod. Technol.* **2022**, *12*, 207–213. [CrossRef]
12. Hoelscher, K.P.; De Stefano, G.; Riley, M.; Young, S. Application of nanotechnology in drilling fluids. In Proceedings of the SPE International Oilfield Nanotechnology Conference and Exhibition, Noordwijk, The Netherlands, 12–14 June 2012; SPE: Richardson, TX, USA, 2012; p. SPE-157031.
13. Cheraghian, G.G.; Afrand, M. Nanotechnology for drilling operations. In *Emerging Nanotechnologies for Renewable Energy*; Elsevier: Amsterdam, The Netherlands, 2021; pp. 135–148.

14. Vargas, J.; Roldán, L.J.; Lopera, S.H.; Cardenas, J.C.; Zabala, R.D.; Franco, C.A.; Cortés, F.B. Effect of silica nanoparticles on thermal stability in bentonite free water-based drilling fluids to improve its rheological and filtration properties after aging process. In Proceedings of the Offshore Technology Conference Brasil, Rio de Janeiro, Brazil, 29–31 October 2019; OTC: Boca Raton, FL, USA, 2019; p. D011S007R003.
15. Rowan, N.J. The role of digital t1chnologies in supporting and improving fishery and aquaculture across the supply chain–Quo Vadis? *Aquac. Fish.* **2023**, *8*, 365–374. [CrossRef]
16. Li, M.-C.; Wu, Q.; Song, K.; De Hoop, C.F.; Lee, S.; Qing, Y.; Wu, Y. Cellulose nanocrystals and polyanionic cellulose as additives in bentonite water-based drilling fluids: Rheological modeling and filtration mechanisms. *Ind. Eng. Chem. Res.* **2015**, *55*, 133–143. [CrossRef]
17. Mao, H.; Qiu, Z.; Shen, Z.; Huang, W. Hydrophobic associated polymer based silica nanoparticles composite with core–shell structure as a filtrate reducer for drilling fluid at ultra-high temperature. *J. Pet. Sci. Eng.* **2015**, *129*, 1–14. [CrossRef]
18. Abdollahi, M.; Pourmahdi, M.; Nasiri, A.R. Synthesis and characterization of lignosulfonate/acrylamide graft copolymers and their application in environmentally friendly water-based drilling fluid. *J. Pet. Sci. Eng.* **2018**, *171*, 484–494. [CrossRef]
19. Nooripoor, V.; Nazemi, R.; Hashemi, A. Employing nano-sized additives as filtration control agent in water-based drilling fluids: Study on barium sulfate, bentonite, surface-modified bentonite, titanium oxide, and silicon oxide. *Energy Sources Part A Recover. Util. Environ. Eff.* **2024**, *46*, 13806–13822. [CrossRef]
20. Nooripoor, V.; Hashemi, A. Effect of a modified nano clay and nano graphene on rheology, stability of water-in-oil emulsion, and filtration control ability of oil-based drilling fluids: A comparative experimental approach. *Oil Gas Sci. Technol. d'IFP Energies Nouv.* **2020**, *75*, 40. [CrossRef]
21. *GB/T 16783.1-2006*; Petroleum and Natural Gas Industries—Field Testing of Drilling Fluids—Part 1: Water-Based Fluids. SAC: Shenzhen, China, 2016.
22. Gao, J.; Zhang, G.; Wang, L.; Ding, L.; Shi, H.; Lai, X.; Wen, X.; Ma, S.; Huang, C. Rheological properties of an ultra-high salt hydrophobic associated polymer as a fracturing fluid system. *RSC Adv.* **2019**, *9*, 15246–15256. [CrossRef]
23. Mohamed, A.; Elkatatny, S.; Al-Majed, A. Removal of calcium carbonate water-based filter cake using a green biodegradable acid. *Sustainability* **2020**, *12*, 994. [CrossRef]
24. Koh, J.K.; Lai, C.W.; Johan, M.R.; Gan, S.S.; Chua, W.W. Recent advances of modified polyacrylamide in drilling technology. *J. Pet. Sci. Eng.* **2022**, *215*, 110566. [CrossRef]
25. Liu, J.; Dai, Z.; Xu, K.; Yang, Y.; Lv, K.; Huang, X.; Sun, J. Water-based drilling fluid containing bentonite/poly (sodium 4-styrenesulfonate) composite for ultrahigh-temperature ultradeep drilling and its field performance. *SPE J.* **2020**, *25*, 1193–1203. [CrossRef]
26. Growcock, F.; Harvey, T. Drilling fluids. In *Drilling Fluids Processing Handbook*; Elsevier: Burlington, MA, USA, 2005; Volume 1.
27. Bergane, C.; Hammadi, L. Impact of organophilic clay on rheological properties of gasoil-based drilling muds. *J. Pet. Explor. Prod. Technol.* **2020**, *10*, 3533–3540. [CrossRef]
28. Li, J.; Sun, J.; Lv, K.; Ji, Y.; Ji, J.; Liu, J. Nano-modified polymer gels as temperature-and salt-resistant fluid-loss additive for water-based drilling fluids. *Gels* **2022**, *8*, 547. [CrossRef] [PubMed]
29. Cai, J.; Chenevert, M.E.; Sharma, M.M.; Friedheim, J. Decreasing Water Invasion into Atoka Shale Using Non-modified Silica Nanoparticles. *SPE Drill. Compl.* **2011**, *27*, 103–112. [CrossRef]
30. Haruna, M.A.; Pervaiz, S.; Hu, Z.; Nourafkan, E.; Wen, D. Improved rheology and high-temperature stability of hydrolyzed polyacrylamide using graphene oxide nanosheet. *J. Appl. Polym. Sci.* **2019**, *136*, 47582. [CrossRef]
31. Bavoh, C.B.; Ofei, T.N.; Lal, B.; Sharif, A.M.; Shahpin, M.H.B.A.; Sundramoorthy, J.D. Assessing the impact of an ionic liquid on NaCl/KCl/polymer water-based mud (WBM) for drilling gas hydrate-bearing sediments. *J. Mol. Liq.* **2019**, *294*, 111643. [CrossRef]
32. Lalji, S.M.; Ali, S.I.; Ahmed, R.; Hashmi, S.; Awan, Z.U.H.; Qazi, S.W.A. Influence of graphene oxide on salt-polymer mud rheology and Pakistan shale swelling inhibition behavior. *Arab. J. Geosci.* **2022**, *15*, 612. [CrossRef]
33. Chen, Y.; Yin, Q.; Zhang, X.; Xue, X.; Jia, H. The crystallization behaviors and rheological properties of polypropylene/graphene nanocomposites: The role of surface structure of reduced graphene oxide. *Thermochim. Acta* **2018**, *661*, 124–136. [CrossRef]
34. Huang, X.; Sun, J.; Lv, K.; Liu, J.; Shen, H.; Zhang, F. Application of core-shell structural acrylic resin/nano-SiO2 composite in water based drilling fluid to plug shale pores. *J. Nat. Gas Sci. Eng.* **2018**, *55*, 418–425. [CrossRef]
35. Akhtarmanesh, S.; Shahrabi, M.J.A.; Atashnezhad, A. Improvement of wellbore stability in shale using nanoparticles. *J. Pet. Sci. Eng.* **2013**, *112*, 290–295. [CrossRef]
36. Morgani, M.S.; Saboori, R.; Sabbaghi, S. Hydrogen sulfide removal in water-based drilling fluid by metal oxide nanoparticle and ZnO/TiO2 nanocomposite. *Mater. Res. Express* **2017**, *4*, 75501. [CrossRef]
37. Naheed, Z.; Nasreen, S.; Tiwari, A.K.; Arsh, S. Analyzing the effectiveness of energy aid for driving the transition towards energy decarbonization; Evidence from Asian developing countries. *J. Environ. Manag.* **2024**, *360*, 121125. [CrossRef]

38. Aftab, A.; Ismail, A.R.; Khokhar, S.; Ibupoto, Z.H. Novel zinc oxide nanoparticles deposited acrylamide composite used for enhancing the performance of water-based drilling fluids at elevated temperature conditions. *J. Pet. Sci. Eng.* **2016**, *146*, 1142–1157. [CrossRef]
39. Ruan, Y.; Wang, W.; Abubakar, M.; Ahmad, N. Global Economic Resilience: Developing Green Growth Strategies, Renewable Energy Integration, and Environmental Economics for Sustainability. *Renew. Energy* **2025**, *251*, 123341. [CrossRef]
40. Luo, Y.; Lin, L.; Luo, P.; Li, X.; Ren, W.; Yi, T. Polymer-laponite composites as filtrate reducer for high temperature and salt resistant drilling fluid: Characterization and performance evaluation. *Colloids Surf. A Physicochem. Eng. Asp.* **2024**, *688*, 133679. [CrossRef]
41. Ezell, R.G.; Ezzat, A.M.; Turner, J.K.; Wu, J.J. New filtration-control polymer for improved brine-based reservoir drilling-fluids performance at temperatures in excess of 400 F and high pressure. In Proceedings of the SPE International Conference and Exhibition on Formation Damage Control, Lafayette, LA, USA, 10–12 February 2010; SPE: Richardson, TX, USA, 2010; p. SPE-128119.
42. Tavakkoli, O.; Kamyab, H.; Shariati, M.; Mohamed, A.M.; Junin, R. Effect of nanoparticles on the performance of polymer/surfactant flooding for enhanced oil recovery: A review. *Fuel* **2022**, *312*, 122867. [CrossRef]
43. Peng, H.; Chen, H.; Sun, Y.; Zhang, X.; Abubakar, M. Environmental economics of green recovery in mineral extraction: Balancing production and conservation. *Resour. Policy* **2024**, *96*, 105202. [CrossRef]
44. Needaa, A.-M.; Pourafshary, P.; Hamoud, A.-H.; Jamil, A. Controlling bentonite-based drilling mud properties using sepiolite nanoparticles. *Pet. Explor. Dev.* **2016**, *43*, 717–723. [CrossRef]
45. Davoodi, S.; Al-Shargabi, M.; Wood, D.A.; Rukavishnikov, V.S.; Minaev, K.M. Synthetic polymers: A review of applications in drilling fluids. *Pet. Sci.* **2024**, *21*, 475–518. [CrossRef]
46. Hayati, M.M.; Aminlou, A.; Abapour, M.; Shafie-khah, M.; Shahinzadeh, H.; Gharehpetian, G.B. 18 The Power-to-X (PtX) Effect in Ancillary Service Markets. In *Power-to-X in Regional Energy Systems*; CRC Press: Boca Raton, FL, USA, 2025; p. 432.
47. Li, M.-C.; Tang, Z.; Liu, C.; Huang, R.; Koo, M.S.; Zhou, G.; Wu, Q. Water-redispersible cellulose nanofiber and polyanionic cellulose hybrids for high-performance water-based drilling fluids. *Ind. Eng. Chem. Res.* **2020**, *59*, 14352–14363. [CrossRef]
48. Aftab, A.; Ali, M.; Sahito, M.F.; Mohanty, U.S.; Jha, N.K.; Akhondzadeh, H.; Azhar, M.R.; Ismail, A.; Keshavarz, A.; Iglauer, S. Environmental friendliness and high performance of multifunctional tween 80/ZnO-nanoparticles-added water-based drilling fluid: An experimental approach. *ACS Sustain. Chem. Eng.* **2020**, *8*, 11224–11243. [CrossRef]
49. Pandey, A.; Chaturvedi, K.R.; Sharma, T. Assessment of polymer based carbonation in weak/strong alkaline media for energy production and carbon storage: An approach to address carbon emissions. *J. Clean. Prod.* **2021**, *328*, 129628. [CrossRef]
50. Chaturvedi, K.R.; Trivedi, J.; Sharma, T. Evaluation of polymer-assisted carbonated water injection in sandstone reservoir: Absorption kinetics, rheology, and oil recovery results. *Energy Fuels* **2019**, *33*, 5438–5451. [CrossRef]
51. Chen, Y.; Shull, K.R. Thermothickening behavior of self-stabilized colloids formed from associating polymers. *Macromolecules* **2019**, *52*, 4926–4933. [CrossRef]
52. Bayat, A.E.; Harati, S.; Kolivandi, H. Evaluation of rheological and filtration properties of a polymeric water-based drilling mud in presence of nano additives at various temperatures. *Colloids Surf. A Physicochem. Eng. Asp.* **2021**, *627*, 127128. [CrossRef]
53. Dong, X.; Sun, J.; Huang, X.; Lv, K.; Zhou, Z.; Gao, C. Nano-laponite/polymer composite as filtration reducer on water-based drilling fluid and mechanism study. *R. Soc. Open Sci.* **2022**, *9*, 220385. [CrossRef] [PubMed]
54. Awaysheh, A.; Heron, R.A.; Perry, T.; Wilson, J.I. On the relation between corporate social responsibility and financial performance. *Strateg. Manag. J.* **2020**, *41*, 965–987. [CrossRef]
55. Belu, C.; Manescu, C. Strategic corporate social responsibility and economic performance. *Appl. Econ.* **2013**, *45*, 2751–2764. [CrossRef]
56. Aramburu, I.A.; Pescador, I.G. The effects of corporate social responsibility on customer loyalty: The mediating effect of reputation in cooperative banks versus commercial banks in the Basque country. *J. Bus. Ethics* **2019**, *154*, 701–719. [CrossRef]
57. Ahamed, M.M.; Luintel, K.B.; Mallick, S.K. Does local knowledge spillover matter for firm productivity? The role of financial access and corporate governance. *Res. Policy* **2023**, *52*, 104837. [CrossRef]
58. Ai, Y.; Chi, Z.; Sun, G.; Zhou, H.; Kong, T. The research on non-linear relationship between enterprise digital transformation and stock price crash risk. *N. Am. J. Econ. Financ.* **2023**, *68*, 101984. [CrossRef]
59. Liu, Y.; Bian, Y.; Zhang, W. How does enterprises' digital transformation impact the educational structure of employees? Evidence from China. *Sustainability* **2022**, *14*, 9432. [CrossRef]
60. Liu, X.; Wang, X.; Yu, W. Opportunity or challenge? Research on the influence of digital finance on digital transformation of agribusiness. *Sustainability* **2023**, *15*, 1072. [CrossRef]
61. Lu, Z.; Li, H. Does environmental information disclosure affect green innovation? *Econ. Anal. Policy* **2023**, *80*, 47–59. [CrossRef]

62. Liao, F.; Hu, Y.; Chen, M.; Xu, S. Digital transformation and corporate green supply chain efficiency: Evidence from China. *Econ. Anal. Policy* **2024**, *81*, 195–207. [CrossRef]
63. Li, S.; Wang, Q. Green finance policy and digital transformation of heavily polluting firms: Evidence from China. *Financ. Res. Lett.* **2023**, *55*, 103876. [CrossRef]
64. Li, R.; Rao, J.; Wan, L. The digital economy, enterprise digital transformation, and enterprise innovation. *Manag. Decis. Econ.* **2022**, *43*, 2875–2886. [CrossRef]
65. Li, M.; Jiang, A.; Ma, J. Digital transformation and income inequality within enterprises–Evidence from listed companies in China. *Pac. Basin Financ. J.* **2023**, *81*, 102133. [CrossRef]
66. Krishnamurti, C.; Shams, S.; Pensiero, D.; Velayutham, E. Socially responsible firms and mergers and acquisitions performance: Australian evidence. *Pac. Basin Financ. J.* **2019**, *57*, 101193. [CrossRef]
67. Ko, K.-C.; Nie, J.; Ran, R.; Gu, Y. Corporate social responsibility, social identity, and innovation performance in China. *Pac. Basin Financ. J.* **2020**, *63*, 101415. [CrossRef]
68. Guo, L.; Xu, L. The effects of digital transformation on firm performance: Evidence from China's manufacturing sector. *Sustainability* **2021**, *13*, 12844. [CrossRef]
69. Eller, R.; Alford, P.; Kallmünzer, A.; Peters, M. Antecedents, consequences, and challenges of small and medium-sized enterprise digitalization. *J. Bus. Res.* **2020**, *112*, 119–127. [CrossRef]
70. Cui, L.; Wang, Y. Can corporate digital transformation alleviate financial distress? *Financ. Res. Lett.* **2023**, *55*, 103983. [CrossRef]
71. Cui, D.; Ding, M.; He, Z.; Shen, M. Restricted shares and CSR: Evidence from foreign strategic investors in China. *Pac. Basin Financ. J.* **2022**, *76*, 101865. [CrossRef]
72. El-Khayat, H.M.M.; Gaber, H.S.; Helal, N.S.; Zayed, K.M.; Flefel, H.E.; Mohammed, W.A.; Sayed, S.S.M. Comparative study on Lake Manzala in Egypt after and before development and purifications: Ecosystem and biomarkers; biochemical, histopathological and immunohistochemical. *Environ. Pollut.* **2023**, *339*, 122741. [CrossRef]
73. Li, X.; Gao, Y.; Zhang, M.; Zhang, Y.; Zhou, M.; Peng, L.; He, A.; Zhang, X.; Yan, X.; Wang, Y.; et al. In vitro lung and gastrointestinal bioaccessibility of potentially toxic metals in Pb-contaminated alkaline urban soil: The role of particle size fractions. *Ecotoxicol. Environ. Saf.* **2020**, *190*, 110151. [CrossRef]
74. Lu, X.; Yang, Q.; Wang, H.; Zhu, Y. A global meta-analysis of the correlation between soil physicochemical properties and lead bioaccessibility. *J. Hazard. Mater.* **2023**, *453*, 131440. [CrossRef]
75. Ağbulut, Ü. Turkey's electricity generation problem and nuclear energy policy. *Energy Sources Part A Recover. Util. Environ. Eff.* **2019**, *41*, 2281–2298. [CrossRef]
76. Roy, S.; Singha, N.; Bose, A.; Basak, D.; Chowdhury, I.R. Multi-influencing factor (MIF) and RS–GIS-based determination of agriculture site suitability for achieving sustainable development of Sub-Himalayan region, India. *Environ. Dev. Sustain.* **2023**, *25*, 7101–7133. [CrossRef]
77. Liu, G.N.; Wu, Y.; Wang, S.C.; Liao, J.; Yu, S.W. The Main Water Environmental Issues and Assessment-zonation of Eco-environmental Geology Risks for Typical River-lake-wetland Systems in the Central Yangtze River. *S. China Geol.* **2022**, *38*, 226–239.
78. Jin, X.; Wei, L.; Wang, Y.; Lu, Y. Construction of ecological security pattern based on the importance of ecosystem service functions and ecological sensitivity assessment: A case study in Fengxian County of Jiangsu Province, China. *Environ. Dev. Sustain.* **2021**, *23*, 563–590. [CrossRef]
79. Pata, U.K.; Kartal, M.T.; Adebayo, T.S.; Ullah, S. Enhancing environmental quality in the United States by linking biomass energy consumption and load capacity factor. *Geosci. Front.* **2023**, *14*, 101531. [CrossRef]
80. Wang, J.; Zhang, S.; Zhang, Q. The relationship of renewable energy consumption to financial development and economic growth in China. *Renew. Energy* **2021**, *170*, 897–904. [CrossRef]
81. Zhang, Q.; Shah, S.A.R.; Yang, L. Modeling the effect of disaggregated renewable energies on ecological footprint in E5 economies: Do economic growth and R&D matter? *Appl. Energy* **2022**, *310*, 118522. [CrossRef]
82. Tkalec, Ž.; Antignac, J.-P.; Bandow, N.; Béen, F.M.; Belova, L.; Bessems, J.; Le Bizec, B.; Brack, W.; Cano-Sancho, G.; Chaker, J.; et al. Innovative analytical methodologies for characterizing chemical exposure with a view to next-generation risk assessment. *Environ. Int.* **2024**, *186*, 108585. [CrossRef]
83. Ma, Q.; Liu, Y.; Yang, X.; Guo, Y.; Xiang, T.; Wang, Y.; Yan, Y.; Li, D.; Nie, T.; Li, Z.; et al. Effect-directed analysis for revealing aryl hydrocarbon receptor agonists in sediment samples from an electronic waste recycling town in China. *Environ. Pollut.* **2022**, *308*, 119659. [CrossRef]
84. Ma, Q.; Yang, X.; Guo, Y.; Wang, Y.; Liu, Y.; Zhang, S.; Xie, H.Q.; Xiang, T.; Li, Z.; Nie, T.; et al. Effect-directed analysis of estrogenic chemicals in sediments from an electronic-waste recycling area. *Environ. Pollut.* **2022**, *306*, 119369. [CrossRef]

85. Shui, Y.; Zhong, J.; Yang, Y.; Zeng, Y.; Liu, S. Employment Transfer of Rural Female Labor and Family Welfare Effect in Mountainous Areas: An Empirical Analysis Based on Panel Data. *Land* **2022**, *11*, 2134. [CrossRef]
86. Xin-gang, Z.; Jin, Z. Industrial restructuring, energy consumption and economic growth: Evidence from China. *J. Clean. Prod.* **2022**, *335*, 130242. [CrossRef]
87. Su, Y.; Fan, Q. Renewable energy technology innovation, industrial structure upgrading and green development from the perspective of China's provinces. *Technol. Forecast. Soc. Change* **2022**, *180*, 121727. [CrossRef]
88. Shang, J.; Wang, Z.; Li, L.; Chen, Y.; Li, P. A study on the correlation between technology innovation and the new-type urbanization in Shaanxi province. *Technol. Forecast. Soc. Change* **2018**, *135*, 266–273. [CrossRef]
89. Xu, H.; Jiao, M. City size, industrial structure and urbanization quality—A case study of the Yangtze River Delta urban agglomeration in China. *Land Use Policy* **2021**, *111*, 105735. [CrossRef]
90. Zhang, H.; Feng, C. Innovative low-carbon city: Carbon emission performance evaluation of innovative city construction. *S. China J. Econ.* **2021**, *3*, 36–53.
91. Zhang, C.; Yao, Y.; Zhou, H. External technology dependence and manufacturing TFP: Evidence from China. *Res. Int. Bus. Financ.* **2023**, *64*, 101885. [CrossRef]
92. Gao, S.; Lin, D.; Li, A.; Deng, L.; Dong, A.; Zhang, J. A zwitterionic polymer filtrate reducer with enhanced anti-polyelectrolyte effect for high-temperature and high-salinity conditions. *Chem. Eng. J.* **2025**, *508*, 161087. [CrossRef]
93. Nathaniel, S.; Anyanwu, O.; Shah, M. Renewable energy, urbanization, and ecological footprint in the Middle East and North Africa region. *Environ. Sci. Pollut. Res.* **2020**, *27*, 14601–14613. [CrossRef]
94. Cheraghian, G. Application of nano-particles of clay to improve drilling fluid. *Int. J. Nanosci. Nanotechnol.* **2017**, *13*, 177–186.
95. Movahedi, H.; Jamshidi, S.; Hajipour, M. New insight into the filtration control of drilling fluids using a graphene-based nanocomposite under static and dynamic conditions. *ACS Sustain. Chem. Eng.* **2021**, *9*, 12844–12857. [CrossRef]
96. Aghayan, S.K. Moghadas hosseinzadeh F. The impact of traveling on infectious diseases transmission with a focus on air travel: A Narrative Review. *Int. J. Travel. Med. Glob. Health* **2025**, *13*, 33–41.
97. Ma, J.; Pang, S.; Zhang, Z.; Xia, B.; An, Y. Experimental study on the polymer/graphene oxide composite as a fluid loss agent for water-based drilling fluids. *ACS Omega* **2021**, *6*, 9750–9763. [CrossRef]
98. Medhi, S.; Chowdhury, S.; Gupta, D.K.; Mazumdar, A. An investigation on the effects of silica and copper oxide nanoparticles on rheological and fluid loss property of drilling fluids. *J. Pet. Explor. Prod. Technol.* **2020**, *10*, 91–101. [CrossRef]
99. Biswas, M.C.; Chowdhury, A.; Hossain, M.M.; Hossain, M.K. Applications, drawbacks, and future scope of nanoparticle-based polymer composites. In *Nanoparticle-Based Polymer Composites*; Elsevier: Amsterdam, The Netherlands, 2022; pp. 243–275.
100. Davoodi, S.; SA, A.R.; Soleimanian, A.; Jahromi, A.F. Application of a novel acrylamide copolymer containing highly hydrophobic comonomer as filtration control and rheology modifier additive in water-based drilling mud. *J. Pet. Sci. Eng.* **2019**, *180*, 747–755. [CrossRef]
101. Dufresne, A. Cellulose nanomaterial reinforced polymer nanocomposites. *Curr. Opin. Colloid Interface Sci.* **2017**, *29*, 1–8. [CrossRef]
102. Elochukwu, H.; Gholami, R.; Dol, S.S. An approach to improve the cuttings carrying capacity of nanosilica based muds. *J. Pet. Sci. Eng.* **2017**, *152*, 309–316. [CrossRef]
103. Kafashi, S.; Rasaei, M.; Karimi, G. Effects of sugarcane and polyanionic cellulose on rheological properties of drilling mud: An experimental approach. *Egypt. J. Pet.* **2017**, *26*, 371–374. [CrossRef]
104. Villada, Y.; Gallardo, F.; Erdmann, E.; Casis, N.; Olivares, L.; Estenoz, D. Functional characterization on colloidal suspensions containing xanthan gum (XGD) and polyanionic cellulose (PAC) used in drilling fluids for a shale formation. *Appl. Clay Sci.* **2017**, *149*, 59–66. [CrossRef]
105. Agwu, O.E.; Okon, A.N.; Akpanika, O.I. Activation of local bentonitic clays for use as viscosifiers in water-based drilling fluids. *J. Sci. Res. Rep.* **2016**, *12*, 1–11. [CrossRef]
106. Chimanowsky, J.P., Jr.; Neto, R.P.C.; Tavares, M.I.B. NMR evaluation of polystyrene nanocomposites degradated by repeated extrusion processing. *Polym. Degrad. Stab.* **2015**, *118*, 178–187. [CrossRef]
107. Alizadeh, S.; Sabbaghi, S.; Soleymani, M. Synthesis of Alumina/Polyacrylamide nanocomposite and its influence on viscosity of drilling fluid. *Int. J. Nano Dimens.* **2015**, *6*, 271–276.
108. Aghdam, S.B.; Moslemizadeh, A.; Kowsari, E.; Asghari, N. Synthesis and performance evaluation of a novel polymeric fluid loss controller in water-based drilling fluids: High-temperature and high-salinity conditions. *J. Nat. Gas Sci. Eng.* **2020**, *83*, 103576. [CrossRef]
109. Lin, Z.; Li, F.; Liu, X.; Su, J. High temperature degradable pullulan microsphere/polymer composite for filtration loss control of water-based drilling fluids. *Geoenergy Sci. Eng.* **2024**, *243*, 213362. [CrossRef]
110. Zhao, D.; Gao, Y.; Wu, Z.; Shabaz, M. Harnessing robotics for environmental economics and energy strategy: Advancing global economic policies towards green growth. *Energy Strateg. Rev.* **2024**, *55*, 101504. [CrossRef]

111. Sun, L.; Wang, J.; Wang, S.; Li, H. Artificial intelligence of robotics and green transformation: Evidence from Chinese manufacturing firms. *Environ. Dev. Sustain.* **2025**, 1–27. [CrossRef]
112. Elaf, R.; Ali, A.B.; Saad, M.; Hussein, I.A.; Bai, B. Development of eco-friendly chitosan-g-polyacrylamide preformed particle gel for conformance control in high-temperature and high-salinity reservoirs. *Geoenergy Sci. Eng.* **2023**, *230*, 212136. [CrossRef]
113. Sun, J.; Peng, Y.; Liang, Z.; Zhou, Y.; Li, L. AI-Driven Synergies: The Role of Artificial Intelligence in Enhancing Green Product Innovation for Environmental and Organizational Performance in China's Petrochemical Industry. *Open Access Libr. J.* **2025**, *12*, 1–15. [CrossRef]
114. Wang, M.; Wang, Y.; Feng, C. Artificial intelligence, institutional environment, and corporate green transformation: Evidence from China's resource-based sector. *Int. Rev. Econ. Financ.* **2025**, *103*, 104473. [CrossRef]
115. Gao, J.; Zheng, M.; Yuan, H.; Yang, Y.; Chen, H.; Fan, Y.; Shen, W. Nano-blade structured hematite/magnetite heterojunction with ZnO cocatalyst and plasmon Au adornment for improvement in solar water splitting. *J. Alloys Compd.* **2025**, *1037*, 182416. [CrossRef]
116. Barry, M.M.; Jung, Y.; Lee, J.-K.; Phuoc, T.X.; Chyu, M.K. Fluid filtration and rheological properties of nanoparticle additive and intercalated clay hybrid bentonite drilling fluids. *J. Pet. Sci. Eng.* **2015**, *127*, 338–346. [CrossRef]
117. Mohammed, R.K.; Farzaneh, H. Power-to-X in Southern Iraq: Techno-economic assessment of solar-powered hydrogen electrolysis combined with carbon capture and storage for sustainable energy solutions. *Energy Convers. Manag. X* **2025**, *26*, 100918. [CrossRef]
118. Obobisa, E.S.; Ahakwa, I. Stimulating the adoption of green technology innovation, clean energy resources, green finance, and environmental taxes: The way to achieve net zero CO_2 emissions in Europe? *Technol. Forecast. Soc. Change* **2024**, *205*, 123489. [CrossRef]
119. Kanugrahan, S.P.; Hakam, D.F.; Nugraha, H. Techno-Economic Analysis of Indonesia Power Generation Expansion to Achieve Economic Sustainability and Net Zero Carbon 2050. *Sustainability* **2022**, *14*, 9038. [CrossRef]
120. Zhang, D.; Chen, X.H.; Lau, C.K.M.; Cai, Y. The causal relationship between green finance and geopolitical risk: Implications for environmental management. *J. Environ. Manag.* **2023**, *327*, 116949. [CrossRef] [PubMed]
121. Zheng, N.; Zhu, J.; Zhu, H.; Xuan, J.; Xu, H.; Ni, M. Toward a circular carbon economy: Production of green C1 compounds through high-temperature CO2 electrolysis. *Nexus* **2025**, *2*, 100053. [CrossRef]
122. Afrane, S.; Ampah, J.D.; Yusuf, A.A.; Jinjuan, Z.; Yang, P.; Chen, J.L.; Mao, G. Role of negative emission technologies in South Africa's pathway to net zero emissions by 2050. *Energy Sustain. Dev.* **2024**, *79*, 101401. [CrossRef]
123. Peecock, A.; Hull-Bailey, B.; Hastings, A.; Martinez-Felipe, A.; Wilcox, L.B. Techno-economic assessment of liquid carrier methods for intercontinental shipping of hydrogen: A case study. *Int. J. Hydrogen Energy* **2024**, *94*, 971–983. [CrossRef]
124. Zhang, J.; Raza, S.M.F.; Huang, Y.; Wang, C. What affect energy poverty in China? A path towards sustainable development. *Econ. Res. Istraz.* **2023**, *36*, 2111316. [CrossRef]
125. Ahmad, H.M.; Kamal, M.S.; Al-Harthi, M.A.; Elkatatny, S.M.; Murtaza, M.M. Synthesis and experimental investigation of novel CNT-polymer nanocomposite to enhance borehole stability at high temperature drilling applications. In Proceedings of the SPE Kingdom of Saudi Arabia Annual Technical Symposium and Exhibition, Dammam, Saudi Arabia, 23–26 April 2018; SPE: Richardson, TX, USA, 2018; p. SPE-192352.
126. Oseh, J.O.; Mohd, N.M.N.A.; Gbadamosi, A.O.; Agi, A.; Blkoor, S.O.; Ismail, I.; Igwilo, K.C.; Igbafe, A.I. Polymer nanocomposites application in drilling fluids: A review. *Geoenergy Sci. Eng.* **2023**, *222*, 211416. [CrossRef]
127. Bayat, A.E.; Moghanloo, P.J.; Piroozian, A.; Rafati, R. Experimental investigation of rheological and filtration properties of water-based drilling fluids in presence of various nanoparticles. *Colloids Surf. A Physicochem. Eng. Asp.* **2018**, *555*, 256–263. [CrossRef]
128. Rana, A.; Arfaj, M.K.; Saleh, T.A. Graphene grafted with glucopyranose as a shale swelling inhibitor in water-based drilling mud. *Appl. Clay Sci.* **2020**, *199*, 105806. [CrossRef]
129. Blkoor, S.O.; Norddin, M.N.A.M.; Ismail, I.; Oseh, J.O.; Agi, A.; Gbadamosi, A.O.; Okoli, N.O.; Onyejekwe, I.M.; Risal, A.R. Amphipathic anionic surfactant modified hydrophilic polyethylene glycol-nanosilica composite as effective viscosifier and filtration control agent for water-based drilling muds. *Arab. J. Chem.* **2022**, *15*, 103741. [CrossRef]
130. Ikram, R.; Jan, B.M.; Vejpravova, J.; Choudhary, M.I.; Chowdhury, Z.Z. "Recent advances of graphene-derived nanocomposites in water-based drilling fluids. *Nanomaterials* **2020**, *10*, 2004. [CrossRef]
131. Ahmad, H.M.; Iqbal, T.; Al Harthi, M.A.; Kamal, M.S. Synergistic effect of polymer and nanoparticles on shale hydration and swelling performance of drilling fluids. *J. Pet. Sci. Eng.* **2021**, *205*, 108763. [CrossRef]

materials

MDPI

Article

A Dual-Dynamic Crosslinked Polysaccharide-Based Hydrogel Loaded with Exosomes for Promoting Diabetic Wound Healing

Ding Lin [1], Zhenhao Li [1], Jianying Hao [1], Xiaobo Xu [1], Xiuqiang Li [1,2], Yuan Feng [1], Xiaochen Lu [1], Fanglian Yao [1,3,4], Hong Zhang [1,3,4] and Junjie Li [1,3,4,*]

1 School of Chemical Engineering and Technology, Tianjin University, Tianjin 300350, China; abc120010_@tju.edu.cn (D.L.)
2 School of Science, Tianjin University, Tianjin 300350, China
3 State Key Laboratory of Synthetic Biology, Tianjin University, Tianjin 300350, China
4 Frontiers Science Center for Synthetic Biology (Ministry of Education), Tianjin University, Tianjin 300350, China
* Correspondence: li41308@tju.edu.cn

Abstract

Diabetic wounds are often accompanied by severe inflammation, which is unfavorable for vascular growth and wound repair. Therefore, promoting the healing of diabetic wounds is of great significance. In this study, carboxymethyl chitosan (CMCS) was grafted with 4-formylphenylboronic acid (FPBA) and then crosslinked with oxidized sodium alginate (OAlg) to form a dual-dynamic covalent hydrogel (CPOA) based on borate ester bond and Schiff base bonds. Mesenchymal stem cells' exosomes (Exos) were incorporated into the CPOA to construct CPOA@Exos for diabetic wound healing. Owing to the dual-dynamic covalent crosslinking network, the CPOA hydrogel showed good injectability and self-healing ability. In addition, the hydrogel displayed reactive oxygen species (ROS) responsive properties, enabling both scavenging of multiple free radicals and on-demand release of Exos in the ROS-rich wound microenvironment. A diabetic wound model was established on C57 mice, and treatment with CPOA@Exos demonstrated that it could promote the polarization of macrophages toward the M2 phenotype, enhance cellular proliferation in the wounded area, and thereby accelerate the healing of diabetic wounds. In conclusion, this study provides a new hydrogel wound dressing that can inhibit inflammation for the management of diabetic wounds.

Keywords: hydrogel; diabetic wound; reactive oxygen species (ROS); exosomes

1. Introduction

Diabetes mellitus has affected over 400 million individuals worldwide [1], and the global prevalence of diabetes among adults is projected to reach 7.7% by 2030 [2]. Diabetic wounds, a common complication of diabetes, affect approximately 25% of diabetic patients [3], posing a significant public health threat to human health.

Normal wound healing involves four sequential phases: hemostasis, inflammation, proliferation, and remodeling [4]. Diabetic wounds, however, present pathological hallmarks (hyperglycemia, chronic inflammation and excessive reactive oxygen species [ROS] [5,6]), which trap the wounds in a prolonged inflammatory state, impeding healing progression. Excessive ROS induces damage to cellular components, proteins, and DNA, exacerbating tissue injury [7], while sustained inflammation suppresses extracellular matrix formation, delaying wound closure [8]. Notably, local immune dysfunction is the core driver of excessive inflammation [9,10], with macrophages serving as key regulators [11]. In normal healing, macrophages polarize to the pro-inflammatory M1 phenotype (early phase [12]) and the pro-healing M2 phenotype (late phase [13]). However, in diabetic wounds, the persistence of M1 macrophages impedes M2 activation, resulting in chronic inflammation [14]. Thus, modulating the M1/M2 phenotypic polarization of macrophages and scavenging excessive ROS are critical therapeutic targets for diabetic wound treatment [15].

Current clinical management of diabetic wounds primarily relies on conventional dressings such as gauze. While these provide a basic physical barrier, their repeated removal may cause adhesive trauma to the wound [16]. Hydrogel dressings, by contrast, are promising alternatives, as they possess physiological water content [17], good biocompatibility, and the ability to maintain a moist wound microenvironment—all of which favor healing [18]. To adapt to the irregular topography of diabetic wounds, hydrogels must exhibit robust adhesive properties [19]. The Schiff base reaction (between aldehyde and amino groups) is a widely used hydrogel crosslinking strategy [20], endowing hydrogels with self-healing capabilities [21] and strong tissue adhesion—mediated by the Schiff reaction between aldehyde groups and tissue-derived amino groups [22,23]. However, diabetic wounds are often accompanied by severe oxidative stress [24], and the hydrogels, despite maintaining a moist environment, fail to effectively promote wound repair. To address this limitation, integrating stimulus-responsive moieties into hydrogel networks has emerged as a promising approach, enabling the hydrogels to respond to disease-specific microenvironments, achieve controlled release of loaded drugs, and optimize therapeutic efficacy [25]. Specifically, the incorporation of dynamic borate ester bonds into the gel network confers ROS-responsive properties, as these bonds can dissociate upon exposure to ROS [26], and even scavenge a portion of ROS, thereby reducing oxidative stress at the wound site [27]. Combining Schiff base (adhesion/self-healing) and borate ester (ROS responsiveness) bonds fabricates multifunctional hydrogels ideal for diabetic wounds [28,29].

Drug loading in hydrogels is an effective strategy to promote diabetic wound repair [30]. Exosomes from mesenchymal stem cells (Exos) are nanosized vesicles (30–150 nm in diameter) secreted by cells, containing various proteins, lipids, and nucleic acids, which play a crucial role in intercellular communication [31]. Numerous studies have demonstrated the Exos enriched with bioactive growth factors can induce M2 macrophage polarization through multiple pathways, such as the integrin β3/SOCS3/STAT3 pathway [32] and S1P/SK1/S1PR1 signaling pathway [33], thereby regulating tissue inflammation and immune responses. However, direct application of Exos to wounds suffers from rapid clearance, limiting their therapeutic efficacy [34]. In contrast, loading Exos into hydrogels can enhance their stability and enable sustained release, thereby improving their therapeutic effects in diabetic wound treatment [35,36].

Dual-dynamic crosslinking systems based on Schiff base and borate ester chemistry have been reported in the literature to enhance the therapeutic efficacy of diabetic wounds [5]. In contrast, this study is distinguished by the innovative design of a modified polysaccharide-based dual-crosslinked network hydrogel (CPOA) composed of carboxymethyl chitosan-grafted phenylboronic acid (CP) and oxidized sodium alginate (OAlg) based on Schiff base and borate ester bonds and its synergistic combination with Exos (CPOA@Exos) to accelerate wound healing (Scheme 1). Specifically, 4-formylphenylboronic acid (FPBA) was grafted onto carboxymethyl chitosan (CMCS) to obtain a series of FPBA-grafted CMCS derivatives. The CP was then mixed with OAlg to form the CPOA hydrogel. In a weakly alkaline microenvironment, the boronic acid groups of CP reacted with cis-

 https://doi.org/10.3390/ma19020445

diol units on OAlg to form borate ester bonds. Simultaneously, the amino groups in CP formed Schiff base bonds with the aldehyde groups in OAlg, collectively resulting in a dual dynamic covalent crosslinked hydrogel network. This modified polysaccharide matrix provides excellent biocompatibility, making it an ideal carrier for Exos loading. Moreover, the borate ester bonds impart ROS-responsive properties to the hydrogel, allowing on-demand release mechanism that enables the Exos to fully exert their anti-inflammatory and wound-repair-promoting effects. Beyond this, this system integrates the intrinsic antioxidant capacity of the hydrogel with the biological functions of Exos, resulting in a synergistic therapeutic effect that effectively promotes diabetic wound repair. Additionally, this study evaluated the mechanical properties, self-healing behavior, and adhesion performance of the CPOA hydrogels, and the *in vivo* repair efficacy was assessed using a diabetic wound model in C57 mice.

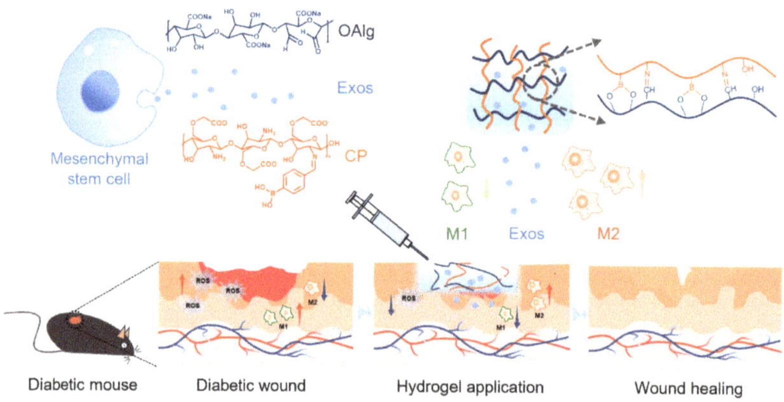

Scheme 1. Schematic of the dual-dynamic crosslinked hydrogel structure and its repair effect on diabetic wounds (OAlg: oxidized sodium alginate, Exos: exosomes, CP: carboxymethyl chitosan-grafted phenylboronic acid, M1: M1 macrophages, M2: M2 macrophages).

2. Materials and Methods

2.1. Materials

Sodium alginate (Alg), hydrogen peroxide (H_2O_2) and 1,1-diphenyl-2-trinitrophenylhydrazine (DPPH) were purchased from Aladdin Co., Ltd. (Shanghai, China). Carboxymethyl chitosan (CMCS) was supplied by Yuanye Bio-Technology Co., Ltd. (Shanghai, China). 4-formylphenylboronic acid (FPBA), Sodium hydroxide (NaOH) and sodium periodate ($NaIO_4$) were sourced from Meryer Chemical Technology Co., Ltd. (Shanghai, China). Hydrochloric acid (HCl) was obtained from Fengchuan Chemical Reagent Co., Ltd. (Tianjin, China). Ethylene glycol was purchased from Heowns Biochem Technologies LLC (Tianjin, China). Exosomes (Exos) were acquired from Mingyang New Materials Co., Ltd. (Tianjin, China). The ABTS assay kit was provided by Beijing Bio-Ocean Technology Co., Ltd. (Beijing, China). RPMI-1640 basal medium and the CCK-8 cell proliferation assay kit were purchased from Beijing Solarbio Technology Co., Ltd. (Beijing, China). All chemicals were used as received without further purification unless otherwise stated.

2.2. Preparation of Oxidized Sodium Alginate

First, 5 g of Alg was dissolved in 25 mL of anhydrous ethanol, while 1 g of $NaIO_4$ was dissolved in 25 mL of water. The two solutions were then mixed under stirring for 4 h. The reaction was then terminated by adding 2 mL of ethylene glycol. The resulting

https://doi.org/10.3390/ma19020445

mixture was dialyzed against water for 3 days and subsequently freeze-dried to yield OAlg. The oxidation degree (*OD*) of the resulting OAlg was quantified by hydroxylamine hydrochloride titration. Specifically, 0.1 g of OAlg was dissolved in 25 mL of hydroxylamine hydrochloride-methyl orange solution. After stirring for 2 h, the mixture solution was titrated with a 10 mmol/L NaOH solution, and the released HCl was measured using potentiometric titration. The *OD* of the OAlg was calculated according to Equation (1):

$$OD(\%) = \frac{0.001 \times c(NaOH) \times \Delta V/2}{w(OAlg)/198.11} \times 100 \quad (1)$$

Whereas, $c(NaOH)$ is the concentration of NaOH solution, ΔV represents the volume of NaOH solution consumed, and $w(OAlg)$ is the mass of the OAlg.

2.3. Preparation of 4-Formylphenylboronic Acid Grafted Carboxymethyl Chitosan

CMCS was dissolved in 80 mL of water. Varying amounts of FPBA (0.08, 0.16, 0.24 and 0.32 g) were then added into the CMCS solution, respectively. The mixtures were stirred for 6 h, and the resulting solutions were purified by precipitation with anhydrous ethanol. The collected solids were dried to obtain a series of FPBA-grafted CMCS derivatives, which were designated as CP1, CP2, CP3 and CP4, respectively, based on the substitution degree of FPBA on the CMCS chain. The chemical structure of the prepared CP was confirmed by nuclear magnetic resonance (^{1}H NMR) spectroscopy (Bruker, Billerica, MA, USA, Avance NEO, 800 MHz) and Fourier transform infrared (FTIR) spectroscopy (FEI, Hillsboro, OR, USA, iS50).

2.4. Preparation of Dual-Dynamic Crosslinked Hydrogel Loaded with Exosomes

First, 6% (w/v) OAlg, CMCS, CP1, CP2, CP3 and CP4 solutions in PBS were prepared separately. To investigate how the CP2-to-OAlg volume ratio affects hydrogel properties and to identify the optimal formulation, OAlg and CP2 solutions were mixed in a centrifuge tube at a series of volume ratios (3:7, 4:6, 5:5, 6:4 and 7:3) using vortex oscillation. This preparation method yielded hydrogels with varying compositions. Adjusting this ratio allows direct modulation of the crosslinking degree of both Schiff base and borate ester bonds, which in turn regulates the hydrogel's structure and overall properties (particularly its mechanical properties), a critical factor for subsequent applications. Then OAlg solution was mixed with CMCS, CP1 and CP2 separately at a 1:1 volume ratio to fabricate a dually dynamically cross-linked injectable hydrogel, designated COA, CP1OA and CP2OA. The specific formulations are listed in Table S1. The network of the CPOA hydrogel was investigated via Scanning Electron Microscope (SEM) (FEI, Hillsboro, OR, USA, Apreo S LoVac). The effect of the FPBA substitution degree on gelation formation was thoroughly investigated. Based on these results, exosomes derived from mesenchymal stem cells were incorporated into the CPOA hydrogel, yielding an exosome-loaded dually dynamically cross-linked hydrogel (CPOA@Exos). The morphology of exosomes was observed using transmission electron microscopy (TEM) (JEOL, Tokyo, Japan, JEM-2100F).

2.5. Rheological Properties Test

The rheological properties of the CPOA hydrogels were evaluated using a rotational rheometer (Anton Paar, Graz, Austria, MCR 302e). Approximately 400 μL of the hydrogel sample was placed on the instrument plate, and the measuring gap was set to 1.0 mm at 25 °C. The test sequence was as follows: first, a time sweep was performed at a fixed frequency of 1.0 Hz and a strain of 1% to characterize the storage modulus (G') and loss modulus (G''). Subsequently, a strain amplitude sweep from 1% to 1000% was conducted at a constant

https://doi.org/10.3390/ma19020445

frequency of 1.0 Hz to determine the critical strain at which the hydrogel failed. Finally, the hydrogel's viscosity was examined under shear rates ranging from 0.001 s^{-1} to 10 s^{-1}.

2.6. Injectability and Self-Healing Property Evaluation

The injection force of the hydrogels was evaluated using a 1 mL dual-barrel syringe with a 21-gauge needle mounted on a universal testing machine (SPAI, Jinan, China, WDW-T05). The syringe was compressed at a rate of 10 mm/min, and the changes in force-displacement were recorded.

The self-healing capability of the CPOA hydrogels was evaluated through macroscopic and rheological assessments. For the macroscopic test, two CPOA hydrogels with different colors (pink and colorless) were prepared, each cut into two halves. The segments from different hydrogels were then tightly merged and incubated at room temperature for 1 h to assess the interfacial integration. Subsequently, a rotational rheometer (MCR 302e) was employed to conduct a cyclic strain test, alternating between low strain (1%, 50 s) and high strain (400%, 50 s) at a fixed frequency of 1.0 Hz. The G' and G'' modulus were recorded to quantify the healing behavior.

2.7. Adhesion Performance Test

The adhesion properties of the prepared CPOA hydrogels were evaluated via lap shear tests. A total of 200 μL of hydrogel was applied onto a rectangular pig skin with a width of 20 mm, resulting in an overlap area of 2 cm^2. Another piece of pig skin was then placed over it to ensure complete contact. The skin-hydrogel samples were clamped to the jaws of a universal tensile testing machine and subjected to tensile separation at a crosshead speed of 20 mm/min to measure the adhesive strength.

2.8. Degradation In Vitro Test

To investigate the degradation behavior, the hydrogels were immersed in PBS and H_2O_2 (500 μmol/L) solutions at 37 °C. The immersion solutions were refreshed, and the hydrogels were taken out to record their morphological changes every 24 h. After 7 days, all the liquids were removed, and the remaining hydrogels were freeze-dried and weighed. The mass retention rate at different time points was calculated using Equation (2),

$$Retention\ rate\ (\%) = \frac{w_t}{w_0} \times 100 \quad (2)$$

Whereas, w_0 and w_t refer to the initial mass and the freeze-dried mass at time t, respectively.

2.9. Measurement of Exosomes Release Profile

To minimize interference from hydrogel degradation products during exosomes release measurement, 100 μL of CPOA@Exos was loaded into a dialysis bag (MWCO: 8000–14,000). The bag was then immersed in 500 μL PBS or H_2O_2 (500 μmol/L) solutions and incubated at 37 °C. At predetermined time intervals, 20 μL of the supernatant was collected and replaced with an equal volume of the corresponding fresh solution. The released Exos were quantified using a BCA protein assay kit (Beyotime Biotechnology, Shanghai, China). Briefly, reagents A and B from the BCA kit were mixed at a ratio of 50:1. Subsequently, 200 μL of the prepared BCA working solution was added to each well, and the plate was incubated at 37 °C for 1 h. The absorbance at 562 nm was measured using a microplate reader (Tecan, Männedorf, Switzerland). The standard curve was generated

https://doi.org/10.3390/ma19020445

using the Exos standard data, and the Exos release curve was derived from this. The release rate of Exos from the CPOA hydrogel was calculated according to Equation (3)

$$Exos\ release\ rate(\%) = \frac{A_t - b}{A_1 - b} \times 100 \tag{3}$$

Whereas b represents the intercept of the standard curve, A_t represents the absorbance of 20 μL of supernatant mixed with 200 μL of BCA solution at time t, and A_1 represents the absorbance of Exos ($1/6 \times 10^9$ particles/mL) mixed with 200 μL of BCA solution.

2.10. Measurement of Antioxidant Properties

The antioxidant properties of hydrogels were evaluated by measuring their free radical scavenging activities against $ABTS\cdot^+$, DPPH· and hydroxyl radicals. For the $ABTS\cdot^+$ scavenging assay, a stock solution was prepared by mixing 7 mmol/L $ABTS\cdot^+$ with an equal volume of 2.45 mmol/L potassium persulfate solution. The mixture was allowed to react in the dark at 4 °C for 12 h before use. The stock solution was then diluted with water to achieve an absorbance of 0.7 ± 0.02 at 734 nm, thus obtaining the $ABTS\cdot^+$ working solution. Subsequently, 200 μL of CPOA hydrogel was added to 1 mL of $ABTS\cdot^+$ working solution. The mixture was incubated in the dark for 10 min, and the absorbance (A_1) was measured at 734 nm. For the control group, an equal volume of water was used instead of the hydrogel, and its absorbance was recorded as A_0. The $ABTS\cdot^+$ scavenging effect was calculated using Equation (4)

$$ABTS\cdot^+ scavenging\ effect\ (\%) = \frac{A_0 - A_1}{A_0} \times 100 \tag{4}$$

For the DPPH· scavenging assay, 200 μL of CPOA hydrogel was added into 1 mL of 0.1 mmol/L DPPH ethanol solution. The mixture was incubated in the dark for 30 min, and the absorbance was measured at 517 nm (A_1). Two controls were established: in the blank group, the DPPH solution was replaced with ethanol, while in the negative control, the hydrogel was replaced with water. Their absorbances were designated as A_2 and A_0, respectively. The DPPH· scavenging effect was determined according to Equation (5).

$$DPPH\cdot scavenging\ effect\ (\%) = (1 - \frac{A_1 - A_2}{A_0}) \times 100 \tag{5}$$

For the ·OH scavenging assay, a reaction mixture was prepared in a test tube by adding 0.3 mL of 10 mmol/L salicylic acid solution, 0.3 mL of 10 mmol/L ferrous sulfate solution and 0.3 mL of 8.8 mmol/L hydrogen peroxide solution. Then, 200 μL of CPOA hydrogel was soaked in the mixture solution and incubated in the dark at 37 °C for 30 min. After incubation, the absorbance of the solution was measured at 510 nm (A_1). In the blank control, salicylic acid solution was replaced with an equal volume of water, and in the negative control, the hydrogel was replaced with an equal volume of water. The corresponding absorbance values were recorded as A_2 and A_0, respectively. The ·OH scavenging effect was determined according to Equation (6).

$$\cdot OH\ scavenging\ effect\ (\%) = (1 - \frac{A_1 - A_2}{A_0}) \times 100 \tag{6}$$

2.11. Hemocompatibility Test

Red blood cells (RBCs) were isolated from anticoagulated rabbit blood (Solarbio, Beijing, China) by centrifugation at 1500 rpm for 10 min and subsequently diluted to 10%

https://doi.org/10.3390/ma19020445

(v/v) suspension. Meanwhile, 400 μL of CPOA hydrogel was immersed in 4 mL of PBS at 37 °C for 24 h to prepare the hydrogel extract. Then, 0.9 mL of the extract was mixed with 0.1 mL of RBC suspension and incubated at 37 °C for 4 h. After centrifugation at 1500 rpm for 5 min, the absorbance of the supernatant was measured at 541 nm. In the control, PBS and water were used to replace the extract as the negative and positive controls, respectively. The hemolysis rate was calculated using Equation (7), where $A_{positive}$, $A_{negative}$ and A_{sample} represent the absorbance of the positive control, negative control, and hydrogel group, respectively.

$$Hemolysis\ rate\ (\%) = \frac{A_{sample} - A_{negative}}{A_{positive} - A_{negative}} \times 100 \quad (7)$$

2.12. Cytotoxicity Assay

The compatibility of the hydrogel was evaluated using L929 cells via the CCK-8 assay and acridine orange (AO) (Baiao Leibo, Beijing, China) fluorescence staining. The CPOA hydrogel extract was obtained by immersing 1 mL of hydrogel in 10 mL of RPMI 1640 medium for 24 h. Meanwhile, L929 cells were seeded in 96-well plates at a density of 5000 cells/well. After being cultured for 12 h, the culture medium was replaced with hydrogel extract, and the cells were incubated for another 24 h or 48 h. Cells incubated with RPMI 1640 medium instead of the extract served as the negative control. Cell viability was determined using a CCK-8 kit according to the manufacturer's protocol and was calculated using Equation (8), where $A_{control}$, A_{blank}, and A_{sample} represent the absorbance of the control group, the blank group (medium without cells) and the hydrogel group, respectively.

$$Cell\ viability\ (\%) = \frac{A_{sample} - A_{blank}}{A_{control} - A_{blank}} \times 100 \quad (8)$$

In addition, AO staining was employed to assess cell viability and cell morphology. After washing with PBS, the cells were incubated with 100 μL of AO solution at 37 °C for 20 min. Following incubation, the cells were washed three times with PBS to remove excess dye. Cell morphology was observed under a fluorescence microscope (ECLIPSE Ts2, Nikon, Tokyo, Japan).

2.13. Scratch Wound Healing Assays

L929 cells were seeded in a 24-well plate and incubated for 24 h. A straight scratch wound was then created on the monolayer using a pipette tip. After washing with PBS, 500 μL of culture medium was added to each well, followed by the addition of 100 μL of CPOA or CPOA@Exos hydrogel. The cells were further incubated for 24 h, after which phase-contrast images were captured. The wound area was measured using ImageJ (version 1.54r) software, and the cell migration rate was calculated according to Equation (9), where S_0 represents the initial scratch area and S_a denotes the scratch area after 24 h.

$$Migration\ rate\ (\%) = \frac{S_0 - S_a}{S_0} \times 100 \quad (9)$$

2.14. Cell Proliferative Behaviors

To evaluate the cell proliferation-promoting effect of the CPOA@Exos hydrogel, a cell proliferation assay was performed using L929 cells. Specifically, the cells were seeded in 48-well plates at a density of 3000 cells/well. After culturing for 12 h, 50 μL of CPOA, 50 μL of CPOA@Exos, or 5 μL of Exos with 45 μL PBS were added to the corresponding wells, respectively. The cells were then co-cultured with these materials for 48 h and

https://doi.org/10.3390/ma19020445

120 h. At each time point, the morphology of cells was observed under a microscope first; subsequently, the cell viability was detected using a CCK-8 cell proliferation kit, and was calculated using Equation (8), where $A_{control}$, A_{blank}, and A_{sample} represent the absorbance of the control group at 48 h, the blank group (medium without cells) and the hydrogel group, respectively.

2.15. Establishment of a Diabetic Skin Wound Model

All animal experiments were approved by Animal Ethical and Welfare of Tianjin University (Approval No.: TJUE-2024-138). Type 1 diabetes was induced in C57BL/6 mice (male, weighing 20–25 g) through selective destruction of pancreatic β-cells using streptozotocin (STZ). Briefly, after a 16 h fast, the mice received a single intraperitoneal injection of STZ. Following the injections, the drinking water was replaced with a 7% glucose solution for 2 h. Three days post-injection of STZ, blood glucose levels were monitored daily for consecutive days. Mice with blood glucose levels exceeding 16.8 mmol/L were considered diabetic. Subsequently, an 8 mm full-thickness skin defect model was created on the back of the prepared diabetic mice. The wounds were treated with 200 μL of PBS, CPOA, Exos (10^9 particles/mL) and CPOA@Exos, respectively. Wound healing progression was documented using digital photography on days 0, 3, 7 and 14. The wound areas were recorded and quantified using ImageJ software, and the percentage of wound closure was calculated according to Equation (10), where S_0 represents the initial wound area and S_t denotes the wound area at day 3, 7 and 14.

$$Healing\ rate\ (\%) = (1 - \frac{S_t}{S_0}) \times 100 \tag{10}$$

2.16. Histochemical and Immunofluorescence Staining

At designated time points post-treatment, the mice were euthanized. Full-thickness skin tissues were harvested and fixed in 4% paraformaldehyde. For hematoxylin and eosin (H&E) staining (Solarbio, Beijing, China), deparaffinized sections were treated sequentially with hematoxylin (2 min), differentiation (10 s), water rinse, eosin (15 s), dehydration in ethanol, and mounting. For Masson's trichrome staining (Solarbio, Beijing, China), sections were stained with Weigert's hematoxylin (5 min), differentiated, blued (3 min), ponceau acid fuchsin (4 min), phosphomolybdic acid (1 min), aniline blue (10 s), rinsed between steps, dehydrated, and mounted. Images were acquired using a microscope (ECLIPSE Ts2, Nikon, Japan).

Additionally, immunofluorescence staining was performed to assess inflammation and angiogenesis through the healing process. Briefly, on days 3 and 7, the tissue sections were stained for CD86 and CD206 (1:400, Affinity Biosciences, Cincinnati, OH, USA) to observe macrophage polarization; On day 14, sections were stained for PCNA (1:400, Affinity Biosciences) to evaluate cell proliferation activity. Furthermore, immunohistochemical staining for IL-10 and TNF-α (1:400, Affinity Biosciences) was conducted on tissue sections on day 7 to assess the inflammatory cytokine expression levels during wound healing. Finally, the sections were examined under a fluorescence microscope (ECLIPSE Ts2, Nikon, Japan), and images were captured for further analysis. The ImageJ software was adopted for analyzing and calculating the cell fluorescence intensity.

2.17. Statistical Analysis

Each experiment was repeated at least three times ($n \geq 3$). The experimental data are presented as mean ± standard deviation. The experimental data were statistically analyzed

https://doi.org/10.3390/ma19020445

using Origin (version 2018) and Excel software. Statistical significance is indicated as follows: * $p < 0.05$, ** $p < 0.01$, *** $p < 0.001$, **** $p < 0.0001$.

3. Results and Discussion

3.1. Formation and Property of the Dual Dynamic Cross-Linked Hydrogel

To prepare a dual-crosslinked network hydrogel with ROS-responsive properties, OAlg was synthesized by periodate oxidation of alginate using sodium periodate. As shown in Figure 1a, new chemical shifts at 5.4 and 5.6 ppm appeared in the ^{1}H NMR spectra of OAlg compared with Alg, indicating the formation of a hemiacetal structure from the aldehyde group. In the FT-IR spectra, OAlg exhibited a new C=O stretching vibration at 1720 cm^{-1} (Figure 1b), which was absent in the spectrum of Alg. These results confirmed the successful synthesis of OAlg. The oxidation degree of OAlg was 14.74 $\pm$ 2.52%, measured by hydroxylamine hydrochloride titration. Meanwhile, FPBA was grafted onto CMCS chain via the amino groups of CMCS and the formyl groups of FPBA, yielding FPBA-grafted CMCS derivatives (CP). The substitution degree of FPBA could be modulated by varying the amount of FPBA. As shown in Figure 1c, compared to CMCS, the FT-IR spectra of CP exhibited new absorption bands at 1500 cm^{-1} and 1380 cm^{-1}, which are assigned to the C=C stretching vibration of the benzene ring and the B-O stretching vibration from the boric acid moiety, respectively. Furthermore, by comparing the ^{1}H NMR spectra of CMCS with CP1, new chemical shifts were observed in the range of 7.6–7.9 ppm (Figure 1d), which corresponded to the proton resonances of the benzene ring, indicating the successful synthesis of CP. The relative ratio of the integral area of the proton peaks (corresponding to carbons b, c, d and e) (Figure 1e) at 7.6–7.9 ppm to that of the proton peak associated with carbon a in CMCS at 4.31–4.61 ppm was used to calculate the substitution degree of FPBA, which were 6.00%, 16.50%, 24.75% and 31.75% for CP1, CP2, CP3 and CP4, respectively.

When CP and OAlg solutions were mixed, a dynamic cross-linked network was formed through the formation of Schiff base bonds between amino and aldehyde groups, as well as the creation of borate ester bonds between the adjacent hydroxyl groups on the G units of OAlg and the boronic acid structures in CP [37] (Figure 1e). The pH of the OAlg solution and CP2 solution was measured using a calibrated pH meter and determined to be 6.74 and 8.53, respectively. Upon mixing, the solution rapidly transitioned from a flowable liquid to a gel state. When the prepared hydrogel was immersed in a culture medium (pH 7.4), no detectable change in the medium pH was observed. The formation of dynamic covalent bonds was evidenced by the FT-IR spectra in Figure 1f, in which both COA and CP2OA hydrogels exhibited a decreased intensity in the C=O vibrational absorption compared with OAlg, suggesting the formation of Schiff base bonds. In addition, the spectrum of CP2OA revealed the asymmetric stretching vibration of B-O-C bond at 1090 cm^{-1}, indicating the formation of borate ester bonds. Moreover, hydrogen bonding among the boric acid moieties of CP was also a key contributor to the hydrogel network formation. Especially, with the increase in the substitution degree of FPBA, both the G' and G'' increased. When the substitution degrees of FPBA on CMCS exceeded 24.75%, the 6% CP3 and CP4 solutions spontaneously formed hydrogels (Figure 1g), with their G' exceeding their G'' (Figure 1h). This gelation is likely attributable to the fact that once the substitution degrees of boronic acid groups from the grafted FPBA reached a critical level, they readily formed an extensive hydrogen-bond network with other functional groups such as amino and hydroxyl groups [38]. Based on these results, CP1OA and CP2OA were selected to fabricate injectable hydrogels with a dual dynamic crosslinked network. The substitution degree of FPBA was the main factor governing hydrogel formation. CP2OA

https://doi.org/10.3390/ma19020445

exhibited the fastest gelation, requiring only 5.00 ± 0.71 s, compared to 1034.00 ± 137.64 s for COA (Figure 1i).

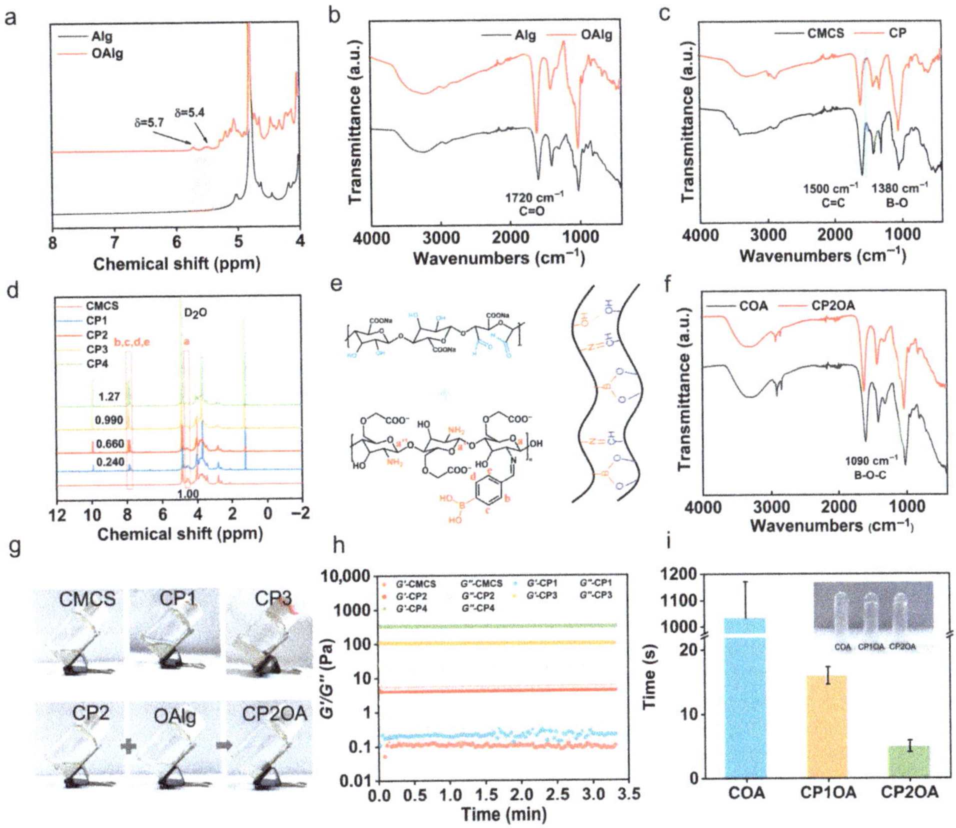

Figure 1. Formation and property of the dual dynamic cross-linked CPOA hydrogel. (**a**) ^{1}H NMR spectra of Alg and OAlg; (**b**) FT-IR spectra of Alg and OAlg; (**c**) FT-IR spectra of CMCS and CP; (**d**) ^{1}H NMR spectra of CMCS, CP1, CP2, CP3 and CP4; (**e**) The chemical structure and crosslinked structures of OAlg and CP; (**f**) FT-IR spectra of COA and CP2OA; (**g**) Spontaneous gelation of CP3 and hydrogel formation process of CP2OA; (**h**) Time sweep test (frequency = 1 Hz, strain = 1%) of CMCS, CP1, CP2, CP3 and CP4; (**i**) Gelation time of COA, CP1OA and CP2OA.

3.2. The Rheological and Adhesive Properties

As the primary structural components, the dynamic Schiff base and borate ester bonds directly dictated the mechanical properties of the CPOA hydrogels. The CP/OAlg ratio and the FPBA substitution degree were thus identified as critical parameters for modulating the network structure and its mechanical properties. As shown in Figure 2a, the G' of all the CPOA hydrogels exceeded their G'', confirming the formation of a stable gel network. The G' value was only 136.38 ± 14.88 Pa at a CP2/OAlg ratio of 3:7, but increased with the OAlg content, reaching a maximum of 451.72 ± 28.82 Pa at a ratio of 5:5. Furthermore, variations in the substitution degree of FPBA also significantly influenced the structure and properties of CPOA hydrogels. SEM images showed that the hydrogel's network structure became denser as the FPBA substitution degree in the CPs increased (Figure S1). Consistent with the trend, time sweep tests showed a corresponding increase

https://doi.org/10.3390/ma19020445

in G', with values of 97.07 ± 15.02, 269.54 ± 37.61 and 451.72 ± 28.82 Pa for COA, CP1OA and CP2OA, respectively (Figure 2b), further confirming the enhancement of mechanical properties with a higher FPBA substitution degree. In the strain amplitude sweeps, the G' of different hydrogels was greater than the G'' under low strain conditions. With the increase in strain, the curves of G' and G'' intersected, indicating the destruction of the hydrogel network structure. Notably, all hydrogel groups could withstand a strain exceeding 100% (Figure S2a,b), which is conducive to their tight adhesion to the wound site.

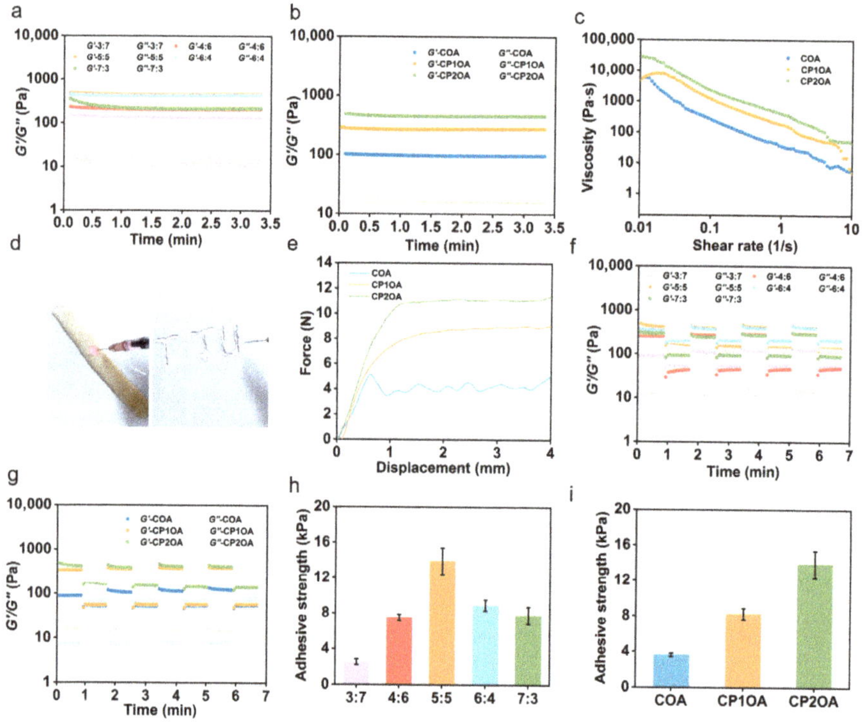

Figure 2. Rheological and adhesive properties of the hydrogels: (**a**) Time sweep test of hydrogels with different ratios; (**b**) Time sweep test of COA, CP1OA and CP2OA; (**c**) Shear-thinning behavior of COA, CP1OA and CP2OA; (**d**) Injectability demonstration; (**e**) Injection force of COA, CP1OA and CP2OA; (**f**) Cyclic strain test of hydrogels with different ratios (low strain = 1%, high strain = 400%); (**g**) Cyclic strain test of COA, CP1OA and CP2OA (low strain = 1%, high strain = 400%); (**h**) Lap shear test of hydrogels with different ratio; (**i**) Lap shear test of COA, CP1OA and CP2OA.

Owing to the dual dynamic crosslinked network, the CPOA hydrogel showed good self-healing property and injectability. As shown in Figure 2c and Figure S2c, the CPOA hydrogels exhibited good shear thinning properties regardless of the composites, with viscosity decreasing as shear rate increased. Therefore, the hydrogel can be injected through a syringe fitted with a 21-gauge needle and subsequently shaped into a stable "TJU"-shaped gel (Figure 2d). The injection force was dependent on the FPBA substitution degree. The COA hydrogel exhibited an injection force of 4.22 ± 0.43 N, which increased with the formation of borate ester crosslinks. Notably, the injection force of CP2OA reached 10.98 ± 0.20 N (Figure 2e). Despite these values, they allowed for easier handling during manual operation and helped minimize the risk of surgical damage [39]. The dynamic crosslinking nature of borate esters and Schiff bases also endowed the CPOA hydrogel with good self-healing properties. As demonstrated by continuous

https://doi.org/10.3390/ma19020445

alternating strain tests, under high strain (400%), the hydrogel structure underwent disrupted ($G'' > G'$), whereas returning to low strain (1%), the dynamic crosslinked network recovered ($G' > G''$). Notably, after three consecutive cycles, G' exhibited no significant change (Figure 2f,g). Moreover, after being cut and physically reconnected, the hydrogel rapidly restored its original structural integrity, owing to the reversible nature of the dynamic bonds (Figure S3a).

Upon injection into the wound surface, the aldehyde groups and phenylboronic acid groups present in CPOA hydrogels formed Schiff base bonds and hydrogen bonds with amino groups on the skin surface, resulting in strong adhesion to pig skin and various animal organs (Figure S3b). Bonding tests indicated that the CPOA hydrogel experienced cohesive failure upon peel-off (Figure S4), indicating that its mechanical properties were the primary determinant of adhesion strength. In the lap shear test, as the ratio of CP2/OAlg increased, the adhesive strength of the CPOA hydrogels first increased and then decreased, reaching a maximum when the ratio of CP2/OAlg was 5:5 (Figure 2h). Meanwhile, as the substitution degree of FPBA increased, the mechanical properties of the hydrogels improved; the adhesive strength measured for COA, CP1OA, and CP2OA was 3.65 ± 0.19, 8.24 ± 0.65, and 13.89 ± 1.52 kPa, respectively (Figure 2i), leading to a corresponding enhancement in adhesion performance.

3.3. ROS Responsiveness and Exos Release Behavior

3.3.1. ROS Responsive Properties

The prepared CPOA hydrogel exhibits good ROS responsiveness owing to the presence of borate ester bonds [27]. The COA hydrogel exhibited no significant difference in its retention ratio between PBS and H_2O_2 solution over 7 days. Specifically, its retention rates in PBS and H_2O_2 solution at day 7 were 50.28 ± 3.07% and 48.61 ± 2.15%, respectively. Upon introduction of a borate ester crosslinked structure, the retention rate of the CPOA hydrogel decreased in H_2O_2 solution, the values of CP1OA and CP2OA decreased from 43.00 ± 2.43% and 51.96 ± 3.19% in PBS to 27.45 ± 2.92% and 15.23 ± 6.92% in H_2O_2 solution, respectively (Figure 3a), demonstrating distinct ROS-responsive behavior. Meanwhile, the CP2OA hydrogel exhibited a faster volume reduction rate in the H_2O_2 solution owing to the ROS-responsive property of CP2OA (Figure S5).

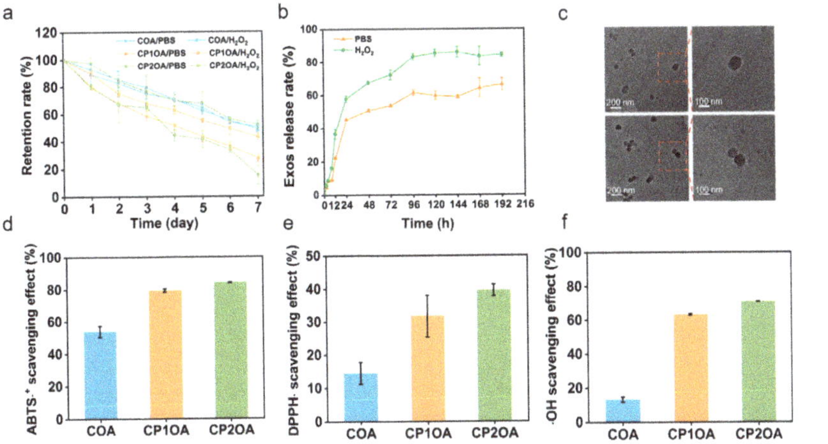

Figure 3. ROS responsiveness and Exos release behavior. (**a**) Degradation behaviors of CPOA hydrogels in PBS and H_2O_2 solution; (**b**) Exos release from CPOA hydrogels in PBS and H_2O_2 solution; (**c**) The Exos under TEM; Scavenging efficiency against (**d**) ABTS radical (ABTS·$^+$), (**e**) DPPH radical (DPPH·) and (**f**) hydroxyl radical (·OH).

https://doi.org/10.3390/ma19020445

3.3.2. Exos Release Behavior

The higher density of borate ester bonds in CP2OA resulted in a more pronounced ROS response, thereby facilitating the responsive release of Exos. As shown in Figure 3b, the cumulative release of Exos from CP2OA hydrogel calculated through the standard curve (Figure S6a) reached 84.37 ± 1.39% in H_2O_2 microenvironment, significantly higher than the 66.27 ± 3.75% released in PBS over 7 days. Additionally, the Exos a spherical morphology with diameters around 100 nm (Figure 3c), and the measured nanoparticle size also falls within the range of 30–150 nm (Figure S6b), enabling their encapsulation within the hydrogel while maintaining stability.

3.3.3. ROS Scavenging Ability

Given the ROS-responsive properties of the prepared CPOA hydrogels, we further evaluated their antioxidative activity. As shown in Figure 3d–f, the COA hydrogel exhibited significantly lower scavenging effect against $ABTS\cdot^+$, DPPH·, and ·OH compared with CPOA hydrogels. Furthermore, the scavenging efficiency for all three radicals was enhanced with an increase in the FPBA substitution degree, which raised the borate ester bond content within the hydrogel. CP2OA demonstrated high scavenging rates of 84.21 ± 0.27%, 39.59 ± 1.76% and 70.63 ± 0.17% against $ABTS\cdot^+$, DPPH·, and ·OH, respectively. These results indicate that CPOA hydrogels have considerable potential for mitigating excessively high ROS levels on diabetic wound surfaces.

3.4. *In Vitro Biological Properties*

Biocompatibility was an essential requirement for the application of CPOA hydrogel in the biomedical field [18]. The CPOA@Exos system was expected to exhibit inherent biocompatibility. This was confirmed by an in vitro hemolysis assay, which showed that the hemolysis rates of all hydrogels were below 5%, indicating no significant hemolytic effect on red blood cells (Figure 4a). The cytotoxicity assay using L929 cells showed similar results. As shown in Figure 4b, after co-culturing the hydrogel extract with cells for 24 and 48 h, CCK-8 assay results suggested that there was no significant difference in cell viability compared to the control group. Similarly, AO staining also demonstrated that all cells were stained green fluorescence, and the number of cells at 48 h significantly increased compared to that at 24 h, indicating the CPOA hydrogel has no cytotoxicity (Figure 4c). Moreover, the effects of CPOA@Exos on cell migration were evaluated using a scratch assay. As shown in Figure 4d,e, the width of the scratch wound decreased noticeably over time in all groups. The cell migration rate was 50.85 ± 3.66% in the control group. A slight increase was observed in the CPOA hydrogel group (54.49 ± 5.19%), while the incorporation of Exos significantly enhanced cell migration. Specifically, the CPOA@Exos group exhibited a cell migration rate of 64.87 ± 4.23%, which was significantly higher than that of the control group (** $p < 0.01$), though slightly lower than that of the free Exos group. To investigate the effects of CPOA@Exos on the proliferation behaviors of cells, cell morphology and viability were examined at 48 h and 120 h. As shown in Figure S7a, the cell number in all groups was significantly higher at 120 h than at 48 h. At each time point, Exos and CPOA@Exos groups exhibited significantly higher cell numbers compared with the control and CPOA groups. Consistent with these observations, the CCK-8 assay confirmed that the highest proliferation rate occurred in CPOA@Exos groups (Figure S7b).

https://doi.org/10.3390/ma19020445

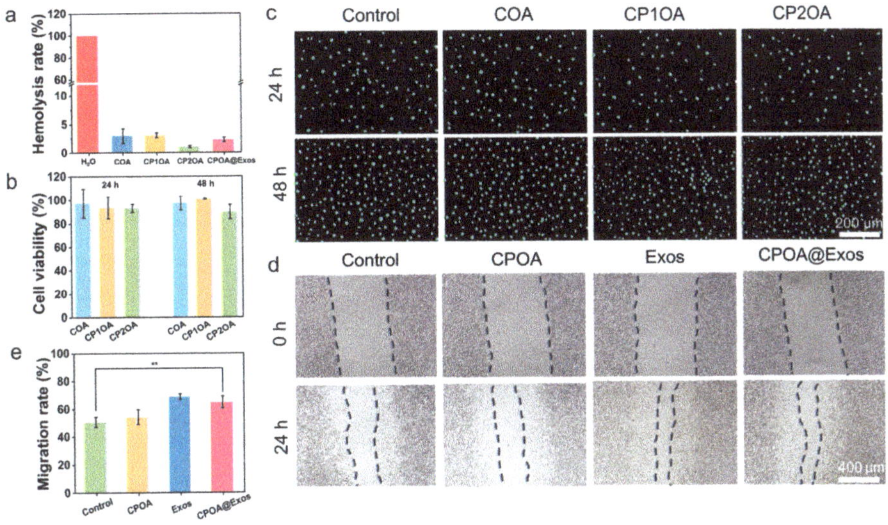

Figure 4. Biocompatibility and cell migratory activity of CPOA@Exos. (**a**) Hemocompatibility; (**b**) Cell viability of L929 cells assessed by CCK-8 assay; (**c**) AO fluorescence staining images of L929 cells after treatment with different hydrogels; (**d**) Cell migration images of L929 cells after co-cultured 24 h with different hydrogels or Exos; (**e**) Statistical analysis of cell migration rate in migration images. (** $p < 0.01$).

3.5. Repair Effects of CPOA@Exos on Diabetic Wound In Vivo

To evaluate the repair effects of CPOA@Exos on diabetic wounds *in vivo*, an STZ-induced diabetic mouse model was established in C57 mice. Three days after STZ injection, all mice developed diabetes, characterized by sustained blood glucose levels exceeding 16.8 mmol/L, which confirmed the establishment of the model (Figure S8). After treated with different materials, the healing process was visually monitored. As shown in Figure 5a,b, the wound area decreased progressively in all groups, with all treatment groups demonstrating markedly enhanced healing rate relative to the control. On day 7, the CPOA and free Exos groups exhibited healing rates of 55.97 ± 2.82% and 65.21 ± 2.67%, respectively. Notably, the CPOA@Exos group showed a higher healing rate of 79.70 ± 2.03% than all other groups (** $p < 0.01$), approaching nearly complete healing (97.97 ± 0.90%) by day 14. This accelerated wound repair can be attributed to the complementary functions of CPOA and Exos, such as wound coverage [17], antioxidant activity [27], and anti-inflammatory activity [40]. Moreover, the ROS-responsive nature of the CPOA hydrogel enables a more controlled and prolonged release of Exos, thereby enhancing their utilization efficiency and therapeutic duration [41]. To further evaluate diabetic wound healing, histological analysis of the injured skin tissue was stained using H&E and Masson's trichrome staining at day 14. As shown in Figure 5c, the CPOA@Exos group displayed abundant newly formed hair follicles near the wound site along with a restored and smooth epidermal surface, collectively indicating advanced tissue maturation and superior healing. Moreover, Masson's trichrome staining (Figure 5d) revealed more collagen deposition in the CPOA@Exos group compared to other groups. During the early phase of wound healing, collagen fibers are characteristically loose and randomly oriented, whereas they gradually assemble into a dense, well-aligned architecture as tissue repair proceeds toward completion. These phenomena were consistent with the previous report [42]. Notably, the collagen fibers in the CPOA@Exos group displayed a distinctly

 https://doi.org/10.3390/ma19020445

denser and more organized arrangement relative to the control group, suggesting improved extracellular matrix remodeling and wound repair.

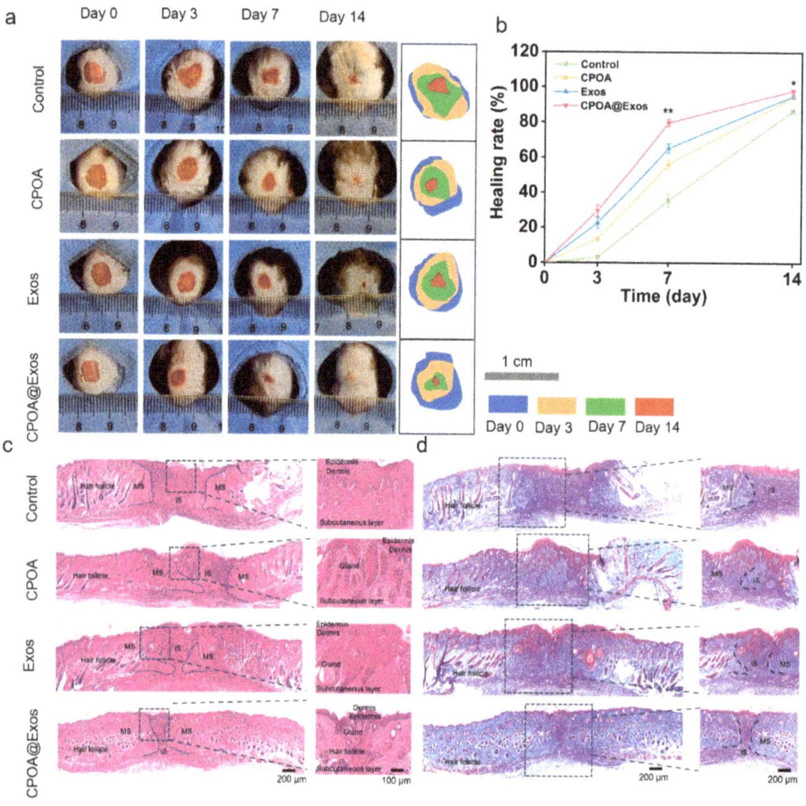

Figure 5. Hydrogel promoted diabetic wound healing: (**a**) Changes in wound size on days 0, 3, 7 and 14; (**b**) Wound healing rate; (**c**) H&E-stained images at day 14, immature skins (IS), mature skins (MS); (**d**) Masson's trichrome-stained images at day 14. (* $p < 0.05$, ** $p < 0.01$).

3.6. Modulation of the Diabetic Wound Microenvironment

The diabetic inflammatory response is a major factor impairing diabetic wound healing [15]. Given this, we evaluated the immunomodulatory effects of CPOA@Exos on the diabetic wound microenvironment via immunofluorescence staining for the macrophage markers CD86 and CD206. As shown in Figure 6a, CD86-labeled pro-inflammatory M1 phenotype macrophages (red) were predominant in the control group on day 3, with only a sparse presence of CD206-labeled anti-inflammatory M2 phenotype macrophages (green). The control group exhibited severe inflammatory responses, with macrophages predominantly polarized into the M1 phenotype characterized by phagocytic and clearance properties, showing a CD86 expression rate of 13.44 ± 1.61%. Notably, M1 macrophages remained the dominant population (7.22 ± 0.97%) even by day 7 (Figure 6b). Furthermore, due to the good antioxidant capacity of the CPOA hydrogel and the immunomodulatory function of Exos, the CD86-positive cell population in both CPOA and Exos groups was significantly decreased compared to the control group on both days 3 and 7. A corresponding increase was observed in the expression levels of CD206-positive cells. As expected, the CPOA@Exos group demonstrated the lowest CD86 expression level of 2.77 ± 0.86%

https://doi.org/10.3390/ma19020445

on day 3 and 0.45 ± 0.22% on day 7 (* $p < 0.05$, ** $p < 0.01$), alongside the highest CD206 expression levels (4.63 ± 0.50% and 8.25 ± 1.18% on days 3 and 7) (Figure 6c). At the same time, CPOA@Exos effectively upregulated the anti-inflammatory cytokine IL-10 (Figure S9a,b) and downregulated the pro-inflammatory cytokine TNF-α (Figure S9c) compared to the control group, thereby suppressing excessive inflammatory responses in the wound microenvironment.

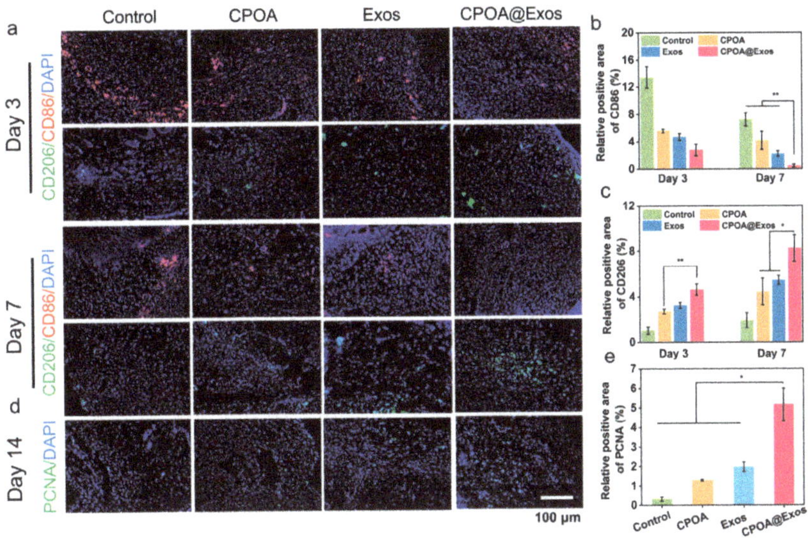

Figure 6. Modulation of the diabetic wound microenvironment in vivo. (**a**) Immunofluorescence staining images of CD86 and CD206 in the wound area on days 3 and 7; Statistical analysis of (**b**) CD86 and (**c**) CD206 fluorescence intensity; (**d**) PCNA immunofluorescence staining images in wound areas at day 14, (**e**) Statistical analysis of PCNA fluorescence intensity in immunofluorescence images. (* $p < 0.05$, ** $p < 0.01$).

These results suggest a phenotypic shift in macrophages within the wound tissue, towards the anti-inflammatory M2-phenotype macrophages, which are known to secrete factors that promote angiogenesis and resolve inflammation. This transition indicates that CPOA@Exos can effectively modulate macrophage polarization to M2 during the inflammatory phase of wound healing. Owing to its modulatory effect on the inflammatory microenvironment, the developed CPOA@Exos hydrogel can further regulate cellular proliferation in the injured area. As shown in Figure 6d,e, the number of PCNA-positive cells in CPOA and Exos increased compared with the control group, and the number of PCNA-positive cells in CPOA@Exos was the highest compared with other groups.

4. Conclusions

In summary, this study fabricated a hydrogel by grafting FPBA onto CMCS and mixed it with OAlg, resulting in a dynamic crosslinked network based on Schiff base and borate ester bonds. The hydrogel can effectively load exosomes. The prepared CPOA hydrogel demonstrated stable mechanical strength, tissue adhesion, and ROS responsiveness. Moreover, the dual-crosslinked hydrogel enhanced multiple hydrogel characteristics, including suitable gelation time, self-healing ability, injectability, mechanical robustness, and adhesiveness. Leveraging dynamic borate ester bonds, it enabled sustained release of exosomes at the diabetic wound site while effectively scavenging excess ROS, thereby facilitating the

https://doi.org/10.3390/ma19020445

healing process. The CPOA@Exos system was shown to modulate inflammatory responses in diabetic wounds, promote macrophage polarization toward the M2 phenotype, and accelerate wound repair. This work presents a promising therapeutic strategy for the management of diabetic wounds.

Supplementary Materials: The following supporting information can be downloaded at: https://www.mdpi.com/article/10.3390/ma19020445/s1, Table S1: The formulations of hydrogels. Figure S1. Morphology of COA, CP1OA and CP2OA hydrogel, Figure S2: Rheological properties of the hydrogels: (a) Strain amplitude sweeps test of hydrogels with different ratios; (b) Strain amplitude sweeps test of COA, CP1OA and CP2OA; (c) Shear-thinning behavior of hydrogels with different ratios, Figure S3: Pictures of self-healing and adhesion of hydrogels: (a) Self-healing process of hydrogels; (b) Adhesive performance of the hydrogels on various substrates, Figure S4: The adhesion stress of COA, CP1OA and CP2OA hydrogel via bonding test, Figure S5: Morphological and pH value changes of CP2OA immersed in PBS and H_2O_2 within 7 days, Figure S6: Exos characterization: (a) The standard curve of Exos assessed by BCA kit; (b) Particle size of the Exos. Figure S7: Cell proliferative behaviors in the presence of CPOA@Exos. (a) Cell morphology at 48 h and 120 h; (b) Cell viability rate at 48 h and 120 h (* $p < 0.05$, ** $p < 0.01$), Figure S8: Changes in blood glucose in mice after injection of STZ, Figure S9: Immunohistochemical staining images of diabetic wounds on day 7 after treatment by CPOA@Exos. (a) Immunohistochemical images of IL-10 and TNF-α; (b) Positive area ratio of IL-10; (c) Positive area ratio of TNF-α (* $p < 0.05$, ** $p < 0.01$).

Author Contributions: Conceptualization, D.L., F.Y., H.Z. and J.L.; methodology, D.L., Z.L., J.H. and X.L. (Xiuqiang Li); investigation, D.L., Z.L., J.H., X.L. (Xiuqiang Li), Y.F. and X.L. (Xiaochen Lu); validation, D.L., Z.L., J.H., X.X., X.L. (Xiuqiang Li), Y.F. and X.L. (Xiaochen Lu); resources, J.L., F.Y. and H.Z.; data curation, D.L.; writing—original draft preparation, D.L. and J.L.; writing—review and editing, D.L. and J.L.; visualization, D.L. and Z.L.; supervision, J.L., F.Y. and H.Z.; funding acquisition, J.L., F.Y. and H.Z. All authors have read and agreed to the published version of the manuscript.

Funding: This study was funded by the State Key Laboratory of Advanced Medical Materials and Devices Research Grant (YGSKL-TJU-2025-KF06).

Institutional Review Board Statement: This study was approved on 28 February 2024 by the Animal Ethical and Welfare Committee of Tianjin University (Approval No.: TJUE-2024-138).

Informed Consent Statement: Not applicable.

Data Availability Statement: The original contributions presented in this study are included in the article/Supplementary Material. Further inquiries can be directed to the corresponding author.

Conflicts of Interest: The authors declare no conflicts of interest.

References

1. Cho, N.H.; Shaw, J.E.; Karuranga, S.; Huang, Y.; da Rocha Fernandes, J.D.; Ohlrogge, A.W.; Malanda, B. IDF Diabetes Atlas: Global estimates of diabetes prevalence for 2017 and projections for 2045. *Diabetes Res. Clin. Pract.* **2018**, *138*, 271–281. [CrossRef]
2. Shaw, J.E.; Sicree, R.A.; Zimmet, P.Z. Global estimates of the prevalence of diabetes for 2010 and 2030. *Diabetes Res. Clin. Pract.* **2010**, *87*, 4–14. [CrossRef]
3. Singh, N.; Armstrong, D.G.; Lipsky, B.A. Preventing foot ulcers in patients with diabetes. *JAMA* **2005**, *293*, 217–228. [CrossRef]
4. Sun, B.K.; Siprashvili, Z.; Khavari, P.A. Advances in skin grafting and treatment of cutaneous wounds. *Science* **2014**, *346*, 941–945. [CrossRef]
5. Zhu, S.; Zhao, B.; Li, M.; Wang, H.; Zhu, J.; Li, Q.; Gao, H.; Feng, Q.; Cao, X. Microenvironment responsive nanocomposite hydrogel with NIR photothermal therapy, vascularization and anti-inflammation for diabetic infected wound healing. *Bioact. Mater.* **2023**, *26*, 306–320. [CrossRef] [PubMed]
6. Hu, N.; Cai, Z.; Jiang, X.; Wang, C.; Tang, T.; Xu, T.; Chen, H.; Li, X.; Du, X.; Cui, W. Hypoxia-pretreated ADSC-derived exosome-embedded hydrogels promote angiogenesis and accelerate diabetic wound healing. *Acta Biomater.* **2023**, *157*, 175–186. [CrossRef] [PubMed]

7. Deng, L.; Du, C.; Song, P.; Chen, T.; Rui, S.; Armstrong, D.G.; Deng, W. The Role of Oxidative Stress and Antioxidants in Diabetic Wound Healing. *Oxidative Med. Cell. Longev.* **2021**, *2021*, 8852759. [CrossRef]
8. Lohmann, N.; Schirmer, L.; Atallah, P.; Wandel, E.; Ferrer, R.A.; Werner, C.; Simon, J.C.; Franz, S.; Freudenberg, U. Glycosaminoglycan-based hydrogels capture inflammatory chemokines and rescue defective wound healing in mice. *Sci. Transl. Med.* **2017**, *9*, eaai9044. [CrossRef]
9. Monaghan, M.G.; Borah, R.; Thomsen, C.; Browne, S. Thou shall not heal: Overcoming the non-healing behaviour of diabetic foot ulcers by engineering the inflammatory microenvironment. *Adv. Drug Deliv. Rev.* **2023**, *203*, 115120. [CrossRef] [PubMed]
10. Mohsin, F.; Javaid, S.; Tariq, M.; Mustafa, M. Molecular immunological mechanisms of impaired wound healing in diabetic foot ulcers (DFU), current therapeutic strategies and future directions. *Int. Immunopharmacol.* **2024**, *139*, 112713. [CrossRef]
11. Jin, N.; Wang, Z.; Tang, X.; Jin, N.; Wang, X. Promoting Diabetic Wound Healing through a Hydrogel-Based Cascade Regulation Strategy of Fibroblast-Macrophage. *Adv. Healthc. Mater.* **2024**, *13*, 2400526. [CrossRef]
12. Tian, D.M.; Wan, H.H.; Chen, J.R.; Ye, Y.B.; He, Y.; Liu, Y.; Tang, L.Y.; He, Z.Y.; Liu, K.Z.; Gao, C.J.; et al. In-situ formed elastin-based hydrogels enhance wound healing via promoting innate immune cells recruitment and angiogenesis. *Mater. Today Bio.* **2022**, *15*, 100300. [CrossRef]
13. Villarreal-Leal, R.A.; Healey, G.D.; Corradetti, B. Biomimetic immunomodulation strategies for effective tissue repair and restoration. *Adv. Drug Deliv. Rev.* **2021**, *179*, 113913. [CrossRef] [PubMed]
14. Hesketh, M.; Sahin, K.B.; West, Z.E.; Murray, R.Z. Macrophage phenotypes regulate scar formation and chronic wound healing. *Int. J. Mol. Sci.* **2017**, *18*, 1545. [CrossRef]
15. Boateng, J.; Catanzano, O. Advanced Therapeutic Dressings for Effective Wound Healing—A Review. *J. Pharm. Sci.* **2015**, *104*, 3653–3680. [CrossRef] [PubMed]
16. Pereira, R.F.; Bártolo, P.J. Traditional Therapies for Skin Wound Healing. *Adv. Wound Care* **2016**, *5*, 208–229. [CrossRef] [PubMed]
17. Zheng, B.D.; Xiao, M.T. Polysaccharide-based hydrogel with photothermal effect for accelerating wound healing. *Carbohydr. Polym.* **2023**, *299*, 120228. [CrossRef]
18. Shao, M.; Shi, Z.; Zhang, X.; Zhai, B.; Sun, J. Synthesis and Properties of Biodegradable Hydrogel Based on Polysaccharide Wound Dressing. *Materials* **2023**, *16*, 1358. [CrossRef]
19. Liang, L.; Zhang, H.; Yao, F.; Li, J. Programmable tissue-adhesive hydrogels with temporal and spatial selectivity. *Chem. Soc. Rev.* **2025**, *54*, 12043–12079. [CrossRef]
20. Wang, X.; Song, R.; Johnson, M.; A, S.; He, Z.; Milne, C.; Wang, X.; Lara-Sáez, I.; Xu, Q.; Wang, W. An Injectable Chitosan-Based Self-Healable Hydrogel System as an Antibacterial Wound Dressing. *Materials* **2021**, *14*, 5956. [CrossRef]
21. Zhao, J.; Zhu, H.; Xu, T.; Xu, S.; Wang, Y.; Hou, Y.; Zhuo, M.; Du, K.; Su, Y.; Han, X.; et al. Self-Healing Oxidized Dextran/Sodium Alginate Hydrogel Dressing with Hemostatic Activity Speeds Up Wound Healing in Burn Injuries. *ACS Appl. Mater. Interfaces* **2025**, *17*, 2940–2951. [CrossRef]
22. Liu, X.; Hu, J.; Hu, Y.; Liu, Y.; Wei, Y.; Huang, D. Multifunctional injectable oxidized sodium alginate/carboxymethyl chitosan hydrogel for rapid hemostasis. *Colloids Surf. B Biointerfaces* **2025**, *245*, 114346. [CrossRef]
23. Xie, M.; Zeng, Y.; Wu, H.; Wang, S.; Zhao, J. Multifunctional carboxymethyl chitosan/oxidized dextran/sodium alginate hydrogels as dressing for hemostasis and closure of infected wounds. *Int. J. Biol. Macromol.* **2022**, *219*, 1337–1350. [CrossRef]
24. Yang, Y.; Zhong, S.; Meng, F.; Cui, X. Multi-Functional hydrogels to promote diabetic wound Healing: A review. *Chem. Eng. J.* **2024**, *497*, 154855. [CrossRef]
25. Gan, Z.; Xiao, Z.; Zhang, Z.; Li, Y.; Liu, C.; Chen, X.; Liu, Y.; Wu, D.; Liu, C.; Shuai, X.; et al. Stiffness-tuned and ROS-sensitive hydrogel incorporating complement C5a receptor antagonist modulates antibacterial activity of macrophages for periodontitis treatment. *Bioact. Mater.* **2023**, *25*, 347–359. [CrossRef]
26. Marco-Dufort, B.; Tibbitt, M.W. Design of moldable hydrogels for biomedical applications using dynamic covalent boronic esters. *Mater. Today Chem.* **2019**, *12*, 16–33. [CrossRef]
27. Miao, Y.; Wei, J.; Chen, X.; Shi, J.; Zhang, L.; Wang, L.; Yang, J.; Ma, L.; Duan, J. Evaluation of living bacterial therapy assisted by pH/reactive oxygen species dual-responsive sodium alginate-based hydrogel for wound infections. *Int. J. Biol. Macromol.* **2024**, *271*, 132536. [CrossRef] [PubMed]
28. Pan, G.; Li, M.; Mu, L.; Huang, Y.; Liang, Y.; Guo, B. Photothermal/Photodynamic Synergistic Antibacterial Hydrogel Dressing with pH/Glucose Dual Responsive Pirfenidone Release for Diabetic Foot Ulcers. *Adv. Funct. Mater.* **2025**, *35*, 2416205. [CrossRef]
29. Li, J.; Du, C.; Yang, X.; Yao, Y.; Qin, D.; Meng, F.; Yang, S.; Tan, Y.; Chen, X.; Jiang, W.; et al. Instantaneous Self-Healing Chitosan Hydrogels with Enhanced Drug Leakage Resistance for Infected Stretchable Wounds Healing. *Small* **2025**, *21*, 2409641. [CrossRef] [PubMed]
30. Xu, Z.; Liang, B.; Tian, J.; Wu, J. Anti-inflammation biomaterial platforms for chronic wound healing. *Biomater. Sci.* **2021**, *9*, 4388–4409. [CrossRef]

 https://doi.org/10.3390/ma19020445

31. Kalluri, R.; LeBleu, V.S. The biology, function, and biomedical applications of exosomes. *Science* **2020**, *367*, eaau6977. [CrossRef]
32. Xie, Z.; Hao, H.; Tong, C.; Cheng, Y.; Liu, J.; Pang, Y.; Si, Y.; Guo, Y.; Zang, L.; Mu, Y.; et al. Human umbilical cord-derived mesenchymal stem cells elicit macrophages into an anti-inflammatory phenotype to alleviate insulin resistance in type 2 diabetic rats. *Stem Cells* **2016**, *34*, 627–639. [CrossRef]
33. Zhao, H.; Shang, Q.; Pan, Z.; Bai, Y.; Li, Z.; Zhang, H.; Zhang, Q.; Guo, C.; Zhang, L.; Wang, Q. Exosomes From Adipose-Derived Stem Cells Attenuate Adipose Inflammation and Obesity Through Polarizing M2 Macrophages and Beiging in White Adipose Tissue. *Diabetes* **2017**, *67*, 235–247. [CrossRef]
34. Gao, Y.; Yuan, Z.; Yuan, X.; Wan, Z.; Yu, Y.; Zhan, Q.; Zhao, Y.; Han, J.; Huang, J.; Xiong, C.; et al. Bioinspired porous microspheres for sustained hypoxic exosomes release and vascularized bone regeneration. *Bioact. Mater.* **2022**, *14*, 377–388. [CrossRef]
35. Shiekh, P.A.; Singh, A.; Kumar, A. Exosome laden oxygen releasing antioxidant and antibacterial cryogel wound dressing OxOBand alleviate diabetic and infectious wound healing. *Biomaterials* **2020**, *249*, 120020. [CrossRef] [PubMed]
36. Tao, S.-C.; Guo, S.-C.; Li, M.; Ke, Q.-F.; Guo, Y.-P.; Zhang, C.-Q. Chitosan Wound Dressings Incorporating Exosomes Derived from MicroRNA-126-Overexpressing Synovium Mesenchymal Stem Cells Provide Sustained Release of Exosomes and Heal Full-Thickness Skin Defects in a Diabetic Rat Model. *Stem Cells Transl. Med.* **2017**, *6*, 736–747. [CrossRef] [PubMed]
37. Li, L.; Bai, G.; Gu, W.; Niu, C.; Feng, Y.; Wei, Z.; Chen, K.; Guo, X. Boronic ester bonds hydrogel with temperature and pH responsiveness for controlled release fertilizer. *Ind. Crops Prod.* **2024**, *221*, 119336. [CrossRef]
38. Liu, W.; Ge, F.; Du, P.; Yan, X.; Jiang, H. High-performance room-temperature phosphorescence materials based on arylboronic acids activated by a dual hydrogen-bonded network. *Chem. Eng. J.* **2025**, *522*, 167561. [CrossRef]
39. Ma, W.; Dong, W.; Zhao, S.; Du, T.; Wang, Y.; Yao, J.; Liu, Z.; Sun, D.; Zhang, M. An injectable adhesive antibacterial hydrogel wound dressing for infected skin wounds. *Biomater. Adv.* **2022**, *134*, 112584. [CrossRef] [PubMed]
40. Xia, W.; Jin, M.; Feng, Z.; Zhang, J.; Rong, Y.; Zhang, Y.; Zhang, S.; Yu, Y.; Yang, H.; Wang, T. Injectable decellularzied extracellular matrix hydrogel derived from human umbilical cord: A novel perspective to deal with refractory wound via medical wastes. *Mater. Des.* **2023**, *229*, 111877. [CrossRef]
41. Jing, Y.; Huang, T.; Zhao, B.; Zhao, L.; Zhang, N.; Zhang, K.; Wang, K.; Wang, J.; Hua, J.; Tu, Q. A ROS/glucose stimulated-responsive ADSCs-derived exosomes-release hydrogel system for diabetic wound healing. *Chem. Eng. J.* **2024**, *487*, 150561. [CrossRef]
42. Xie, H.; Wang, Z.; Wang, R.; Chen, Q.; Yu, A.; Lu, A. Self-Healing, Injectable Hydrogel Dressing for Monitoring and Therapy of Diabetic Wound. *Adv. Funct. Mater.* **2024**, *34*, 2401209. [CrossRef]

https://doi.org/10.3390/ma19020445

materials

Article

Antimicrobial Properties of Polymer-Based Nanocomposites Modified by Nanoparticles Produced by Green Chemistry

Anna Wasilewska [1,2], Magda Bielicka [1,2], Urszula Klekotka [2], Grzegorz Markiewicz [3], Marek Jałbrzykowski [3], Wioleta Lewandowska [1,4], Izabela Swiecicka [4,5] and Beata Kalska-Szostko [2,*]

[1] Doctoral School, University of Bialystok, Ciołkowskiego 1K 1, 15-245 Białystok, Poland
[2] Faculty of Chemistry, University of Białystok, Ciołkowskiego 1K 1, 15-245 Białystok, Poland
[3] Faculty of Mechanical Engineering, Bialystok University of Technology, Wiejska 45C, 15-351 Białystok, Poland
[4] Faculty of Biology, University of Bialystok, Ciolkowskiego 1J, 15-245 Bialystok, Poland
[5] Laboratory of Applied Microbiology, University of Bialystok, Ciolkowskiego 1J, 15-245 Bialystok, Poland
* Correspondence: author: kalska@uwb.edu.pl; Tel.: +48-85-738-8069

Abstract

A significant driving force in nanotechnology development is the environmentally friendly synthesis of nanomaterials using natural extracts as reducing and stabilizing agents. In this study, silver and copper nanoparticles were synthesized and compared using two approaches: (1) a green synthesis pathway employing beetroot extract as a natural bio-reductant and stabilizer, and (2) a conventional chemical reduction method. The resulting nanoparticles were extensively characterized using transmission electron microscopy (TEM), X-ray diffraction (XRD), UV-Vis spectroscopy, and dynamic light scattering (DLS). The study revealed that the green synthesis route produced nanoparticles with well-defined morphology, high stability, and strong antimicrobial potential, outperforming those obtained via conventional chemical synthesis. Copper nanoparticles synthesized using beetroot extract exhibited particularly enhanced fungicidal and bactericidal properties, demonstrating the effectiveness of plant-based reducing agents in producing functional nanostructures. To further evaluate potential applications, the green-synthesized nanoparticles were incorporated into a polypropylene matrix, confirming their integrity and activity within the composite system. This work emphasizes the role of green synthesis in designing high-performance nanomaterials and highlights the promising capabilities of beetroot extract as a sustainable and efficient reducing and stabilizing medium for silver and copper nanoparticle production.

Keywords: Ag NPs; Cu NPs; green chemistry; composites; antimicrobial activity

1. Introduction

In recent years, the subject of polymer-based composites reinforced with nanoparticles has attracted the interest of many scientists [1–3]. Polymer nanocomposites are an innovative class of materials that exhibit improved functional properties due to the addition of very small particles (10^{-7} to 10^{-9} m) to the polymer matrix [2]. These include strength, mechanical resistance, flexibility, abrasion, mega-volt insulation, electrical conductivity, barrier to gas permeability, fire resistance, and scratch resistance, etc. [3]. Introducing a nanofiller changes the properties of the polymer, giving the nanocomposite attractive features that can be used, for example, in packaging design, medical applications, and many others [4]. Nanoparticles introduced into the polymer can have various shapes, e.g., nanofibers, nanotubes, nanorods, nanoplates, or nanograins [5]. Typically, the surfaces of the nanoparticles are modified to improve the adhesion of the polymer to the nanoparticles [6], which is a critical step to obtain new or maintain attractive properties of the nanocomposite [7]. An exciting direction in developing new nanocomposites is related to one possessing antimicrobial properties [8].

One of the types of polymer-inorganic nanocomposites is a system made of a polymer matrix and inorganic nanoparticles embedded in its structure [9]. Polymers used to create hybrid systems are divided in terms of their origin into two main groups: synthetic polymers and natural polymers (biopolymers, e.g., polysaccharides, lipids) [10]. Synthetic polymers can be precisely designed, and their properties are well adapted to the needs. On the other hand, biopolymers have a strictly defined structure, but their great diversity, internal biocompatibility, and biodegradability compensate for the lower possibility of manipulating the properties. Polyvinylpyrrolidone (PVP)-coated nanoparticles are characterized by good solubility in water and some organic solvents, thanks to which phase transfer between these solvents is possible [11]. For example, PVP-coated silver nanoparticles show high stability and biocompatibility [12]. Attention should also be paid to nanoparticles introduced into the polymer matrix. The presented research is focused on nanoparticles fabricated both by traditional and 'green' synthesis methods and checking whether synthesis routine affects the morphological properties and stability of composites, and their antimicrobial potential [13], since an exciting direction concerns usage of nanocomposites as antimicrobial agents [8].

Green synthesis of nanoparticles, based on the principles of green chemistry, aims to reduce hazardous substances, minimize energy consumption, and employ renewable or naturally derived reagents. Plant extracts are increasingly used as reducing and stabilizing agents, offering a non-toxic and low-cost alternative to chemical reduction. However, despite the growing number of studies, many plant sources remain underexplored. Beetroot extract is one such promising but insufficiently studied bioreductant. It contains betalains and phenolic compounds capable of reducing and stabilizing metal nanoparticles. Unlike commonly used extracts (e.g., green tea or neem), beetroot is inexpensive, widely available, and offers potential for developing new eco-friendly synthesis pathways.

Metal/polymer nanocomposites, due to the high dispersion potential of the metal clusters, the large nanoparticle/polymer contact surface ensures high reactivity and, ultimately, good metal-releasing properties. That can be attractive, for example, in plant cultivation and textile production, etc. [14,15]. Although much work has been devoted to synthesizing, researching, and developing metal-based nanocomposite materials, little is known about their antifungal and/or antibacterial properties [16]. This was the motivation to undertake the presented research. Prospective use of the developed nanomaterials embedded in the polymer can be envisaged in the preparation of antifungal coatings for use in, for example, the textile industry, gardening, etc. [17,18]. The dispersion of nanoparticles in the polymer improves the properties of the final nanocomposite. A significant challenge in preparing polymer nanocomposites is obtaining a homogeneous distribution of nanoparticles in the polymer matrix, knowing that better dispersion of nanoparticles allows for more significant matrix-nanoparticle interaction, which can be responsible for improving the biostatic properties of the material. Copper nanoparticles have a solid inhibitory and antifungal effect on a broad spectrum of fungi [19,20]. Nanosilver is known for its antibacterial, antifungal, and antiviral effects [21]. Currently, colloidal silver is used externally, i.e., on the skin and mucous membranes. This compound is often used in ophthalmology to treat bacterial conjunctivitis [22]. It is also possible to modify materials by depositing silver nanoparticles on carriers or

https://doi.org/10.3390/ma19020251

covering various surfaces with them so that they have biocidal, deodorizing, antistatic, and impregnating properties [22–24]. Research on polymer nanocomposites has developed rapidly over the last ten years [25]. Such nanocomposites have attracted the attention of industrial researchers due to their remarkable electrical, thermal, chemical, and biological properties and potential applications in various industrial sectors [24,25]. Despite rising interest in green synthesis, the current literature lacks systematic comparisons between nanoparticles produced by different green reducing agents and their chemically synthesized analogues. In particular, there is limited information on how beetroot extract influences nanoparticle formation, stability, and subsequent performance in polymer matrices.

Copper and silver nanoparticles mixed with a polymer can lead to more efficient antimicrobial properties [26]. Surface modification generally reduces the surface energy of nanoparticles, improving their affinity to the polymer matrix, which affects the processing of such a composite [27–29]. The present study compares the developed method of adding the nanoparticles produced by the 'green chemistry' routine and those synthesized by the traditional method (chemical reduction using borohydride) to the polymer. This work aimed to explore underused or less commonly studied natural reducing agents. Beetroot is not as widely used as, for example, green tea or neem, so it offers a novel contribution and comparative potential. The resulting product is assumed to be characterized by antifungal or antibacterial activity. The product's safety and durability are critical parameters that allow application in the integrated cultivation of vegetables or fruits, e.g., when creating agro textiles [30], or can be used as an additive to any textile products that require antibacterial or antifungal properties. Therefore, the research results can increase the environmental sustainability of nanoparticle production processes, which, in turn, suggests many possible applications, including medicine or engineering [31].

Synthesized by two methods, the nanoparticles were embedded into a polymer and then used for the measurements. The described method consists of adding to the machine at the same time polymer (polypropylene (PP)) and a liquid with nanoparticles produced by 'green chemistry' and those fabricated by chemical reduction. Overall, using PP allows us to test the integration of green-synthesized nanoparticles in a realistic and industrially relevant polymer matrix, highlighting the potential for commercial application of the developed nanomaterials. Despite growing interest in green synthesis of nanoparticles for incorporation into polymer composites, significant research gaps remain. In particular, existing studies often lack detailed comparisons of the advantages of specific plant extracts, such as beetroot, as reducing and stabilizing agents. This results in an incomplete understanding of how these bio-sources influence nanoparticle properties and their subsequent effects on the composite performance. Therefore, this study aims to address these gaps by systematically evaluating the use of beetroot extract for the green synthesis of silver and copper nanoparticles, and assessing their incorporation into polymer matrices. By thoroughly comparing chemically synthesized counterparts and analyzing the mechanistic aspects of nanoparticle formation and antimicrobial activity, the present work strengthens the motivation for applying green-synthesized nanoparticles in polymer composites.

2. Experimental

2.1. Materials

Silver nitrate ($AgNO_3$, ≥99% purity) was purchased from POCH S.A. (Gliwice, Poland). Copper(II) nitrate ($Cu(NO_3)_2{\cdot}3H_2O$, ≥99% purity) was obtained from Chempur (Piekary Śląskie, Poland). Sodium dodecyl sulfate (SDS, ≥98% purity) and sodium borohydride ($NaBH_4$, ≥98% purity) were supplied by Sigma-Aldrich (St. Louis, MO, USA) and used without further purification. Polypropylene (PP, isotactic grade) was provided by Thermo

https://doi.org/10.3390/ma19020251

Scientific (Waltham, MA, USA). Fresh beetroot (*Beta vulgaris* L.) was sourced from a certified local supplier and processed immediately to prepare the natural reducing and stabilizing extract. All aqueous solutions were prepared using deionized water (18.2 MΩ·cm) produced by a Merck Millipore Direct-Q 3UV-R system (Merck KGaA, Darmstadt, Germany).

2.2. Materials and Characterization

Particle structure, morphology, diameter, and size distribution were identified by transmission electron microscopy (FEI Tecnai G2 X-TWIN at 200 kV, Hillsboro, OR, USA). Samples were prepared by dropping freshly prepared nanoparticle solution on the carbon-covered 400 mesh Cu grid and left for drying. Polymer homogeneity and particle incorporation into PP were verified by scanning electron microscopy (SEM—FEI INSPECT S60, Hillsboro, OR, USA) equipped with energy dispersive X-ray spectroscopy (EDX). The crystal structure and the resulting crystallites' size were measured by X-ray diffraction using an XRD diffractometer Agilent Technologies SuperNova (Santa Clara, CA, USA) with a Mo microfocused source, $K_{\alpha 2}$ = 0.71367. A small number of nanoparticles was placed on a nylon loop covered with highly viscous oil for each measurement. Dynamic Light Scattering estimated the hydrodynamic size using a DLS spectrometer, Nanoplus (Gerbrunn, Germany), which performed experiments in water at 20 ± 2 °C. A UV-Vis spectral analysis was performed to confirm the presence of nanoparticles in the obtained suspensions. The Able&Jasco V-670 spectrometer (JASCO Corporation, Tokyo, Japan) with a spectral range of 200–800 nm was used to detect the spectrum of particles dispersed in water.

2.3. Preparation of Beetroot Extract

Beets were purchased from a local store to prepare the beetroot extract. First, the beetroot was washed twice with distilled water to remove any residue (soil, pesticides, etc.). The beetroot was then peeled, cut into several pieces, and added to the water, which was heated to 80 °C for 1 h. After cooling, the resulting mixture was filtered to obtain a clear liquid. The pH of the obtained juice was tested to be 4.2. Figure 1 shows a scheme for obtaining nanoparticles using the described "green chemistry" method.

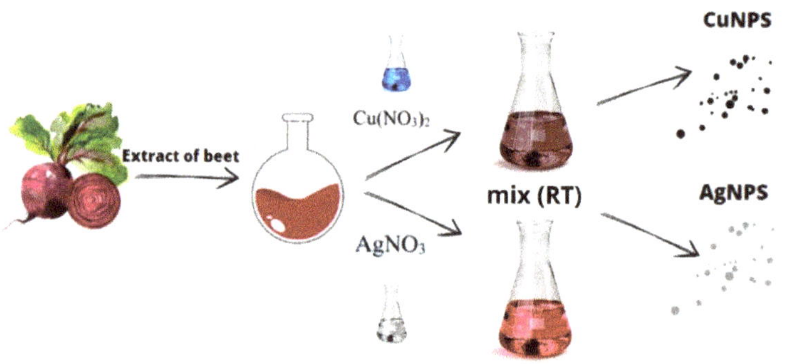

Figure 1. Schematic presentations of "green chemistry" synthesis in the presence of beetroot extract.

2.4. Synthesis

Two synthesis methods were used to obtain silver and copper nanoparticles. The first was based on the chemical reduction of Ag and Cu particles using sodium borohydride and sodium lauryl sulfate (SLS). The second is a green synthesis that uses beetroot extract as a reducing and stabilizing agent for the obtained nanoparticles. The salt used as a precursor of silver ions was silver nitrate ($AgNO_3$), and for the formation of copper nanoparticles, it was copper (II) nitrate ($Cu(NO_3)_2$). In both cases, the salt concentration used for synthesis was 0.1 mol/L.

https://doi.org/10.3390/ma19020251

Two nanoparticle formation processes for both silver and copper were followed:

(1) Sodium borohydride was used as a reducing agent, and sodium dodecyl sulfate was applied as a surface stabilizer. 25 mL of a salt solution and 0.015 g of sodium borohydride were mixed with 0.03 g of SDS-sodium dodecyl sulfate. The synthesis was carried out at room temperature. The solutions were stirred for 1 h. Acetone was used as a filtering solvent to obtain a clear solution. The product was centrifuged at 13,000 rpm for 30 min and then dried to a powder form. The synthesis was carried out at room temperature.

(2) Beetroot extract was used as a stabilizer and reducer. 100 g of beetroot was placed in 250 mL of distilled water. The whole process was carried out at 80 °C for 1 h. After heating, the entire thing was filtered and centrifuged twice to remove the beetroot residue. 25 mL of salt solution and 5 mL of beetroot extract were mixed for 1 h. The final synthesis step was carried out at room temperature. Acetone was used as a filtering solvent to obtain a clear solution. The produced liquid was centrifuged at 13,000 rpm for 30 min, and then the sediment was dried to a powder form.

The summary of the synthesis is presented in Table 1.

Table 1. Representation of sample names.

Sample	Synthesis
C-AgNPs	chemical reduction ($NaBH_4$ + SDS)
G-AgNPs	'green chemistry' (beetroot extract)
C-CuNPs	chemical reduction ($NaBH_4$ + SDS)
G-CuNPs	'green chemistry' (beetroot extract)

2.5. Composite Preparation

The following procedure was used to obtain polymer composites. An appropriate amount (15 g) of polypropylene (PP) was placed in a crusher (Zamak, Mercator, Viscomix VM-30, Skawina, Poland) and then crushed and kneaded at 160 °C for 2 min. After this time, an appropriate amount (7 mL) of the concentrated solution was added dropwise to the already plasticized PP, and the material was still kneaded to mix the components for 2 min. After this time, the formed composite was placed on a press table (Fentijne Press, LabEcon 300, Utrecht, The Netherlands) and kneaded at 200 kN for 10 s at 160 °C to create a thin foil. The developed thin film was cooled down. The procedure was used for all synthesized solutions.

2.6. Antimicrobial Activity

The antibacterial and antifungal activities of the samples of composites made of polymer—polypropylene (PP) and nanoparticles prepared through chemical synthesis and 'green chemistry' against Gram-positive (*Staphylococcus aureus* strain ATCC 6538 and *Bacillus cereus* strain ATCC 10987) and Gram-negative (*Escherichia coli* strain ATCC 11229) bacteria, and fungi (*Candida krusei* strain ATCC 30135) were assessed by measuring the inhibition zones obtained from the disk diffusion assay. Microorganisms stored at −80 °C in LB broth supplemented with glycerol (a ratio of 1:1) were inoculated onto nutrient agar (bacteria) or Sabouraud agar (fungi) and incubated overnight at 37 °C. One colony of the microorganisms was then reinoculated into LB broth and cultured overnight at 37 °C with gentle shaking. The cultures were suspended to a final optical density of 0.2–0.3 at 600 nm using a V-670 spectrophotometer (Jasco International Co., Tokyo, Japan). A volume of 100 μL of the suspension was placed onto agar and spread evenly with a smoothing plunger. Composites formed as foil were aseptically cut into discs with a diameter of 8 mm and placed on the agar surface. The microorganisms were incubated at

https://doi.org/10.3390/ma19020251

37 °C overnight. The antibacterial and antifungal activity of the composites was assessed visually after incubation based on the inhibition of microbial growth around and under the discs. Microorganism cultures with foil discs without composites on the medium surface were considered as a control. All the experiments were carried out in quadruplicate. The microbiological media used in the study were supplied by Oxoid Ltd. (Basingstoke, UK).

3. Results and Discussion

3.1. Transmission Electron Microscopy (TEM)

Details of the morphology and size of the synthesized nanoparticles were recorded using transmission electron microscopy and presented in Figure 2A. TEM images confirm the presence of nanoparticles of sizes and shapes dependent on the methods used during their synthesis. In frame B, a set of representative histograms is plotted. Based on the qualitative and quantitative analysis of the TEM images, the average diameters of the obtained nanostructures were calculated, their degree of agglomeration was determined, and the size distribution was assessed. For details, see Figure 2A and Table 2.

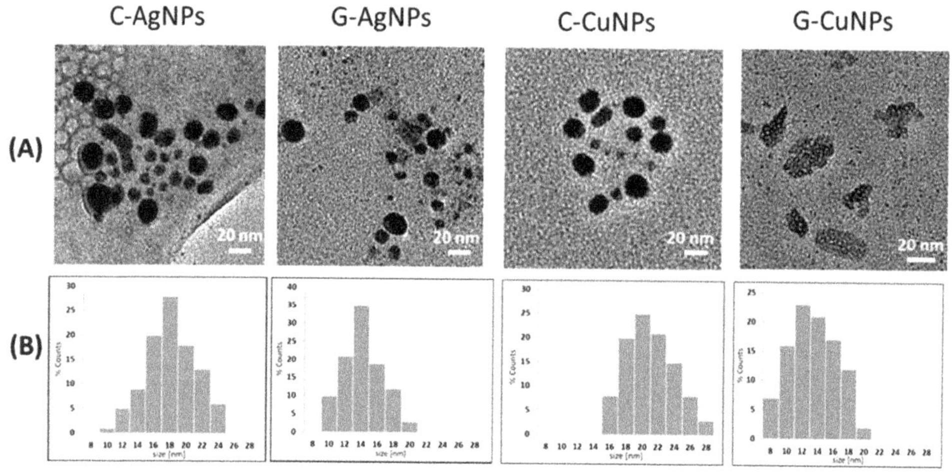

Figure 2. TEM images of the obtained nanoparticles (**A**); Respective histograms (**B**).

Table 2. Numerical values of the obtained nanoparticles.

Sample	Size (TEM) [nm]	UV-Vis Absorbance Maximum ± 5 [nm]	DLS Size [nm]
C-AgNPs	18 ± 3	405	18 ± 4
G-AgNPs	14 ± 2	425	32 ± 5
C-CuNPs	20 ± 3	320	15 ± 3
G-CuNPs	13 ± 3	321	30 ± 4

The TEM analysis revealed that the chemically synthesized silver nanoparticles (C-AgNPs) have an average size of 18 ± 3 nm. In contrast, the green-synthesized silver nanoparticles (G-AgNPs) are smaller, with an average size of 14 ± 2 nm. In the case of Ag, the homogeneity of nanoparticles is not perfect, but the particles are much more similar in size. The most homogeneous nanoparticles were obtained by chemical reduction of copper nanoparticles (see Figure 2A, C-CuNPs). 'Green chemistry' based synthesis of

 https://doi.org/10.3390/ma19020251

Cu results in uneven objects (see Figure 2A, G-CuNPs). For copper nanoparticles, the chemically synthesized CuNPs (C-CuNPs) showed an average size of 20 ± 3 nm, whereas the green-synthesized CuNPs (G-CuNPs) were slightly larger, averaging 13 ± 4 nm. The average diameter of the nanoparticles was estimated based on the TEM images using the commercial ImageJ IJ 1.46r program [32]. The numerical values obtained for each solution are summarized in Table 2.

Copper nanoparticles obtained in the presence of beetroot extract have particles consisting of leaf-like shapes with a rough surface and internal voids. According to the literature, similar polycrystalline images were also seen in TEM images in other Cu nanoparticles [10].

TEM measurements were performed to prove composite fabrication and successful incorporation of nanoparticles into the polymer (Figure 3).

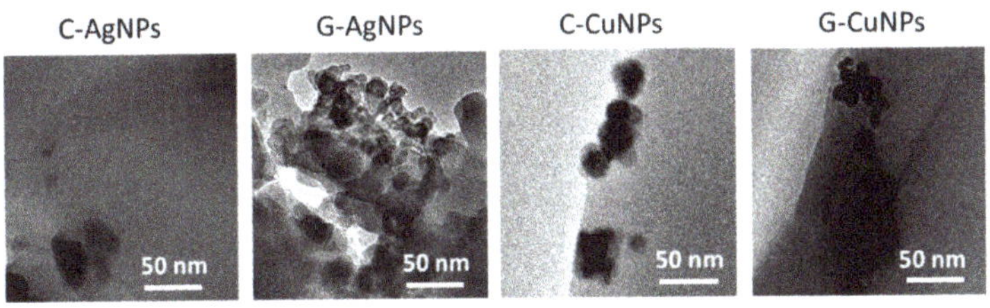

Figure 3. TEM images of a polymer with embedded silver and copper nanoparticles.

The TEM images in Figure 3 show that C-AgNPs, G-AgNPs, C-CuNPs, and G-CuNPs are in the bulk of the polymer. Silver nanoparticles tend to agglomerate much more strongly (regardless of the synthesis method), whereas individual copper nanoparticles are more easily stabilized and remain better dispersed.

3.2. Scanning Electron Microscopy (SEM)

Details of the morphology of resultant polymer-based nanoparticles-modified composites were registered using scanning electron microscopy. The obtained images are depicted in Figure 4 together with EDX spectra verifying the presence of nanoparticles.

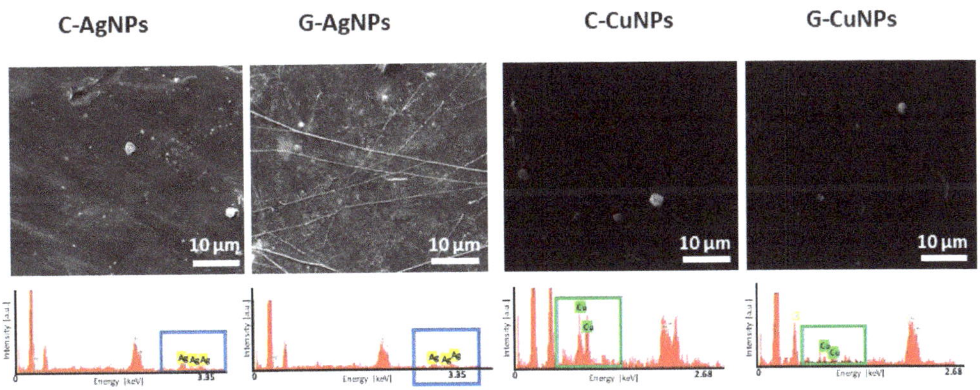

Figure 4. SEM images and EDX spectra of the polymer with embedded silver and copper nanoparticles, respectively.

SEM analysis reveals a relatively smooth and compact polymer matrix with randomly distributed brighter contrast regions, which can be attributed to agglomerates or clusters of embedded metallic nanoparticles. The brighter spots observed in the micrographs arise

https://doi.org/10.3390/ma19020251

from the higher atomic number of silver and copper compared to the polymeric matrix composition, resulting in increased electron backscattering.

Presented SEM images show a random distribution of particle clusters in the matrix (brighter spots). Verification of elemental composition done by EDX confirms the composite warehouse (boxes highlight, respectively, Ag and Cu signals).

SEM analysis proves the concept of obtaining a composite of a polymer matrix with distributed particles.

3.3. X-Ray Diffraction (XRD)

The crystal structure and size of the synthesized nanoparticles were identified using X-ray diffraction. The symmetry of the crystal structure of the investigated nanoparticles was determined based on diffraction patterns. The qualitative analysis of the diffraction patterns allowed us to conclude that silver and copper nanoparticles were formed in the respective reactions in forms observed in bulk [33]. Qualitative XRD patterns analysis presented in Figure 5 of the synthesized nanoparticles allows us to identify diffraction peaks from two crystallographic systems: metallic silver and metallic copper.

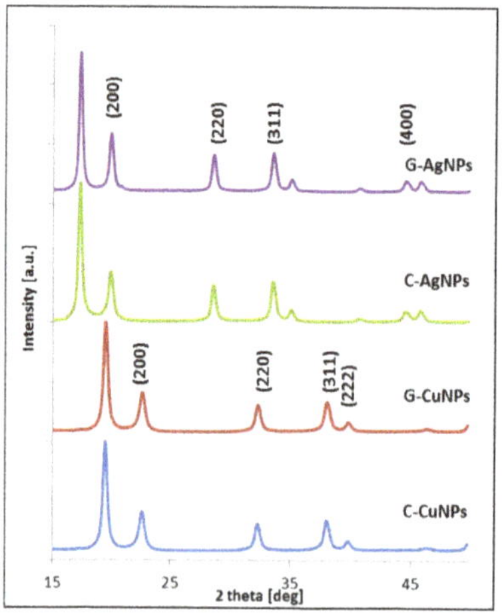

Figure 5. Diffractograms of synthesized Ag and Cu nanoparticles, respectively (C—chemical reduction, G—'green chemistry' reaction).

Quantitative evaluation of the obtained diffractograms can be done with the Williamson–Hall Equation (1), confirming the synthesis of nanoparticles of various sizes [34].

$$\beta cos\theta = \left(\frac{0.9\lambda}{D}\right) + (4\varepsilon sin\theta) \quad (1)$$

where

D—grain size [Å],

λ—wavelength (for Mo source is 0.7136 Å),

β—full width at half maximum intensity of the peak [rad],

ε—strain,

θ—diffraction angle [rad].

The structure of metallic silver has fcc symmetry with the 225 space group number [35]. The experimentally determined silver lattice parameter oscillates between 4.062 and 4.072 ± 0.002 Å, in good agreement with the literature values [36]. The diffraction

https://doi.org/10.3390/ma19020251

peaks from chemical reduction synthesis are located on the two theta axis in positions 17.01°, 19.93°, 28.35°; 33.35°, 34.53°, and 44.21°, which corresponds to the crystal hkl planes according to Miller's nomenclature: (111), (200), (220), (311), (400) [33].

X-ray diffraction pattern analysis also revealed typical patterns for the metallic Cu fcc crystal structure, which can be indexed as (111), (200), (220), (311), (222), (400), and (331) [37]. The experimentally determined copper lattice parameter oscillates between 3.599 and 3.611 ± 0.002 Å, in good agreement with the literature values [36]. This confirms the high crystallinity and structural purity of the synthesized copper nanoparticles. The patterns' angle positions are, respectively, equal: 19.5°, 22.43°, 32.03°, 37.78°, and 39.74°. That data is consistent with published results [11].

The crystallites' average size of the synthesized nanoparticles and the lattice parameter and phase fraction were determined by fitting the diffraction data using the Williamson-Hall Equation (1) [34]. Where the analysis of width of the diffraction peaks at half-maximum allows the size of the nanoparticles' crystallites to be determined. The obtained quantitative data are summarized in Table 3.

Table 3. The size of crystallites and lattice parameters were calculated from XRD data.

Sample	Size (XRD) [nm] ±2	Lattice Parameter (Ag) Å ± 0.002	Lattice Parameter (Cu) Å ± 0.02
C-AgNPs	19	4.072	-
G-AgNPs	13	4.069	-
C-CuNPs	18	-	3.611
G-CuNPs	15	-	3.599

The particle sizes obtained from XRD are in good agreement with those estimated by TEM analysis, showing comparable average values for all investigated samples.

3.4. UV-Vis Spectroscopy

UV-Vis spectroscopy measures the extinction (scattering + absorption) of light passing through a sample of Ag or Cu. Nanoparticles have unique optical properties sensitive to size, shape, concentration, state of aggregation, or refractive index near the nanoparticle surface, making UV-Vis a valuable tool for identifying and characterizing nanomaterials. In Figure 6, the UV-Vis spectra of each solution are presented in series Ag or Cu.

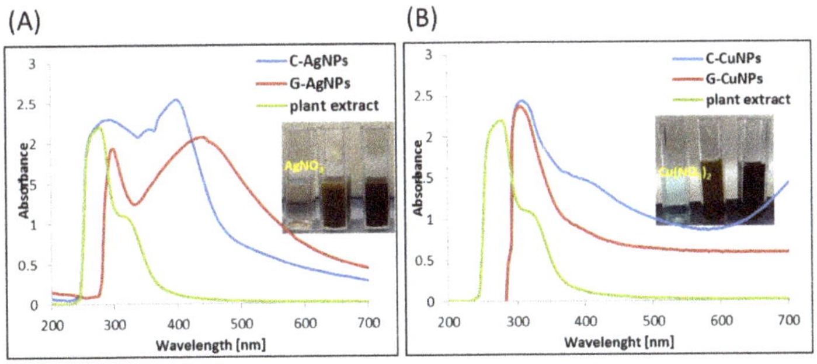

Figure 6. UV-Vis spectra of silver (**A**) and copper (**B**) nanoparticles obtained by chemical reduction (blue line), in the presence of plant extract (red line), and the plant extract.

The characteristic absorption band associated with the surface plasmon resonance (SPR) of AgNPs is observed above 400 nm (Figure 6A), confirming the formation of silver

 https://doi.org/10.3390/ma19020251

nanoparticles [38]. According to the literature, the SPR band of metallic silver typically appears above 390 nm [39]. Therefore, absorption features observed below 390 nm cannot be attributed to the plasmon resonance of AgNPs. Importantly, similar signals are also present in chemically synthesized nanoparticles, indicating that these bands are not specific to the use of plant extract. The absorption observed below 390 nm is more likely related to the presence of organic impurities, such as residual solvents, unreacted precursors, or stabilizing agents, rather than to the nanoparticles themselves. In the case of green synthesis using beetroot extract, these signals may additionally originate from phytochemicals such as betalains, flavonoids, polyphenols, or sugars, which can act as reducing and capping agents during nanoparticle formation. These compounds may remain adsorbed on the nanoparticle surface or dispersed in the colloidal solution, contributing to absorption in the UV region. The shift in the SPR band above 400 nm toward longer wavelengths confirms nanoparticle formation. Moreover, a red shift is commonly associated with an increase in particle size [40].

The UV-Vis absorption spectra shown in Figure 6B display characteristic SPR features of CuNPs. A maximum absorbance observed around 320 nm indicates the presence of copper nanoparticles in both samples [41], which is consistent with previously reported data for CuNPs [42]. Minor shifts in the absorption maximum can be attributed to differences in particle size. The broad band observed around 400 nm is likely associated with organic compounds and/or solvent residues present in the samples rather than with the SPR of CuNPs.

3.5. Dynamic Light Scattering (DLS)

The DLS measurements were conducted to determine the average hydrodynamic size obtained in a given solution of nanoparticles, their size distribution, and their tendency to aggregate. In the solutions of both silver and copper nanoparticles obtained in the presence of beetroot extract, narrow peaks were recorded, indicating a small size distribution of the obtained nanostructures (see Figure 7). To correctly interpret the results of DLS analyses, choosing the appropriate way of presenting them is necessary. Inadequate processing may cause the presence of a few volume percent of larger nanoparticles to overshadow the number of smaller ones completely [43].

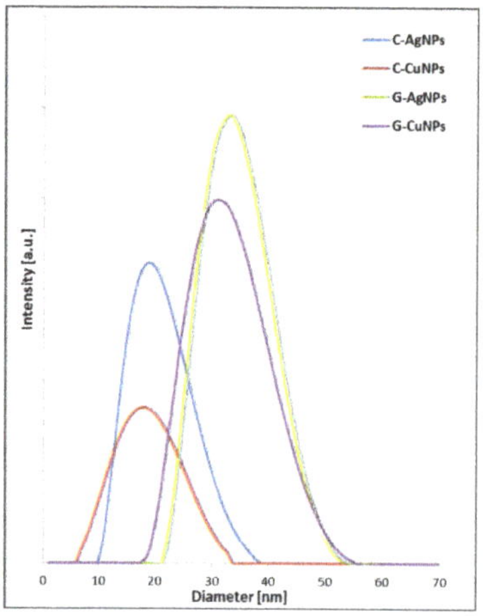

Figure 7. DLS data of the nanoparticles fabricated with chemical reagents (chemical reduction) and beetroot extract.

https://doi.org/10.3390/ma19020251

Qualitative analysis of the DLS spectra allows us to conclude that the nanoparticles obtained due to chemical reduction are consistent with those obtained in TEM imaging. The size of particles registered for green synthesis is in both cases larger than what is related to the presence of an organic corona on the particle's surface. The size distribution of the identified objects in individual samples is associated with the variable hydrodynamic radius, which is related to other organic layers formed around the inorganic core [44].

The estimated size by simulation is defined for a sphere model with the same diffusion coefficient as the measured nanostructure. As a result, all this may cause nanoparticles' diameter to differ significantly from that determined using other methods, such as TEM. Therefore, the DLS measurement identifies more than the core of nanoobjects. It identifies the structure as a whole. In addition, suppose the particles tend to agglomerate, i.e., the system is dynamic, it is recorded in subsequent measurements that there are more and more particles with increasing hydrodynamic radius [45]. Both series of samples, Ag and Cu, show the same tendency (Figure 7). The discrepancy between particle sizes measured by TEM and DLS for green-synthesized AgNPs and CuNPs is due to the presence of organic molecules from the beetroot extract capping the nanoparticles [46]. TEM measures the core size, while DLS measures the hydrodynamic diameter, including the organic layer and solvent shell [45]. Chemically synthesized nanoparticles lack this organic coating, resulting in better agreement between TEM and DLS sizes [44,46].

Smaller nanoparticles exhibit higher antibacterial activity due to larger surface area, enhanced ion release, and stronger membrane interaction. DLS data additionally show dispersion behaviour in aqueous conditions relevant to antibacterial assays. XRD confirms crystalline phase and purity. Since different crystallographic phases of Ag, and Cu, nanoparticles release metal ions at different rates, XRD provides essential structural information explaining variations in antibacterial performance.

3.6. Antimicrobial Effect of Polymers Containing Nanoparticles

Disk diffusion tests were performed to check the antimicrobial properties of the obtained nanocomposites containing polymer and nanoparticles (Figure 8). Each composite type was subjected to the pathogens, and after 24 h, the inhibitory effect was checked visually. The microorganisms were selected as standard Gram-positive, Gram-negative and fungal reference strains widely used in antimicrobial nanomaterial testing.

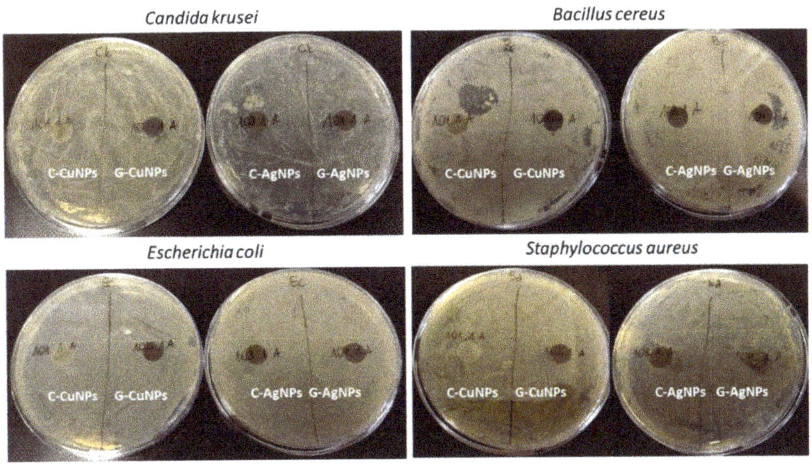

Figure 8. Disc diffusion for a polymer with embedded nanoparticles.

https://doi.org/10.3390/ma19020251

As an additional control, we also tested the polypropylene matrix without nanoparticles. The pure PP material did not exhibit any antibacterial or antifungal activity, confirming that the polymer itself has no intrinsic antimicrobial properties. In the case of a composite with both copper and silver nanoparticles obtained with the 'green chemistry' (right side of the plates), a zone of inhibition of the microorganism tested growth under the discs was noted. Similarly, the composites with silver nanoparticles obtained through the chemical synthesis inhibited the microbial growth under the discs. However, the composites with copper nanoparticles synthesized similarly (chemical synthesis) showed no inhibitory effect on the microorganisms tested.

The obtained polymers with copper nanoparticles can be applied to materials such as agro textiles. Due to its valuable properties, polyvinylpyrrolidone (PVP) is often used in biomedical applications [47]. For instance, adding this polymer to drug carriers significantly increases the efficiency of their assimilation by organisms and extends their resistance time in the bloodstream [48]. Moreover, magnetic nanoparticles coated with poly(ethylene glycol) are also used in anticancer heat therapy, during which cancerous cells are destroyed in a controlled manner [49]. Polymer nanocomposites containing metal nanoparticles are a group of materials with high application potential and extend the scope of use of plastics as materials [50].

Copper ions interact with bacterial cells by binding to negatively charged cell wall and membrane components, disrupting membrane integrity. Furthermore, Cu^{2+} can penetrate bacterial cells and catalyze the formation of reactive oxygen species (ROS) through Fenton-like reactions, leading to oxidative damage of DNA, proteins, and lipids. The generation of oxidative stress ultimately compromises vital cellular processes, leading to bacterial cell death [49]. The second mechanism involves direct contact between Cu NPs and bacterial surfaces, where nanoparticles may adhere to the membrane, induce physical damage or destabilization, and increase permeability [50]. In some cases, internalization of nanoparticles may occur, further disrupting intracellular function.

Research shows that green-chemistry-prepared copper nanoparticles have better results than chemically reduced ones when testing their inhibitory activity. Gram-positive bacteria have one membrane with a multi-layered peptidoglycan polymer and a thick cell wall. In contrast, Gram-negative bacteria have two membranes with a fragile polymer. In our study, copper nanoparticles exhibited antibacterial activity against Gram-negative *E. coli* (ATCC 25922), producing a clear inhibition zone, whereas the copper salt solution (Cu^{2+} ions), used as a non-nanoparticle control, showed significantly lower activity.

According to the literature, small-sized particles penetrate the bacterial cell well and have a better inhibitory effect than on Gram-positive bacteria [51]. The distinct activity is due to differences in particle size and phytochemicals in the plant extract. When the synthesized nanoparticles have a negative charge that strongly binds to the negatively charged microorganism, the concentration of oxygen increases as a result of this excitation [52]. Singlet oxygen is more reactive and enhances the antibacterial effect [52]. In the presented case, we must also consider the specific particle behavior related to the composite form. Copper nanoparticles are widely incorporated into polymeric films due to their excellent antimicrobial and functional properties. However, copper nanoparticles may exhibit cytotoxic or ecotoxic effects at elevated concentrations, primarily due to the release of copper ions (Cu^{2+}). Therefore, evaluating both the extent and the rate of nanoparticle or ion leaching from the films is crucial under realistic conditions. Such assessment is essential to ensure the safety of the final material, particularly in applications involving direct contact with food, biological tissues, or the environment. According to the European Food Safety Authority (EFSA), the tolerable upper intake level (UL) for copper is

https://doi.org/10.3390/ma19020251

5 mg/day for adults. At the same time, the U.S. Environmental Protection Agency (EPA) has set the maximum contaminant level goal (MCLG) for copper in drinking water at 1.3 mg/L. These values provide reference thresholds against which the leaching behavior of Cu-containing films should be evaluated. Ensuring that the release of copper remains well below these limits is critical to minimize potential toxicological risks and to comply with regulatory requirements.

To support the mechanistic interpretation presented in the antimicrobial activity section, it is important to emphasize that the behaviour observed in our study is consistent with well-established mechanisms reported in the literature [53,54]. Numerous studies have demonstrated that green-synthesized metal nanoparticles exhibit enhanced antimicrobial effects due to the combined influence of their small crystallite size, reduced agglomeration, and the presence of phytochemical residues acting as stabilizers and electron donors [53–55]. These phytochemicals (e.g., phenolics, flavonoids, betalains present in beetroot extract) are known to facilitate ROS generation and increase the rate of metal-ion release, which directly correlates with enhanced antimicrobial potency [56]. Furthermore, previous reports have shown that copper nanoparticles produced via green chemistry generate higher levels of reactive oxygen species than chemically reduced particles of similar nominal size, confirming the size and surface chemistry-dependent nature of their activity [57]. In addition, studies on polymer-nanoparticle composites indicate that antimicrobial effects depend strongly on nanoparticle accessibility and surface exposure [58]. The inhibition zones observed for PP composites containing green Ag and Cu nanoparticles align with principle: the nanoparticles produced via the plant-extract route retain higher surface reactivity and interaction potential even after incorporation into the polypropylene matrix [59]. In contrast, chemically synthesized copper nanoparticles, known to exhibit lower redox activity and stronger agglomeration, show diminished or absent antimicrobial effects when embedded in polymers, which is consistent with our findings [60]. Green-synthesized nanoparticles showed stronger effects due to their smaller size, reduced agglomeration, and phytochemical capping layers, which enhanced Cu^{2+}/Ag^{+} release and ROS generation. Gram-negative *E. coli* was particularly sensitive to green Cu NPs, while chemically synthesized Cu NPs showed no inhibition. PP without nanoparticles produced no activity, confirming that antimicrobial effects originate solely from the incorporated nanoparticles. Antibacterial activity can be quantified using MIC/MBC or inhibition-zone measurements.

In this study, green-synthesized Ag and Cu nanoparticles showed smaller and more uniform particle sizes, higher stability (TEM/DLS/UV-Vis), and stronger bactericidal/fungicidal activity compared to chemically synthesized nanoparticles, confirming their superior antimicrobial performance.

4. Conclusions

This study demonstrates that copper and silver nanoparticles synthesized via a green chemistry approach using beetroot extract can be successfully incorporated into a polypropylene polymer matrix while preserving their physicochemical integrity and antimicrobial functionality. Notably, the composites containing green-synthesized Cu nanoparticles exhibited significantly enhanced bactericidal and fungicidal properties compared to silver-based composites and copper nanoparticles obtained by conventional chemical reduction. These results confirm that the synthesis routine plays a decisive role in dictating the biological activity of the resulting polymer nanocomposites. The superior antimicrobial efficiency of green-synthesized Cu nanoparticles highlights their potential as effective, sustainable alternatives to conventionally produced nanomaterials in applications where resistance to bacterial or fungal growth is critical. By maintaining structural

https://doi.org/10.3390/ma19020251

stability within the polymer matrix, these nanoparticles retain their active functionality and open promising avenues for developing antimicrobial packaging, agricultural films, and biomedical materials. Taken together, our findings validate the feasibility of introducing green-synthesized nanoparticles into polymer systems and provide strong evidence that the choice of synthesis method directly influences the biological performance of the resulting nanocomposites. Future work should focus on scaling the synthesis approach, assessing long-term stability, and exploring application-specific optimization, thereby bridging laboratory advances with industrial and clinical applications.

Author Contributions: Conceptualization, B.K.-S.; Validation, M.J. and I.S.; Investigation, A.W., M.B., U.K., G.M., W.L. and B.K.-S.; Data curation, A.W., M.B., U.K., G.M., M.J., W.L., I.S. and B.K.-S.; Writing—original draft, A.W., M.B., U.K., G.M., M.J., W.L., I.S. and B.K.-S.; Writing—review and editing, B.K.-S.; Supervision, B.K.-S. All authors have read and agreed to the published version of the manuscript.

Funding: The research was carried out thanks to the project funded by the Ministry of Education and Science, Poland under Agreement No. NdS 548575/2022/2022. The EU funds partially supported the work (equipment) via a projects with contract numbers POPW.01.03.00-20-034/09-00, and POPW.01.03.00-20-004/11-00. The antimicrobial activity of the nanoparticles was tested in the Laboratory of Applied Microbiology, maintained from the grant of the Ministry of Education and Science in Poland (Decision no. 8636/E-342/SPUB/2021).

Data Availability Statement: The original contributions presented in this study are included in the article. Further inquiries can be directed to the corresponding author.

Acknowledgments: The research is carried out thanks to the project under the competition "Science in young hands for the socio-economic environment." The funds were paid from the Ministry of Education and Science under Agreement No. NdS 548575/2022/2022 of 11 August 2022. EU funds partially financed the work via a project with contract numbers POPW.01.03.00-20-034/09-00, POPW.01.03.00-20-004/11-00. The antimicrobial activity of the nanoparticles was tested in the Laboratory of Applied Microbiology, maintained from the 'Specific Scientific Equipment' grant of the Ministry of Education and Science in Poland (Decision no. 8636/E-342/SPUB/2021). The funders had no role in the study design, data collection, interpretation, or the decision to submit the work for publication.

Conflicts of Interest: The authors declare no conflicts of interest.

References

1. Malik, S.; Muhammad, K.; Waheed, Y. Nanotechnology: A revolution in modern industry. *Molecules* **2023**, *28*, 661. [CrossRef]
2. Todaro, B.; Santi, M. Characterization and Functionalization Approaches for the Study of Polymeric Nanoparticles: The State of the Art in Italian Research. *Micro* **2022**, *3*, 9–21. [CrossRef]
3. Harish, V.; Tewari, D.; Gaur, M.; Yadav, A.B.; Swaroop, S.; Bechelany, M.; Barhoum, A. Review on nanoparticles and nanostructured materials: Bioimaging, biosensing, drug delivery, tissue engineering, antimicrobial, and agro-food applications. *Nanomaterials* **2022**, *12*, 457. [CrossRef] [PubMed]
4. Jałbrzykowski, M.; Oksiuta, Z.; Obidziński, S. Assessment of innovative PLA biopolymer compositions with plant waste fillers. *Eng. Fail. Anal.* **2022**, *139*, 106496. [CrossRef]
5. Jałbrzykowski, M.; Leszczyńska, K.; Obidziński, S. Impact of Nanosilver and Nanocopper on Antifungal Properties of Plastic Elements in the Aspect of Their Use in Agri-Food Processing. *Agric. Eng.* **2018**, *22*, 49–59. [CrossRef]
6. Ying, S.; Guan, Z.; Ofoegbu, P.C.; Clubb, P.; Rico, C.; He, F.; Hong, J. Green synthesis of nanoparticles: Current developments and limitations. *Environ. Technol. Innov.* **2022**, *26*, 102336. [CrossRef]
7. Abbasi, R.; Shineh, G.; Mobaraki, M.; Doughty, S.; Tayebi, L. Structural parameters of nanoparticles affecting their toxicity for biomedical applications: A review. *J. Nanopart. Res.* **2023**, *25*, 43.
8. Velgosova, O.; Mačák, L.; Lisnichuk, M.; Vojtko, M. Synthesis and Analysis of Polymorphic Silver Nanoparticles and Their Incorporation into the Polymer Matrix. *Polymers* **2022**, *14*, 2666. [CrossRef]

https://doi.org/10.3390/ma19020251

9. Baranwal, J.; Barse, B.; Fais, A.; Delogu, G.L.; Kumar, A. Biopolymer: A sustainable material for food and medical applications. *Polymers* **2022**, *14*, 983. [CrossRef]
10. Reddy, M.S.B.; Ponnamma, D.; Choudhary, R.; Sadasivuni, K.K. A comparative review of natural and synthetic biopolymer composite scaffolds. *Polymers* **2021**, *13*, 1105. [CrossRef]
11. Pham, N.D.; Duong, M.M.; Le, M.V.; Hoang, H.A. Preparation and characterization of antifungal colloidal copper nanoparticles and their antifungal activity against *Fusarium oxysporum* and *Phytophthora capsici*. *Comptes Rendus Chim.* **2019**, *22*, 786–793. [CrossRef]
12. Daniel, K.; Vinothini, G.; Subramanian, N.; Nehru, K.; Sivakumar, M. Biosynthesis of Cu, ZVI, and Ag nanoparticles using *Dodonaea viscosa* extract for antibacterial activity against human pathogens. *J. Nanopart. Res.* **2013**, *15*, 1–10.
13. Din, M.I.; Arshad, F.; Hussain, Z.; Mukhtar, M. Green adeptness in the synthesis and stabilization of copper nanoparticles: Catalytic, antibacterial, cytotoxicity, and antioxidant activities. *Nanoscale Res. Lett.* **2017**, *12*, 638. [CrossRef] [PubMed]
14. Adomaviciene, M.; Stanys, S.; Demsar, A.; Godec, M. Insertion of Cu nanoparticles into a polymeric nanofibrous structure via an electrospinning technique. *Fibres Text. East. Eur.* **2010**, *18*, 17–20.
15. Van Deun, J.; Roux, Q.; Deville, S.; Van Acker, T.; Rappu, P.; Miinalainen, I.; Hendrix, A. Feasibility of mechanical extrusion to coat nanoparticles with extracellular vesicle membranes. *Cells* **2020**, *9*, 1797. [CrossRef]
16. Chaudhary, P.; Fatima, F.; Kumar, A. Relevance of nanomaterials in food packaging and its advanced future prospects. *J. Inorg. Organomet. Polym. Mater.* **2020**, *30*, 5180–5192. [CrossRef]
17. Valdeperez, D.; Wang, T.; Eußner, J.P.; Weinert, B.; Hao, J.; Parak, W.J.; Pelaz, B. Polymer-coated nanoparticles: Carrier platforms for hydrophobic water-and air-sensitive metallo-organic compounds. *Pharmacol. Res.* **2017**, *117*, 261–266. [CrossRef]
18. Fatma, S.; Kalainila, P.; Ravindran, E.; Renganathan, S. Green synthesis of copper nanoparticle from *Passiflora foetida* leaf extract and its antibacterial activity. *Asian J. Pharm. Clin. Res.* **2017**, *10*, 79–83. [CrossRef]
19. Mali, S.C.; Raj, S.; Trivedi, R. Biosynthesis of copper oxide nanoparticles using *Enicostemma axillare* (Lam.) leaf extract. *Biochem. Biophys. Rep.* **2019**, *20*, 100699. [CrossRef]
20. Imamura, H.; Kamikoriyama, Y.; Muramatsu, A.; Kanie, K. A mild aqueous synthesis of ligand-free copper nanoparticles for low temperature sintering nanopastes with nickel salt assistance. *Sci. Rep.* **2021**, *11*, 24268. [CrossRef]
21. Klekotka, U.; Rogacz, D.; Szymanek, I.; Malejko, J.; Rychter, P.; Kalska-Szostko, B. Ecotoxicological assessment of magnetite and magnetite/Ag nanoparticles on terrestrial and aquatic biota from different trophic levels. *Chemosphere* **2022**, *308*, 136207. [CrossRef]
22. Kaiser, K.G.; Delattre, V.; Frost, V.J.; Buck, G.W.; Phu, J.V.; Fernandez, T.G.; Pavel, I.E. Nanosilver: An old antibacterial agent with great promise in the fight against antibiotic resistance. *Antibiotics* **2023**, *12*, 1264. [CrossRef]
23. Ghazzy, A.; Naik, R.R.; Shakya, A.K. Metal–Polymer Nanocomposites: A Promising Approach to Antibacterial Materials. *Polymers* **2023**, *15*, 2167. [CrossRef] [PubMed]
24. Kupińska, K.; Michalik, M.; Wilczewska, A.Z.; Jałbrzykowski, M.; Bielicka, M.; Markiewicz, K.H.; Kalska-Szostko, B. Method of polypropylene-zinc composites formation at desired/precise inorganic content. *Polymer* **2023**, *281*, 126102. [CrossRef]
25. Kupińska, K.; Michalik, M.; Krajenta, J.; Bielicka, M.; Markiewicz, K.H.; Kalska-Szostko, B.; Wilczewska, A.Z. An In-Situ Fabrication Method of ZnO and Other Zn (II) Compounds Containing Polypropylene Composites. *Int. J. Mol. Sci.* **2023**, *24*, 2357. [CrossRef]
26. Shakya, A.K.; Al-Sulaibi, M.; Naik, R.R.; Nsairat, H.; Suboh, S.; Abulaila, A. Review on PLGA Polymer Based Nanoparticles with Antimicrobial Properties and Their Application in Various Medical Conditions or Infections. *Polymers* **2023**, *15*, 3597. [CrossRef]
27. Wieszczycka, K.; Staszak, K.; Woźniak-Budych, M.J.; Litowczenko, J.; Maciejewska, B.M.; Jurga, S. Surface functionalization–The way for advanced applications of smart materials. *Coord. Chem. Rev.* **2021**, *436*, 213846.
28. Kango, S.; Kalia, S.; Celli, A.; Njuguna, J.; Habibi, Y.; Kumar, R. Surface modification of inorganic nanoparticles for development of organic–inorganic nanocomposites—A review. *Prog. Polym. Sci.* **2013**, *38*, 1232–1261. [CrossRef]
29. Dufresne, A. Processing of polymer nanocomposites reinforced with polysaccharide nanocrystals. *Molecules* **2010**, *15*, 4111–4128. [CrossRef]
30. Lamberty, A.; Kreyenschmidt, J. Ambient parameter monitoring in fresh fruit and vegetable supply chains using internet of things-enabled sensor and communication technology. *Foods* **2022**, *11*, 1777. [CrossRef]
31. Wasilewska, A.; Klekotka, U.; Zambrzycka, M.; Zambrowski, G.; Święcicka, I.; Kalska-Szostko, B. Physico-chemical properties and antimicrobial activity of silver nanoparticles fabricated by green synthesis. *Food Chem.* **2023**, *400*, 133960. [PubMed]
32. Schneider, C.A.; Rasband, W.S.; Eliceiri, K.W. NIH Image to ImageJ: 25 years of image analysis. *Nat. Methods* **2012**, *9*, 671–675. [CrossRef] [PubMed]
33. Wasilewska, A.; Basa, A.; Zambrzycka, M.; Święcicka, I.; Kalska-Szostko, B. Apple Extract as an Effective Silver Reducer in the Synthesis of Ag Nanoparticles. *Nanomaterials* **2025**, *15*, 595. [CrossRef]

 https://doi.org/10.3390/ma19020251

34. Ali, M.H.; Azad, M.A.K.; Khan, K.A.; Rahman, M.O.; Chakma, U.; Kumer, A. Analysis of Crystallographic Structures and Properties of Silver Nanoparticles Synthesized Using PKL Extract and Nanoscale Characterization Techniques. *ACS Omega* **2023**, *8*, 28133–28142. [CrossRef]
35. Cao, H.; Ni, K.; Du, J.; Wang, Z.; Huang, Y. Allotropic phenomena in silver metals: Search for hexagonal phases, structure determination, and thermodynamic stability. *Acta Mater.* **2025**, *285*, 120701.
36. Kobayashi, M. Review on structural and dynamical properties of silver chalcogenides. *Solid State Ion.* **1990**, *39*, 121–149.
37. Zeng, S.; Shan, S.; Lu, A.; Wang, S.; Caracciolo, D.T.; Robinson, R.J.; Shang, G.; Xue, L.; Zhao, Y.; Zhang, A.; et al. Copper-alloy catalysts: Structural characterization and catalytic synergies. *Catal. Sci. Technol.* **2021**, *11*, 5712–5733. [CrossRef]
38. Castañeda-Aude, J.E.; Morones-Ramírez, J.R.; De Haro-Del Río, D.A.; León-Buitimea, A.; Barriga-Castro, E.D.; Escárcega-González, C.E. Ultra-Small Silver Nanoparticles: A Sustainable Green Synthesis Approach for Antibacterial Activity. *Antibiotics* **2023**, *12*, 574. [CrossRef]
39. Rosman, N.S.R.; Harun, N.A.; Idris, I.; Ismail, W.I.W. Eco-friendly silver nanoparticles (AgNPs) fabricated by green synthesis using the crude extract of marine polychaete, *Marphysa moribidii*: Biosynthesis, characterisation, and antibacterial applications. *Heliyon* **2020**, *6*, e05462.
40. Bélteky, P.; Rónavári, A.; Zakupszky, D.; Boka, E.; Igaz, N.; Szerencsés, B.; Kónya, Z. Are smaller nanoparticles always better? Understanding the biological effect of size-dependent silver nanoparticle aggregation under biorelevant conditions. *Int. J. Nanomed.* **2021**, *16*, 3021–3040. [CrossRef]
41. Nasrollahzadeh, M.; Sajadi, S.M.; Khalaj, M. Green synthesis of copper nanoparticles using aqueous extract of the leaves of *Euphorbia esula* L. and their catalytic activity for ligand-free Ullmann-coupling reaction and reduction of 4-nitrophenol. *RSC Adv.* **2014**, *4*, 47313–47318.
42. Gondwal, M.; Joshi nee Pant, G. Synthesis and catalytic and biological activities of silver and copper nanoparticles using *Cassia occidentalis*. *Int. J. Biomater.* **2018**, *2018*, 6735426. [CrossRef] [PubMed]
43. Jia, Z.; Li, J.; Gao, L.; Yang, D.; Kanaev, A. Dynamic Light Scattering: A Powerful Tool for In Situ Nanoparticle Sizing. *Colloids Interfaces* **2023**, *7*, 15. [CrossRef]
44. Liaqat, N.; Jahan, N.; Anwar, T.; Qureshi, H. Green synthesized silver nanoparticles: Optimization, characterization, antimicrobial activity, and cytotoxicity study by hemolysis assay. *Front. Chem.* **2022**, *10*, 952006. [CrossRef]
45. Dos Santos, V.; da Silveira, N.P.; Bergmann, C.P. In-situ evaluation of particle size distribution of ZrO_2-nanoparticles obtained by sol–gel. *Powder Technol.* **2014**, *267*, 392–397. [CrossRef]
46. Bielicka, M.; Klekotka, U.; Wasilewska, A.; Kalska-Szostko, B. How to determine the correct size of nanoparticles?—A review. *J. Alloys Compd.* **2025**, *1044*, 184302.
47. Domingues, J.M.; Miranda, C.S.; Homem, N.C.; Felgueiras, H.P.; Antunes, J.C. Nanoparticle synthesis and their integration into polymer-based fibers for biomedical applications. *Biomedicines* **2023**, *11*, 1862. [CrossRef]
48. Sultana, A.; Zare, M.; Thomas, V.; Kumar, T.S.; Ramakrishna, S. Nano-based drug delivery systems: Conventional drug delivery routes, recent developments and future prospects. *Med. Drug Discov.* **2022**, *15*, 100134. [CrossRef]
49. Farzin, A.; Etesami, S.A.; Quint, J.; Memic, A.; Tamayol, A. Magnetic nanoparticles in cancer therapy and diagnosis. *Adv. Healthc. Mater.* **2020**, *9*, 1901058. [CrossRef]
50. Gómez-Lázaro, B.; López-Saucedo, F.; Flores-Rojas, G.G.; Camacho-Cruz, A.; Buendía-González, L.; Mendizabal, E.; Bucio, E. Antimicrobial silver-loaded starch-based bioplastic grafted with poly (4-vinylpyridine) as a novel eco-friendly material. *Glob. Health Econ. Sustain.* **2023**, *1*, 1251.
51. More, P.R.; Pandit, S.; Filippis, A.D.; Franci, G.; Mijakovic, I.; Galdiero, M. Silver nanoparticles: Bactericidal and mechanistic approach against drug resistant pathogens. *Microorganisms* **2023**, *11*, 369. [CrossRef] [PubMed]
52. Okkeh, M.; Bloise, N.; Restivo, E.; De Vita, L.; Pallavicini, P.; Visai, L. Gold nanoparticles: Can they be the next magic bullet for multidrug-resistant bacteria? *Nanomaterials* **2021**, *11*, 312. [CrossRef] [PubMed]
53. Karimi, M.; Sadeghi, E.; Bigdeli, S.K.; Zahedifar, M. Optical Properties, Singlet Oxygen, and Free Radical Production Ability with Different UV Irradiations and Antimicrobial Inhibitors against Various Bacterial Species of ZnO:Eu Nanoparticles. *Radiat. Phys. Chem.* **2023**, *212*, 111132.
54. Bae, I.K.; Shin, J.Y.; Son, J.H.; Wang, K.K.; Han, W.S. Antibacterial Effect of Singlet Oxygen Depending on Bacteria Surface Charge. *Photodiagn. Photodyn. Ther.* **2022**, *39*, 102975.
55. Zhang, H.; Du, S.; Wang, Y.; Xue, F. Prevention of Crystal Agglomeration: Mechanisms, Factors, and Impact of Additives. *Crystals* **2024**, *14*, 676. [CrossRef]
56. Ijaz, S.; Iqbal, J.; Abbasi, B.A.; Ullah, Z.; Uddin, S.; Yaseen, T.; Zarshan, F.; Imtiaz, M.; Ali, H.; Jamil, M.; et al. Bioactive Phytochemicals from Beetroot (*Beta vulgaris*) By-Products. In *Bioactive Phytochemicals in By-Products from Leaf, Stem, Root and Tuber Vegetables*; Springer Nature: Cham, Switzerland, 2025; pp. 131–180.

 https://doi.org/10.3390/ma19020251

57. Longobardo, G.; Della Sala, F.; Marino, G.; Barretta, M.; Forte, M.; Paradiso, R.; Borriello, G.; Borzacchiello, A. One-Step Synthesized Folic Acid-Based Carbon Dots: A Biocompatible Nanomaterial for the Treatment of Bacterial Infections in Lung Pathologies. *Nanomaterials* **2025**, *15*, 1657.
58. Harun-Ur-Rashid, M.; Foyez, T.; Krishna, S.B.N.; Poda, S.; Imran, A.B. Recent Advances of Silver Nanoparticle-Based Polymer Nanocomposites for Biomedical Applications. *RSC Adv.* **2025**, *15*, 8480–8505. [CrossRef]
59. Vishnu, D.; Dhandapani, B. A Review on the Synergetic Effect of Plant Extracts on Nanomaterials for the Removal of Metals in Industrial Effluents. *Curr. Anal. Chem.* **2021**, *17*, 260–271. [CrossRef]
60. Fan, X.; Yahia, L.H.; Sacher, E. Antimicrobial Properties of the Ag, Cu Nanoparticle System. *Biology* **2021**, *10*, 137. [CrossRef]

https://doi.org/10.3390/ma19020251

materials

MDPI

Article

Preparation of Particle-Reinforced Resin Using Highly Functional ZnO Particle Filler Driven by Supramolecular Interactions

Haruka Nakagawa [1] and Kohei Iritani [1,2,*]

[1] School of Engineering, Tokyo University of Technology, 1404-1, Katakuramachi, Hachioji 192-0982, Tokyo, Japan; d512500321@edu.teu.ac.jp
[2] Research Center for Advanced Lignin-Based Materials, Tokyo University of Technology, 1404-1, Katakuramachi, Hachioji 192-0982, Tokyo, Japan
* Correspondence: iritanikh@stf.teu.ac.jp

Abstract: The surface modification of zinc oxide nanoparticles (ZnONPs) with organic compounds has been shown to improve their dispersibility. In this study, to develop a highly functional material, ZnONP modified with 6-amino-1-hexanol bearing both amino and hydroxyl functional groups was synthesized. Scanning electron microscopy–energy dispersive X-ray spectroscopy (SEM-EDS) and X-ray photoelectron spectroscopy (XPS) analyses confirmed that functionalized ZnONP was successfully obtained by a hydrothermal synthetic method. The mechanical properties of composite films of polylactic acid (PLA) reinforced with the functionalized ZnONP were then evaluated. The composite containing functionalized ZnONP exhibited a higher maximum stress than that containing unmodified ZnONP. These ZnONP/polymer composites therefore show promise as novel high-performance materials.

Keywords: zinc oxide; bioplastics; surface modification; nanoparticles; hydrothermal synthesis; composites

1. Introduction

Metal oxides [1–4], such as ZnO, TiO_2, and NiO, have a wide range of applications in electronic devices, energy-storage systems, and structural materials. Among them, ZnO has been used in everyday products and biomedical applications because of its UV absorption [5,6], antibacterial properties [7], biocompatibility [8,9], and affordability. To maximize its potential, ZnO nanoparticles (ZnONPs) have attracted intense interest for use in cosmetics, industrial films, and paints due to their UV-shielding ability [10–15]. Surface modification of ZnONP has been employed to prevent aggregation of fine particles and stabilize the suspension [7,11,16,17]. Moreover, functionalizing the ZnONP surface has enabled the creation of high-value-added materials. For example, Kumar and co-workers demonstrated control of oxidative stress in bacteria using ZnONP with surface-adsorbed isopropylamine, and a material for a biosensor was fabricated by chemically modifying ZnONP with triethanolamine to improve stability [18]. In addition, ZnONP has been used as a filler for a particle-reinforcing material. For example, Bao and Liu and co-workers reported an improvement in the dispersibility and antibacterial properties of polymer materials by physically dispersing ZnONP into PLA or PVA matrices [19]. Although high-performance ZnONPs have been produced through surface functionalization, their practical applications remain limited because they often respond selectively to a specific substrate.

Numerous synthetic methods for ZnONP have been developed [20–40]. For example, commonly used methods are sol–gel [20,21], hydrothermal [22–24], wet-chemical [25,26], and spray-pyrolysis [27,28] techniques. The hydrothermal method is particularly attractive because it is sustainable, operates at relatively low temperatures, and requires short processing times. Using this approach, chemical modification of ZnONP with hexanol has also been reported [24]. X-ray photoelectron spectroscopy (XPS) analyses are commonly used to evaluate surface modifications of ZnONP for interfacial bonding characteristics in metal oxide/carbon nanocomposites [41–44]. Nevertheless, comprehensive studies on creating high-value-added ZnONPs through surface functionalization are still scarce.

In this work, we synthesized surface-functionalized ZnONP by the hydrothermal method and explored its use as a reinforcing filler in composite materials. Our goal was to obtain high-strength polymer materials through interactions between the polymer matrix and the functionalized ZnONP. We chose 6-amino-1-hexanol (6AH) as the modifier because its amino group would donate hydrogen bonds, and we selected polylactic acid (PLA) as the polymer matrix because its carbonyl oxygen atom would accept hydrogen bonds (Figure 1). PLA is a biodegradable polymer and therefore offers potential for eco-friendly materials. Because supramolecular interactions are highly versatile, introducing modifiers with diverse chemical structures should open the way to a wide variety of functional composites.

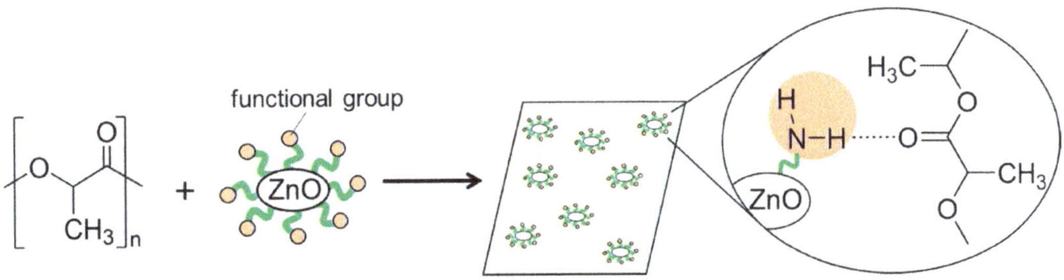

Figure 1. Schematic illustration for fabrication of composite of PLA with functionalized ZnONP; hydrogen-bond formation between amino group on ZnONP and carbonyl group attached with PLA.

2. Materials and Methods

2.1. Reagents

Reagents used for the experiments were purchased from Fujifilm Wako Pure Chemical Industries, Ltd. (Osaka, Japan), and used without further purification.

2.2. Syntheses of Surface-Modified ZnONPs

Syntheses of ZnONPs were conducted according to the previously reported hydrothermal method [24], examining the temperature and time conditions due to the addition of 6AH. $Zn(NO_3)_2{\cdot}6H_2O$ (2.97 g, 10.0 mol) and KOH (1.12 g, 20.0 mol) were added in an autoclave. In the case of surface modifications, 6AH (6.23 g, 53.2 mmol) or 1-hexanol (6.67 mL, 53.2 mmol) was also added as a modifier. The autoclave was kept in an oven at a constant temperature (80 or 120 °C) for a fixed duration (1 or 24 h); subsequently, the product was collected by centrifugation (3500 rpm for 10 min) and washed with water (for the sample without the modifier) or ethanol and water (for that with the modifier).

2.3. Structural Analysis of ZnONP Samples

X-ray diffraction (XRD) measurements were carried out with a SmartLab 3 fully automated multipurpose diffractometer (Rigaku Corp., Tokyo, Japan). The wavelength used for the measurement was CuKα radiation. The crystallite size was averaged at three representative peaks observed within 2θ from 30° to 40°. The average crystallite size D was estimated using the Scherrer equation $D = \frac{K\lambda}{\beta\cos\theta}$, where K is a dimensionless Scherrer's constant, where a value of 0.9 is used as an approximation; λ is the X-ray wavelength, where a value of 0.15418 nm for CuKα is used; β and θ are full width at half maximum and Bragg angle, respectively, in radians. The D values were estimated as averages at three characteristic peaks of θ at 31.78, 34.37, and 36.25.

2.4. Evaluations of Surface Modification

Scanning electron microscopy (SEM) coupled with energy-dispersive X-ray spectroscopy (EDS) was performed on a JSM-IT200 (Keyence Corp., Osaka, Japan). In the case of SEM measurements for organic samples, each sample was Au ion coated by using an IB-2 ion coater (Eiko Corporation, Tokyo, Japan) for charge prevention. For the observation of ZnONP, the sample was prepared by attaching carbon tape to a brass sample stage and mounting the sample on it. For the observation of film, the sample was prepared by attaching carbon tape to the brass sample stage for cross-sectional observation and fixing the film. ZnONPs were analyzed by SEM-EDS using a secondary electron detector (Keyence Corp., Osaka, Japan) at an accelerating voltage of 20.0 kV, at a working distance (WD) of 10.5 mm under high vacuum conditions. Cross-sectional SEM-EDS analysis of the film was performed using a backscattered electron detector (Keyence Corp., Osaka, Japan). at an accelerating voltage of 15.0 kV, at a WD of 10.4 mm under low vacuum conditions. X-ray photoelectron spectroscopy (XPS) data were collected on a JPS-9200 photoelectron spectrometer (JEOL Ltd., Tokyo, Japan). XPS samples were prepared by attaching a paste of In to an Al plate cut to a 50 mm × 50 mm size and then embedding ZnONP samples on top of the paste. The XPS spectrum was measured using monochromatic X-rays (Al). Before XPS measurements, each sample was Au ion coated by using an IB-2 ion coater (Eiko Corporation, Tokyo, Japan). The binding energy scale was calibrated with reference to the Au $4f_{7/2}$ at 84.0 eV [45]. XPS measurements were performed using Al Kα radiation ($h\nu$ = 1486.6 eV) under an accelerating voltage of 10 kV and a discharge current of 10 mA. The energy resolution and pass energy were set to 0.5 eV and 300 eV, respectively, for survey scans, and 0.3 eV and 100 eV, respectively, for O 1s and Zn 2p. Core-level spectra of O 1s and Zn 2p were analyzed after background subtraction using a Shirley method. Peak fitting was carried out using a mixed Gaussian–Lorentzian function, in which the ratio of these functions was also fitted.

2.5. Preparations of PLA Composite Films

ZnONP/PLA composite films were prepared by melting PLA on a hot plate at 220 °C and mechanically mixing ZnONPs of 0, 0.5, 1, 2, and 5 wt%. Subsequently, films were prepared by heating at 200 °C with pressure under 10 MPa. Figure S1 shows the appearance of the films. UV-vis transition spectra were measured with a UV-2600/2700 UV-vis spectrophotometer (Shimadzu Corp., Kyoto, Japan).

2.6. Tensile Testing of Films

Tensile tests were conducted on an EZ-SX tester (EZ Test series, Shimadzu Corp., Kyoto, Japan). To fit into the tensile strength tester, the films were cut to 5 mm × 50 mm using a cutter. The Young's modulus (E) of the film was determined from the equation

$E = (W/A)/(X/L)$, where W is the applied load measured during the tensile test, A is the cross-sectional area of the sample, X is the elongation, and L is the initial gauge length. The load (W) and elongation (X) were obtained from tensile test data. The cross-sectional area (A) was calculated by multiplying the film width (5 mm) with its thickness, measured using a film thickness meter. The gauge length (L) was set to 10 mm for all samples. Stress and strain were defined as W/A and X/L, respectively, and the stress–strain curve was constructed from the tensile test results. The values of the mechanical properties of each film were the averages of 10 test pieces.

2.7. Evaluations of Films Based on FT-IR Measurements

FT-IR spectra were measured using IR Affiniry-1S (Shimadzu Co., Ltd., Kyoto, Japan) by an ATR method. The dried film was placed on a measuring plate at a room temperature under atmosphere and was sandwiched with a fixture to obtain spectra. The range was set to 400–4000 cm^{-1} with 256 scans accumulated. The Happ-Genzel function was used as the apodization function. XRD patterns of the films were measured using the same method described in Section 2.3. Peak fitting was carried out using a mixed Gaussian–Lorentzian function, in which the ratio of these functions was also fitted. Differential scanning calorimetry (DSC) analyses were conducted using a Thermo plus EVO2 DSC8231 (Rigaku Corp., Tokyo, Japan) over a temperature range of 40–200 °C with approximately 10 mg of each sample. Measurements were carried out under dynamic heating conditions at a fixed rate of 5 °C/min. Key thermal transitions, including the glass transition temperature (T_g), cold crystallization temperature (T_{cc}), and melting temperature (T_m), were evaluated from the heating curves of the second heating cycle.

2.8. Water Repellency Test for Films

Static contact angles were measured with a Drop Master system (Kyowa Interface Science Co., Ltd., Saitama, Japan). All pictures were taken within 10 s after dropping water on the film surface. The values of the contact angles of each film were the averages of 5 test pieces.

2.9. Analyses for Particle Distribution in the Composite Film

Image analysis was performed by MATLAB (24.2.0.2871072 (R2024b); updated on 7 February 2025). Five images were analyzed for each sample. The binarized images were obtained using the following procedure: the original image was first converted to grayscale to transform color information into luminance values; a Gaussian filter (σ = 2) was then applied to suppress noise; adaptive thresholding, which determines local thresholds based on the neighborhood intensity distribution, was used to robustly extract target regions under varying illumination conditions; morphological erosion was subsequently applied to remove small artifacts and protrusions; finally, a hole-filling process was performed to close internal cavities within the segmented regions.

3. Results and Discussion

3.1. Structural Analyses of Products

Surface-modified ZnONPs were synthesized under four conditions: 80 or 120 °C for 1 or 24 h. In the paper, we represent the sample as ZnONP-temp.-time, for example, as ZnONP-80-24. XRD patterns confirmed that every sample crystallized as single-phase hexagonal wurtzite-type ZnO [46] (Figure 2 and Figure S2). In Table S1, their crystallite sizes are summarized.

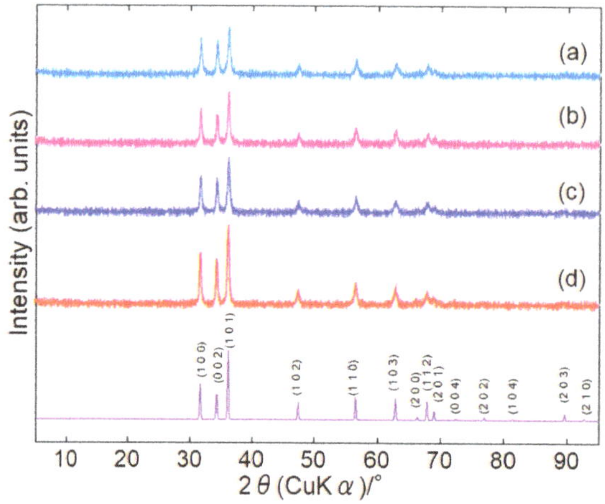

Figure 2. XRD patterns of ZnONP-80-1 (a), ZnONP-120-1 (b), ZnONP-80-24 (c), and ZnONP-120-24 (d).

3.2. Evaluations of Surface Modification

Next, SEM-EDS analyses were carried out to examine particle morphology and elemental distribution. Figure 3a–e and Figure S3a presents SEM images; Figure 3f–j and Figure S3b–d show the corresponding EDS mapping for carbon, zinc, and oxygen; and Figure 4 exhibits EDS spectra of each sample. Although fine primary particles were present in all samples, they formed micrometer-scale agglomerates, likely during drying. In all cases, zinc and oxygen signals were detected. On the other hand, carbon signals appeared only in the samples modified with 6AH, while unmodified ZnO showed no surface carbon. Because the modifier was the only carbon source, these results confirm successful surface modification under all conditions for the 6AH and 1-hexanol modification.

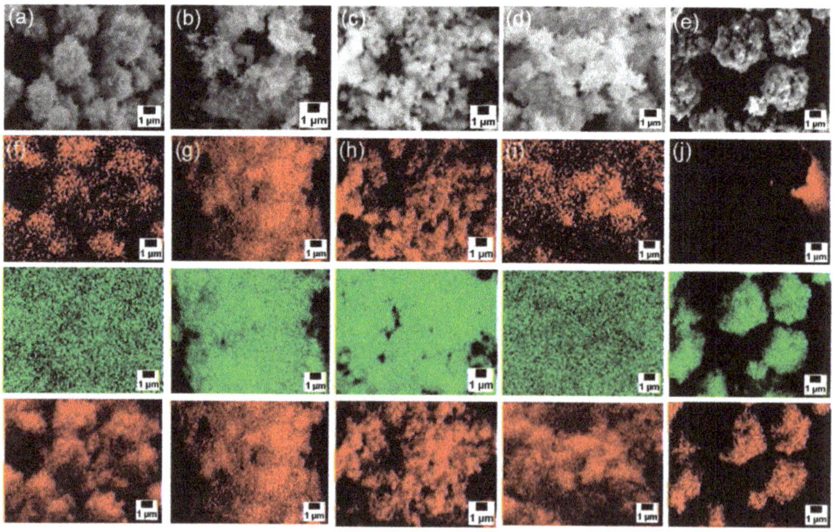

Figure 3. (**a–e**) SEM images and (**f–j**) EDS mappings (top: carbon, middle: zinc, bottom: oxygen) for modified ZnONP-80-1 (**a**,**f**), ZnONP-120-1 (**b**,**g**), ZnONP-80-24 (**c**,**h**), ZnONP-120-24 (**d**,**i**), and unmodified ZnONP-120-1 (**e**,**j**).

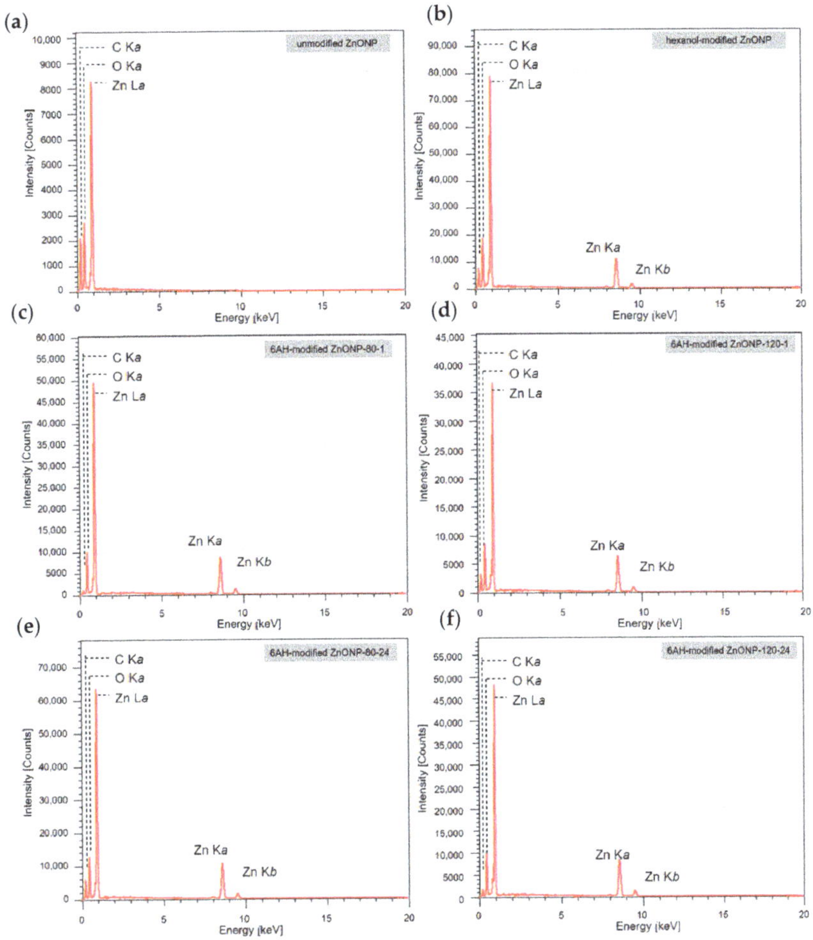

Figure 4. EDS spectra for unmodified ZnONP (**a**), 1-hexanol-modified ZnONP (**b**), 6AH-modified ZnONP-80-1 (**c**), ZnONP-120-1 (**d**), ZnONP-80-24 (**e**), ZnONP-120-24 (**f**).

XPS measurements were conducted to confirm the chemical modification of 6AH on the ZnONP surface. Figures 5–7 present the XPS spectra of the O 1s and Zn 2p regions and survey spectra, respectively. In the survey spectra, peaks corresponding to In and Au used for the fixation of samples and the references of the binding energy, respectively, were observed as well as peaks attributed to ZnO. As shown in Figure 6, the Zn $2p_{1/2}$ and Zn $2p_{3/2}$ peak positions for all samples matched previously reported values [47]. Reported binding energies for bulk oxygen, surface hydroxyl oxygen, and carbon-bonded oxygen are 529.6, 531.2, and 532.7 eV, respectively [47]. Each O 1s spectrum was deconvoluted into three components by Gaussian–Lorentzian fitting using these reference values. The resulting binding energies and peak area ratios are summarized in Table 1. For unmodified ZnONP (Figure 5a), peaks at 530.7 and 531.5 eV correspond to bulk and surface oxygen, respectively, whereas the component attributable to carbon-bonded oxygen was too weak to appear as a distinct peak. In contrast, all 6AH-modified samples exhibited three components assigned to bulk, surface, and carbon-bonded oxygen, indicating successful chemical modification. Quantitative evaluation based on the relative areas of the three peaks showed modification ratios below 10% for every temperature/time combination, with no systematic dependence on either parameter. For the 1-hexanol modification, a similar XPS analysis at the O 1s region was performed (Figure 5b). Its modification ratio was estimated to be 8.8%, which was similar to that obtained with the 6AH modifications.

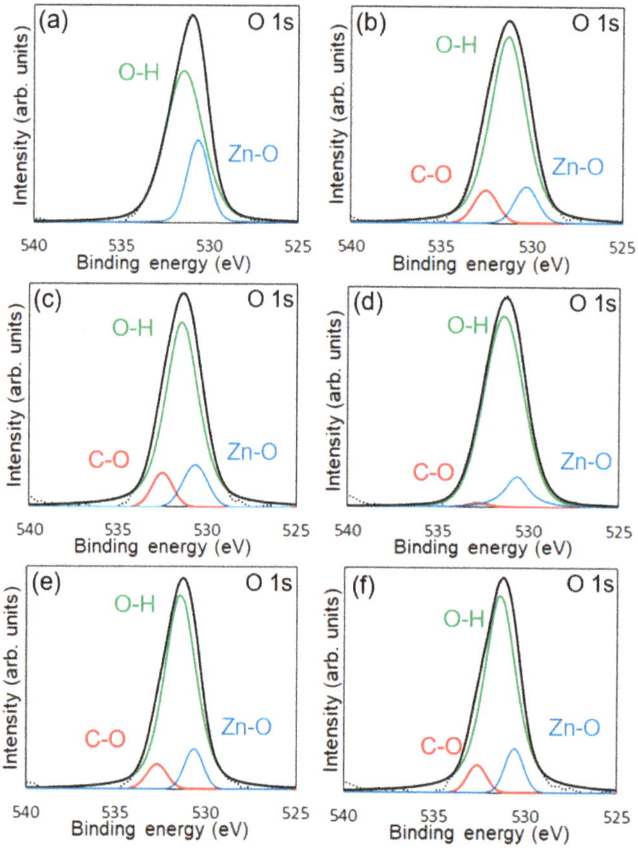

Figure 5. XPS spectra of the O 1s region for unmodified ZnONP (**a**), 1-hexanol-modified ZnONP (**b**), 6AH-modified ZnONP-80-1 (**c**), ZnONP-120-1 (**d**), ZnONP-80-24 (**e**), and ZnONP-120-24 (**f**). The black dotted, black solid, red solid, green solid, and blue solid lines indicate measured spectra, fitting curve, components for carbon-bonded, surface hydroxyl, and bulk oxygen atoms, respectively.

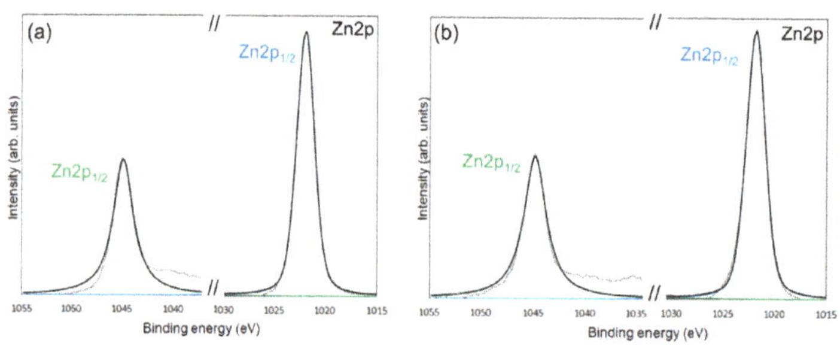

Figure 6. *Cont.*

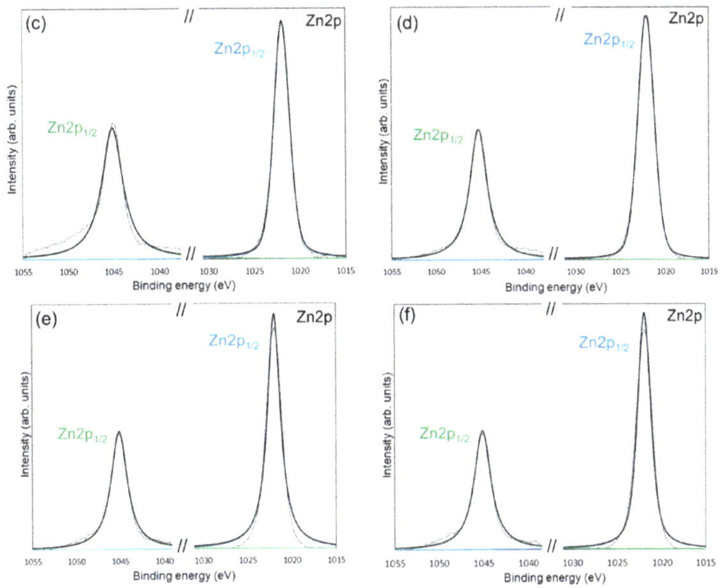

Figure 6. XPS spectra of the Zn $2p_{1/2}$ and Zn $2p_{3/2}$ for unmodified ZnONP (**a**), modified 1-hexanol-ZnONP-120-1 (**b**), 6AH-ZnONP-80-1 (**c**), 6AH-ZnONP-8-24 (**d**), 6AH-ZnONP-120-1 (**e**), 6AH-ZnONP-120-24 (**f**). The black dotted, black solid, green solid, and blue solid lines indicate measured spectra, fitting curve, components for Zn $2p_{1/2}$, and Zn $2p_{3/2}$, respectively.

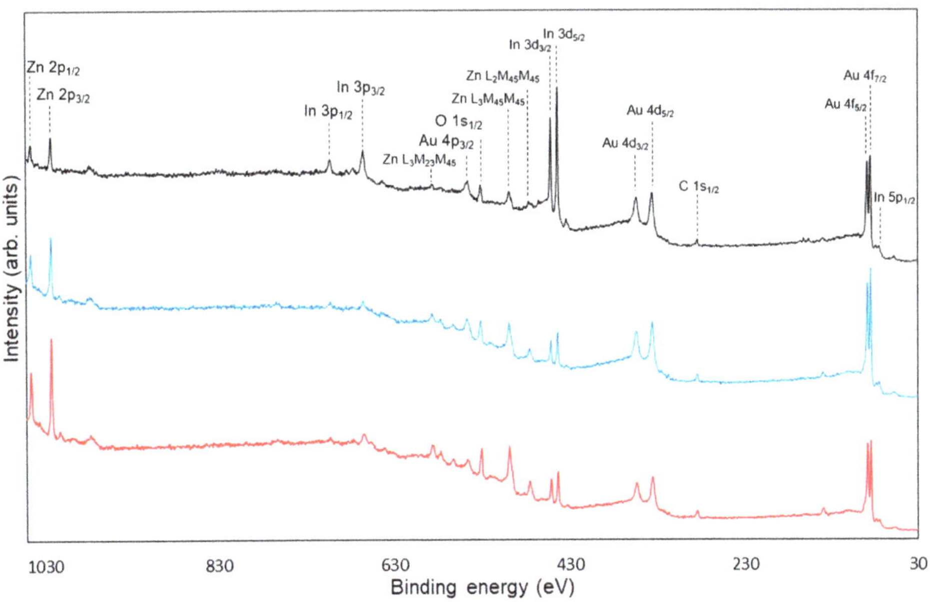

Figure 7. Survey spectra of unmodified ZnONP (black line), 1-hexanol-modified ZnONP (blue line), 6AH-modified ZnONP (red line).

Table 1. Binding energy and peak areas for the O 1s region of each XPS analysis. [a] The component attributable to carbon-bonded oxygen was too weak to appear as a distinct peak.

	Average Binding Energy (eV)			Peak Area (%)	
	Zn-O	Zn-OH	C-O	ZnO + Zn-OH	C-O
Modified ZnONP-80-1	530.7	531.4	532.5	92.2	7.8
Modified ZnONP-120-1	530.6	531.4	532.8	98.1	1.9
Modified ZnONP-80-24	530.6	531.4	532.7	94.2	5.8
Modified ZnONP-120-24	530.6	531.4	532.7	94.3	5.7
Unmodified ZnONP-120-1	530.7	531.5	n. d. [a]	-	-

3.3. Functional Evaluations of Composite Films

The reinforcing potential of the surface-modified particles was assessed in PLA composite films. Figure S1 shows the appearance of the films: the neat PLA was uniform and transparent, whereas the ZnONP-containing films were opaque. In addition, Figure S4 presents the transmittance spectra of the films over the wavelength range of 200–800 nm. The composite film exhibited lower transmittance than the neat PLA film over the visible light range. The transmittance at 400 nm was detected to be 95.8%, 54.5%, and 37.1% for neat PLA, 6AH-ZnONP(1 wt%)/PLA, and 6AH-ZnONP(5 wt%)/PLA films, respectively, indicating that incorporation of ZnONPs induced Mie scattering due to the increased light scattering from the particles with sizes comparable to the wavelengths of visible light. Mechanical properties were measured by tensile testing. Representative stress–strain curves are displayed in Figure S5, and elastic modulus, maximum stress, and elongation at break are plotted against ZnONP content in Figure 8. The incorporation of ZnONPs, whether functionalized or not, generally increased the tensile modulus and reduced elongation at break relative to neat PLA, indicating reduced deformability. Regarding maximum stress, functionalized ZnONP provided higher strength than unmodified ZnONP at loadings $\leq$ 2 wt%. The maximum strength ratios of the composite films (1 wt%) to neat PLA film, $\sigma_{\text{ZnONP}}/\sigma_{\text{PLA}}$, $\sigma_{\text{hexanol-ZnONP}}/\sigma_{\text{PLA}}$, and $\sigma_{\text{6AH-ZnONP}}/\sigma_{\text{PLA}}$, were calculated to be 1.06. 1.63, and 1.89, respectively. The introduction of the amino group on the ZnONP surface significantly improved the composite film strength. Although no clear correlation emerged between the XPS-determined functionalization ratio (Table 1) and film strength, the mechanical trends were consistent across all synthesis conditions without 5 wt% containing films. To investigate the effect of hydrogen bonding [48], FT-IR measurements were performed (Figure 9). In the case of PLA and the composite with unmodified ZnONP, peaks corresponding to a stretching vibration of a carbonyl unit were detected at 1753 cm^{-1}, whereas, for the composite with modified ZnONP, these were low-wavenumber shifts to 1751 cm^{-1}. These results suggest hydrogen bonding between the carbonyl unit of PLA and the functional group of ZnONP, although the shift is not clear, probably due to the small probability of ZnONPs near the surface. The intermolecular interactions between the functional groups and the polymer matrix likely affected the mechanical performance. In the film containing 5 wt%, large variations were observed, and the average values differed significantly depending on the synthetic conditions.

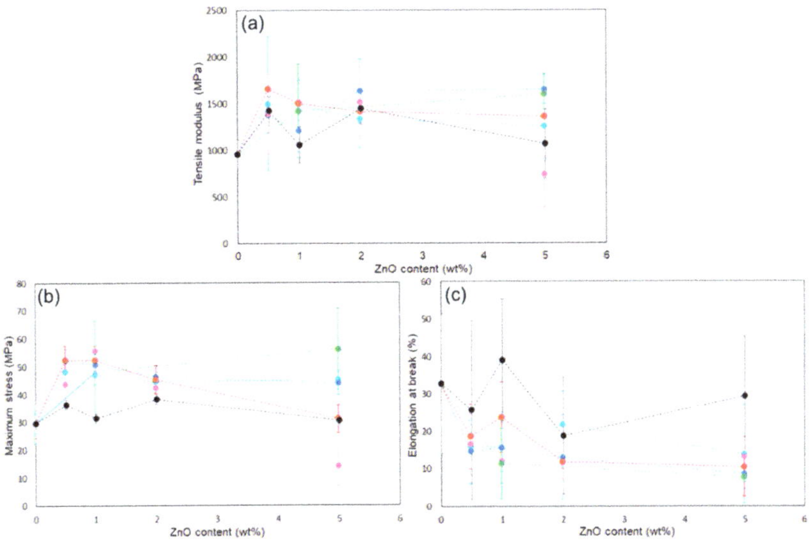

Figure 8. Plots of (**a**) tensile modulus, (**b**) maximum strength, and (**c**) elongation at break of ZnONP-120-1 (black), 6AH-modified ZnONP-80-1 (light blue), 6AH-modified ZnONP-80-24 (navy blue), 6AH-modified ZnONP-120-1 (pink), 6AH-modified ZnONP-120-24 (red), and 1-hexanol-modified ZnONP-120-1 (green).

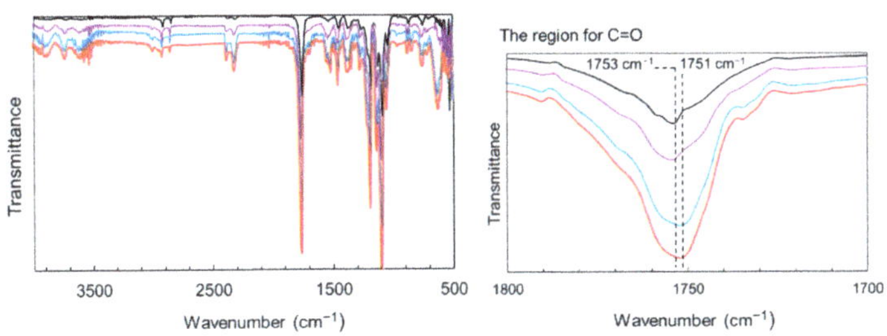

Figure 9. FT-IR spectra of PLA (black line), composite films with unmodified ZnONP (purple line), 1-hexanol-modified ZnONP (blue line), and 6AH-modified ZnONP (red line).

To clarify the mechanism behind the enhancement of the mechanical properties of the composites containing 6AH-modified ZnONP, despite the surface modification ratio of less than 10%, XRD and DSC analyses were conducted (Figures 10 and 11). For XRD patterns, based on curve fitting by the mixed Gaussian–Lorentzian function, two diffraction components were detected at approximately 15° and 20° in 2θ, corresponding to the (203) and (015) lattice plane, respectively [49]. An average value of 2θ and area ratios of diffractions are summarized in Table 2. The area ratio assigned to the (015) lattice plane was 48.4% for 6AH-ZnONP, whereas it remained below 40% for all other ZnONPs. This suggests that functionalization affected the crystal structure of PLA, probably due to interactions between the filler and matrix. In addition, the glass transition temperature (T_g), cold crystallization temperature (T_{cc}), and melt temperature (T_m) were determined by DSC measurements, as summarized in Table 3. The values of the T_g and T_m of composites were comparable to those of neat PLA, indicating that the addition of ZnONPs would have no significant effect on the free volume or the energetic stability of PLA crystals. On the other hand, while T_{cc} of unmodified ZnONP/PLA increased compared to that of neat PLA, T_{cc} of 6AH-modified ZnONP/PLA remained comparable to that of neat PLA. This difference is attributed to the phenomenon that, although ZnONPs suppress polymer chain mobility and thereby elevate T_{cc}, the introduction of the amino group is expected to act as a nucleating agent, facilitating crystallization at a lower temperature. These results indicate that, despite the surface modification ratio being below 10%, it significantly affected the crystalline structure of PLA and the particle sizes, thereby demonstrating the potential to control the mechanical properties of the composite material.

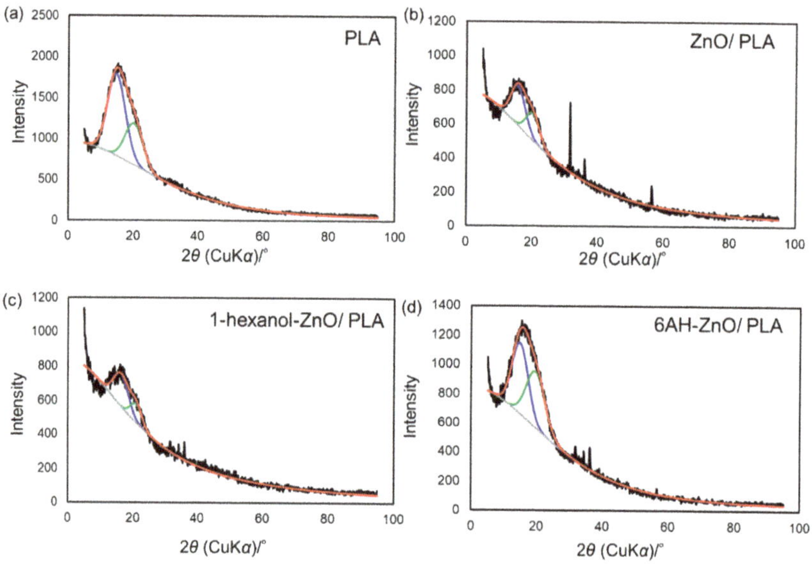

Figure 10. XRD patterns of PLA (**a**), unmodified ZnONP/PLA (**b**), 1-hexanol-modified ZnONP/PLA (**c**), 6AH-modified ZnONP/PLA (**d**). Black, red, blue, green, gray lines show experimental data, fitting curve, two diffraction components, baseline fitting, respectively.

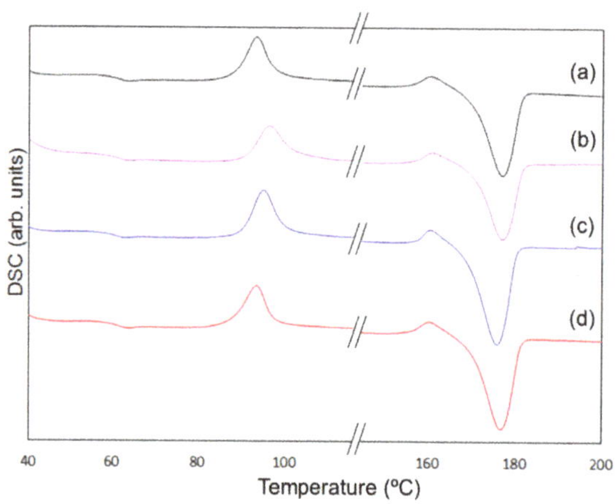

Figure 11. DSC curves of PLA (a), unmodified ZnONP/PLA (b), 1-hexanol-modified ZnONP/PLA (c), 6AH-modified ZnONP/PLA (d).

Table 2. Estimated average and peak area ratio of diffraction components from XRD patterns.

	Diffraction Component 1		Diffraction Component 2	
	Average 2θ/°	Peak Area (%)	Average 2θ/°	Peak Area (%)
PLA	14.8	66.2	20.1	33.8
Unmodified ZnONP/PLA	15.7	60.5	20.3	39.5
1-hexanol-ZnONP/PLA	16.3	68.6	20.9	31.4
6AH-ZnONP/PLA	14.8	51.6	19.6	48.4

Table 3. Evaluated glass transition temperature (T_g), cold crystallization temperature (T_{cc}), and melting temperature (T_m) from DSC measurements.

	T_g (°C)	T_{cc} (°C)	T_m (°C)
PLA	61.2	93.4	177.1
Unmodified ZnONP/PLA	61.2	96.4	177.2
1-hexanol-ZnONP/PLA	60.9	95.1	176.0
6AH-ZnONP/PLA	61.6	93.5	176.9

Subsequently, the composite using 6AH-ZnONP, which was synthesized at 120 °C for 1 h and exhibited distinct mechanical properties, was subjected to a detailed analysis of its film interior to evaluate the dispersibility of ZnONPs. Figure 12a–d and Figure S6a–h show the SEM images and corresponding EDS mappings of Zn from the cross-sections of the composite films with 6AH-ZnONP and unmodified- and 1-hexanol-ZnONP at 1 and 5 wt% ratios, respectively. The regions where Zn was detected were identified as corresponding to particles. The particle size was evaluated by image analyses based on binarization of the data. Details of the binarization method are provided in the Materials and Methods section. A representative binarized EDS mapping image is shown in Figure 12e,f and Figure S6i–l. Histograms of the particle area were fabricated based on the Zn mapping processed by binarization (Figure 13). The estimated particle sizes and ratios of nanometer-sized particles, that is below 1 μm^2, for the composites with 1 and 5 wt% of modified ZnONP are summarized in Table 4. The composites containing unmodified and 1-hexanol-modified ZnONPs exhibited larger average particle sizes and a lower proportion of particles below 1 μm^2, respectively, compared to those containing 6AH-modified ZnONP. These findings suggest that the enhanced compatibility between ZnONPs and PLA is attributed to interactions introduced thorough the surface functionalization. The larger variance observed at 5 wt% contents suggests that, despite the reduction in average particle size, the larger particles probably increased due to aggregation at this concentration.

Table 4. Estimated particle size and ratios of particle below 1 μm^2 based on each binarized image.

		Estimated Particle Size (μm)		Ratios of Particle Below 1 μm^2 (%)
		Average	Standard Deviation	
Unmodified ZnONP/PLA	1 wt%	2.48	4.11	52
	5 wt%	3.36	4.79	40
1-hexanol-ZnONP/PLA	1 wt%	2.37	3.36	48
	5 wt%	2.96	3.84	41
6AH-ZnONP/PLA	1 wt%	0.54	1.04	69
	5 wt%	0.31	1.24	71

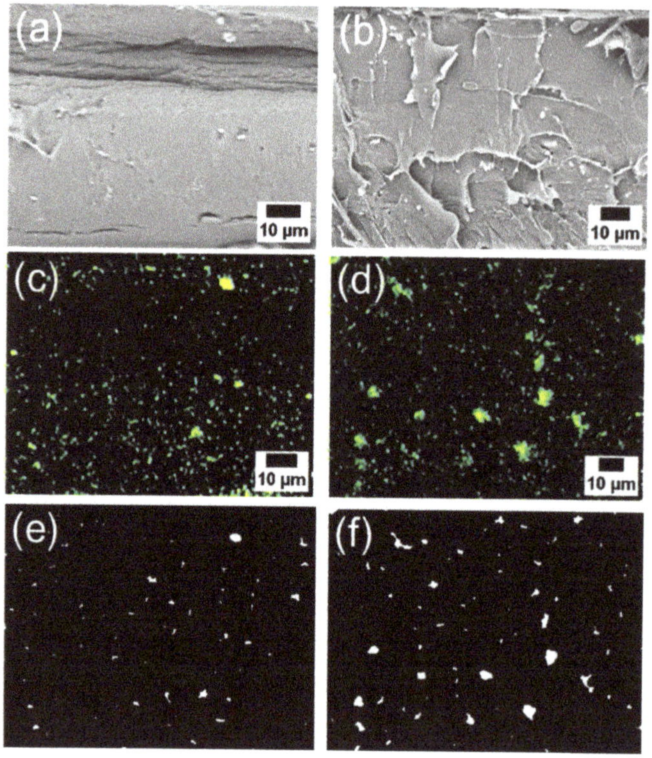

Figure 12. SEM images (**a**,**b**), EDS mappings (**c**,**d**), and binarized images (**e**,**f**) of the cross-sections of the PLA composite films with 6AH-modified ZnONP at 1 wt% (**a**,**c**,**e**) and 5 wt% (**b**,**d**,**f**).

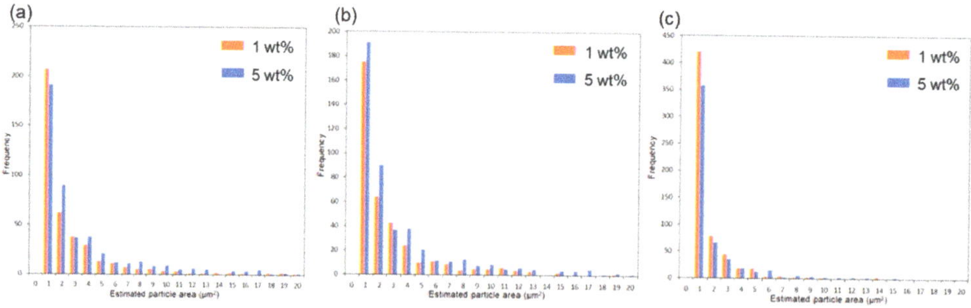

Figure 13. Histograms of particle frequency against estimated particle area from the binarized images of composites containing unmodified (**a**), 1-hexanol-modified (**b**), and 6AH-modified ZnONP (**c**) at 1 and 5 wt%.

Contact angle measurements were carried out to assess the water repellency of the composite surfaces. Side-view images of water droplets on the films and the measured contact angles are shown in Figure 14. Incorporation of surface-modified ZnONP increased the contact angle, indicating reduced surface free energy—presumably as the result of hydrogen bonding between PLA and the functionalized ZnONP.

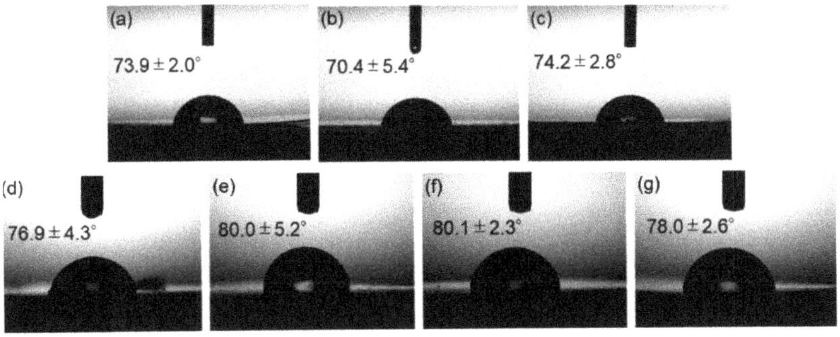

Figure 14. Photographs of neat PLA (**a**) and composite films with unmodified ZnONPs at 1 wt% (**b**) and 5 wt% (**c**); 1-hexanol-modified ZnONPs at 1 wt% (**d**) and 5 wt% (**e**); and 6AH-modified ZnONPs at 1 wt% (**f**) and 5 wt% (**g**). The inserted values indicate contact angles.

4. Conclusions

In this study, ZnO nanoparticles (ZnONPs) surface-functionalized with an amino group were synthesized via a hydrothermal method, and their potential use as a reinforcing filler in a polylatic acid (PLA) composite film was investigated. Within the range of reaction conditions employed, XPS measurements revealed that the surface modification ratio of the ZnONPs remained below 10%. Composite films with PLA were fabricated and subjected to a tensile test to evaluate their mechanical properties. It was found that films containing 6AH-modified ZnONPs exhibited higher tensile strengths than those with unmodified ZnONPs. Furthermore, an increase in water contact angle indicated that the surface modification enhanced the hydrophobicity of the films. In addition, SEM–EDS analyses of the film cross-sections, combined with image analyses, confirmed nanoparticle aggregation; however, no clear dependence on ZnONP content was observed.

This study demonstrated the potential of surface-functionalized ZnONPs as reinforcing fillers, highlighting a promising approach for the development of next-generation sustainable materials. The methodology established here is expected to serve as a foundational technology for the creation of such materials. In particular, due to its low cost and minimal environmental impact, this approach holds significant promise for a wide range of applications, not only in industrial films but also in medical and food-related materials.

Supplementary Materials: The following supporting information can be downloaded at https://www.mdpi.com/article/10.3390/ma18132986/s1, Figure S1: Appearance of the composite films; Figure S2: X-ray diffraction of the synthesized samples; Figure S3: SEM images and EDS mappings for modified 1-hexanol-ZnONP; Figure S4: UV–vis transmittance spectra of neat PLA, 6AH-ZnONP (1 wt%)/PLA, and 6AH-ZnONP (5 wt%)/PLA films, Figure S5: Typical stress–strain curves of PLA film and composites film; Table S1: Crystallite size of the synthesized samples, Figure S6: SEM images, EDS mappings, and binarized images of the cross-sections of the PLA composite films with unmodified ZnONP and 1-hexanol-modified ZnONP.

Author Contributions: Conceptualization, H.N. and K.I.; methodology, H.N.; formal analysis, H.N.; investigation, H.N.; writing—original draft of the manuscript, H.N.; writing—review and editing, supervision, K.I. All authors have read and agreed to the published version of the manuscript.

Funding: This research received no external funding.

Institutional Review Board Statement: Not applicable.

Informed Consent Statement: Not applicable.

Data Availability Statement: The original contributions presented in this study are included in the article/Supplementary Materials. Further inquiries can be directed to the corresponding author.

Acknowledgments: During the preparation of this manuscript and study, the authors used ChatGPT (OpenAI, GPT-4o) for the purposes of language polishing, MATLAB (24.2.0.2871072 (R2024b)) for the purposes of image analyses, respectively. The authors have reviewed and edited the output and take full responsibility for the content of this publication.

Conflicts of Interest: The authors declare no conflict of interest.

References

1. Schilling, K.; Bradford, B.; Castelli, D.; Dufour, E.; Nash, J.F.; Pape, W.; Schulte, S.; Tooley, I.; van den Boschi, J.; Schellauf, F. Human safety review of "nano" titanium dioxide and zinc oxide. *Photochem. Photobiol. Sci.* **2010**, *9*, 495–509. [CrossRef] [PubMed]
2. Falcaro, P.; Ricco, R.; Yazdi, A.; Imaz, I.; Furukawa, S.; Maspoch, D.; Ameloot, R.; Evans, J.D.; Doonan, C.J. Application of metal and metal oxide nanoparticles@MOFs. *Coord. Chem. Rev.* **2016**, *307*, 237–254. [CrossRef]
3. Chavali, M.S.; Nikolova, M.P. Metal oxide nanoparticles and their applications in nanotechnology. *Appl. Sci.* **2019**, *1*, 607. [CrossRef]
4. Shang, G.; Tang, L.; Wu, G.; Yuan, S.; Jia, M.; Guo, X.; Zheng, X.; Wang, W.; Yue, B.; Teng, K.S. High-Performance NiO/TiO_2/ZnO Photovoltaic UV Detector. *Sensors* **2023**, *23*, 2741. [CrossRef] [PubMed]
5. Song, Z.; Kelf, T.A.; Sanchez, W.H.; Roberts, M.S.; Ricka, J.; Frenz, M.; Zvyagin, A.V. Characterization of optical properties of ZnO nanoparticles for quantitative imaging of transdermal transport. *Biomed. Opt. Express* **2011**, *2*, 3321–3333. [CrossRef]
6. Singh, S.; Gade, J.V.; Verma, D.K.; Elyor, B.; Jain, B. Exploring ZnO nanoparticles: UV–visible analysis and different size estimation methods. *Opt. Mater.* **2024**, *152*, 115422. [CrossRef]
7. Tang, E.; Cheng, G.; Ma, X.; Pang, X.; Zhao, Q. Surface modification of zinc oxide nanoparticle by PMAA and its dispersion in aqueous system. *Appl. Surf. Sci.* **2006**, *252*, 5227–5232. [CrossRef]
8. Zhao, L.-H.; Zhang, R.; Zhang, J.; Sun, S.-Q. Synthesis and characterization of biocompatible ZnO nanoparticles. *CrystEngComm* **2012**, *14*, 945. [CrossRef]
9. Liu, J.-H.; Ma, X.; Xu, Y.; Tang, H.; Yang, S.-T.; Yang, Y.-F.; Kang, D.-D.; Wang, H.; Liu, Y. Low toxicity and accumulation of zinc oxide nanoparticles in mice after 270-day consecutive dietary supplementation. *Toxicol. Res.* **2017**, *6*, 134–143. [CrossRef]
10. Zhang, Z.-Y.; Xiong, H.-M. Photoluminescent ZnO Nanoparticles and Their Biological Applications. *Materials* **2015**, *8*, 3101–3127. [CrossRef]
11. Raha, S.; Ahmaruzzaman, M. ZnO nanostructured materials and their potential applications: Progress, challenges and perspectives. *Nanoscale Adv.* **2022**, *4*, 1868–1925. [CrossRef] [PubMed]
12. Yuan, Q.; Hein, S.; Misara, R.D.K. New generation of chitosan-encapsulated ZnO quantum dots loaded with drug: Synthesis, characterization and in vitro drug delivery response. *Acta Biomater.* **2010**, *6*, 2732–2739. [CrossRef] [PubMed]
13. Nie, L.; Gao, L.; Feng, P.; Zhang, J.; Fu, X.; Liu, Y.; Yan, X.; Wang, T. Three-Dimensional Functionalized Tetrapod-like ZnO Nanostructures for Plasmid DNA Delivery. *Small* **2006**, *2*, 621–625. [CrossRef] [PubMed]
14. Zhang, P.; Liu, W. ZnO QD@PMAA-co-PDMAEMA nonviral vector for plasmid DNA delivery and bioimaging. *Biomaterials* **2010**, *31*, 3087–3094. [CrossRef]
15. Zhu, P.; Weng, Z.; Li, X.; Liu, X.; Wu, S.; Yeung, K.W.K.; Wang, X.; Cui, Z.; Yang, X.; Chu, P.K. Biomedical Applications of Functionalized ZnO Nanomaterials: From Biosensors to Bioimaging. *Adv. Mater. Interfaces* **2016**, *3*, 1500494. [CrossRef]
16. Mahajan, M.; Kaur, R.; Gaur, J.; Kumar, S.; Kaushal, S.; Kaur, H.; Bajwa, A.; Nainawat, A.K.; Singh, G. Surface-modified ZnO nanoparticles for enhanced environmental and biomedical performance. *Hybrid Adv.* **2025**, *8*, 100342. [CrossRef]
17. Khan, A.U.H.; Liu, Y.; Naidu, R.; Fang, C.; Shon, H.K.; Zhang, H.; Dharmarajan, R. Changes in the Aggregation Behaviour of Zinc Oxide Nanoparticles Influenced by Perfluorooctanoic Acid, Salts, and Humic Acid in Simulated Waters. *Toxics* **2024**, *12*, 602. [CrossRef]
18. Kumar, A.; Ansari, Z.A.; Fouad, H.; Umar, A.; Ansari, S.G. Oxidative Stress Control in *E. coli* and *S. aureus* Cells Using Amines Adsorbed ZnO. *Sci. Adv. Mater.* **2014**, *6*, 1236–1243. [CrossRef]
19. Duan, L.; Yan, F.; Zhang, L.; Liu, B.; Zhang, Y.; Tian, X.; Liu, Z.; Wang, X.; Wang, S.; Tian, J.; et al. ZnO@Polyvinyl Alcohol/Poly(lactic acid) Nanocomposite Films for the Extended Shelf Life of Pork by Efficient Antibacterial Adhesion. *ACS Omega* **2022**, *7*, 44657–44669. [CrossRef]
20. Mondelaers, D.; Vanhoyland, G.; Van den Rul, H.; D'Haen, J.; Van Bael, M.K.; Mullens, J.; Van Poucke, L.C. Synthesis of ZnO nanopowder via an aqueous acetate-citrate gelation method. *Mater. Res. Bull.* **2002**, *37*, 901–914. [CrossRef]

21. Tokumoto, M.S.; Pulcinelli, S.H.; Santilli, C.V.; Briois, V. Catalysis and Temperature Dependence on the Formation of ZnO Nanoparticles and of Zinc Acetate Derivatives Prepared by the Sol-Gel Route. *J. Phys. Chem. B* **2003**, *107*, 568–574. [CrossRef]
22. Zhang, H.; Yang, D.; Ji, Y.; Ma, X.; Xu, J.; Que, D. Low Temperature Synthesis of Flowerlike ZnO Nanostructures by Cetyltrimethylammonium Bromide-Assisted Hydrothermal Process. *J. Phys. Chem. B* **2004**, *108*, 3955–3958. [CrossRef]
23. Liu, B.; Zeng, H.C. Hydrothermal Synthesis of ZnO Nanorods in the Diameter Regime of 50 nm. *J. Am. Chem. Soc.* **2003**, *125*, 4430–4431. [CrossRef]
24. Mousavand, T.; Ohara, S.; Naka, T.; Umetsu, M.; Takami, S.; Adschiri, T. Organic-ligand-assisted hydrothermal synthesis of ultrafine and hydrophobic ZnO nanoparticles. *J. Mater. Res.* **2010**, *25*, 219–223. [CrossRef]
25. Mahmood, N.B.; Saeed, F.R.; Gbashi, K.R.; Mahmood, U.-S. Synthesis and characterization of zinc oxide nanoparticles via oxalate co-precipitation method. *Mater. Lett. X* **2022**, *13*, 100126. [CrossRef]
26. Ghorbani, H.R.; Mehr, F.P.; Pazoki, H.; Rahmani, B.M. Synthesis of ZnO Nanoparticles by Precipitation Method. *Orient. J. Chem.* **2015**, *31*, 1219–1221. [CrossRef]
27. Ghaffarian, H.R.; Saiedi, M.; Sayyadnejad, M.A.; Rashidi, A.M. Synthesis of ZnO Nanoparticles by Spray Pyrolysis Method. *Iran. J. Chem. Chem. Eng.* **2011**, *30*, 1–6.
28. Wallace, R.; Brown, A.P.; Brydson, R.; Wegner, K.; Milne, S.J. Synthesis of ZnO nanoparticles by flame spray pyrolysis and characterisation protocol. *J. Mater. Sci.* **2013**, *48*, 6393–6403. [CrossRef]
29. Liang, Y.; Wicker, S.; Wang, X.; Erichsen, E.S.; Fu, F. Organozinc Precursor-Derived Crystalline ZnO Nanoparticles: Synthesis, Characterization and Their Spectroscopic Properties. *Nanomaterials* **2018**, *8*, 22. [CrossRef]
30. Primc, G.; Brencic, K.; Mozetic, M.; Gorjanc, M. Recent Advances in the Plasma-Assisted Synthesis of Zinc Oxide Nanoparticles. *Nanomaterials* **2021**, *11*, 1191. [CrossRef]
31. Messai, R.; Ferhat, M.F.; Belmekki, B.; Alam, M.W.; Al-Othoum, M.A.S.; Sadaf, S. GAD plasma-assisted synthesis of ZnO nanoparticles and their photocatalytic activity. *Mater. Res. Express* **2024**, *11*, 015006. [CrossRef]
32. Lu, J.; Ng, K.M.; Yang, S. Efficient, One-Step Mechanochemical Process for the Synthesis of ZnO Nanoparticles. *Ind. Eng. Chem. Res.* **2008**, *47*, 1095–1101. [CrossRef]
33. Otis, G.; Ejgenberg, M.; Mastai, Y. Solvent-Free Mechanochemical Synthesis of ZnO Nanoparticles by High-Energy Ball Milling of ε-$Zn(OH)_2$ Crystals. *Nanomaterials* **2021**, *11*, 238. [CrossRef] [PubMed]
34. Liu, W.; Wang, S.; Wang, J.; Zhang, B.; Liu, L.; Liu, H.; Yang, J. Supercritical hydrothermal synthesis of nano-zinc oxide: Process and mechanism. *Ceram. Int.* **2022**, *48*, 22629–22646. [CrossRef]
35. Sun, P.; Lv, Z.; Sun, C. Study on the Synthesis of Nano Zinc Oxide Particles under Supercritical Hydrothermal Conditions. *Nanomaterials* **2024**, *14*, 844. [CrossRef]
36. Trad, T.M.; Donley, K.B.; Look, D.C.; Eyink, K.G.; Tomich, D.H.; Taylor, C.R. Low temperature deposition of zinc oxide nanoparticles via zinc-rich vapor phase transport and condensation. *J. Cryst. Growth* **2010**, *312*, 3675–3679. [CrossRef]
37. Marin, O.; Real, S.C.; Vega, N.C.; Tirado, M.; Comedi, D. ZnO Nanostructures Synthesized by Vapor Transport and Liquid Phase Synthesis Techniques: Growth and Properties. *Sci. Rev.* **2020**, *1*, 6–23. [CrossRef]
38. López-Cuenca, S.; Pérez Carrillo, L.A.; Rabelero Velasco, M.; Díaz de León, R.; Saade, H.; López, R.G.; Mendizábal, E.; Puig, J.E. High-Yield Synthesis of Zinc Oxide Nanoparticles from Bicontinuous Microemulsions. *J. Nanomater.* **2011**, *2011*, 6. [CrossRef]
39. Yildirim, O.A.; Durucan, C. Synthesis of zinc oxide nanoparticles elaborated by microemulsion method. *J. Alloys Compd.* **2010**, *506*, 944–949. [CrossRef]
40. Usui, H. Electrochemical self-assembly synthesis of zinc oxide nanoparticles and lamellar-structured organic/inorganic hybrids by electrodeposition in surfactant solution. *Electrochim. Acta* **2011**, *56*, 3934–3940. [CrossRef]
41. Rabchinskii, M.K.; Glukhova, O.E.; Sysoev, V.V.; Barkov, P.V.; Ryzhkov, S.A.; Stolyarova, D.Y.; Saveliev, S.D.; Khalturin, B.G.; Varezhnikov, A.S.; Solomatin, M.A.; et al. Delving into the effect of ZnO nanoparticles on the chemistry and electronic properties of aminated graphene: Ab initio and experimental probing. *Surf. Interfaces* **2025**, *65*, 106501. [CrossRef]
42. Rafaela, R.; Dejan, M.; Damjan, B.; Julio, C.; Niksa, K. Synthesis of Silver, Gold, and Platinum Doped Zinc Oxide Nanoparticles by Pulsed Laser Ablation in Water. *Nanomaterials* **2022**, *12*, 3484. [CrossRef] [PubMed]
43. Frankcombe, T.J.; Liu, Y. Interpretation of Oxygen 1s X-ray Photoelectron Spectroscopy of ZnO. *Chem. Mater.* **2023**, *35*, 5468–5474. [CrossRef]
44. Yalovega, G.E.; Brzhezinskaya, M.; Dmitriev, V.O.; Shmatko, V.A.; Ershov, I.V.; Ulyankina, A.A.; Chernysheva, D.V.; Smirnova, N.V. Interfacial Interaction in MeOx/MWNTs (Me–Cu, Ni) Nanostructures as Efficient Electrode Materials for High-Performance Supercapacitors. *Nanomaterials* **2024**, *14*, 947. [CrossRef]
45. Yakimchuk, D.V.; Bundyukova, V.D.; Ustarroz, J.; Terryn, H.; Baert, K.; Kozlovskiy, A.L.; Zdorovets, M.V.; Khubezhov, S.A.; Trukhanov, A.V.; Trukhanov, S.V.; et al. Morphology and Microstructure Evolution of Gold Nanostructures in the Limited Volume Porous Matrices. *Sensors* **2020**, *20*, 4397. [CrossRef]

Various oxygen delivery systems have been developed to address hypoxia, including direct oxygen supply, perfluorocarbon/hemoglobin-based oxygen delivery, and oxygen-generating biomaterials [10]. However, conventional direct oxygen delivery is inherently unsustainable and may cause excessive oxygen accumulation in healthy tissues, which could lead to reactive oxygen species (ROS) production and subsequent cell death [11]. Therefore, precise control of oxygen supply rates is critical to enhancing cell survival and functional performance under hypoxia. Perfluorcarbon (PFC)-based composites are usually employed as oxygen carriers by increasing local dissolved oxygen levels [12]. For instance, Niu et al. developed a methacrylate-poly (ethylene glycol) (PEG)-perfluorooctane hydrogel with a high oxygen-retention capacity to support the survival and proliferation of bone-marrow mesenchymal stem cells under hypoxia [13]. Nevertheless, this approach is constrained by three major limitations: limited oxygen-carrying capacity, uncontrolled release behavior, and potential toxicity [10]. For long-term applications, oxygen-generating materials represent a more suitable alternative. Hydrogen peroxide (H_2O_2) can directly generate an oxygen source through decomposition into water and oxygen [11]. However, H_2O_2-based systems often suffer from rapid oxygen release kinetics due to their high water solubility and decomposition rate [14]. In addition, direct cellular exposure to H_2O_2 may trigger oxidative stress damage.

Calcium peroxide (CaO_2, CPO) can generate O_2 by forming H_2O_2 following its hydrolysis. It is an effective solid-oxygen-generating material that can stably generate O_2 via reacting with water in vivo and in vitro, supporting cell proliferation and differentiation [15], and providing oxygen for tissue-engineered implants before capillary network formation [16]. Some polymers (e.g., PEG [17,18], polydimethylsiloxane [19,20], alginate [12,21–24], gelatin [25–27], etc.) are incorporated with CPO through composite or coating strategies to modulate the oxygen production rate and biocompatibility. For example, in bone tissue engineering [28–31], Hsieh et al. designed CPO/MnO_2-encapsulated poly (lactic-co-glycol acid) (PLGA) microspheres as an oxygen-generating system for local oxygen delivery to promote bone regeneration [32]. However, most reported CPO particles are micron-sized [12,21,26,33–36] and tend to agglomerate and distribute homogeneously in scaffold networks. Although the reaction rate of CPO with water is lower than H_2O_2, it remains too rapid compared to the rate of tissue repair and engineered tissue formation [37]. Adjusting the reaction rate of CPO with water is an effective method to control the rate of oxygen generation and prolong the process of oxygen release. Adding a coating shell layer onto CPO to prepare core–shell nanoparticles is an effective way to modulate the reaction of CPO with water [38]. Recent comprehensive reviews have systematically documented the evolving fabrication strategies for calcium-peroxide-based biomedical platforms, particularly highlighting the structural advantages of core–shell architectures in controlled oxygen release [39]. Silica (SiO_2) is an ideal shell layer that enhances CPO dispersion in aqueous systems and controls the oxygen generation via physical barrier effects [40,41]. Wu et al. designed CaO_2@SiO_2-PAA nanoparticles by encapsulating CaO_2 in hollow mesoporous SiO_2 and coating with polyacrylic acid (PAA). The system could efficiently deliver CaO_2 to the tumor site and release ROS [42]. However, few studies have focused on controlling oxygen generation in CaO_2-based core–shell architectures.

This study aims to develop nanoparticles with tunable oxygen-generating properties to reduce oxidative stress damage in hypoxic cells through sustained oxygen supply. To achieve this goal, we engineered core–shell structured CPO@SiO_2 nanoparticles. The SiO_2 shell was coated on the nano CPO cores to construct these CPO@SiO_2 nanoparticles, which exhibited uniform dispersion. Moreover, the nano CPO core functions as an oxygen source to alleviate hypoxia, while the SiO_2 shell acts as a physical barrier that (1) regulates the oxygen generation rate by controlling the reaction between the CPO core and water; (2) reduces cytotoxicity from excessive H_2O_2 production; and (3) enhances nanoparticle biocompatibility. Therefore, the prepared CPO@SiO_2 could improve cell survival and proliferation under hypoxia and reduce oxidative stress damage. The thickness of the SiO_2 shell layer can be modulated by adjusting the amount of tetraethyl orthosilicate (TEOS). The chemical structure, morphology, size, and oxygen-generating behaviors of the prepared CPO@SiO_2 nanoparticles were systematically characterized. Subsequently, their biocompatibility was evaluated using L929 cells. Finally, the effects of the CPO@SiO_2 nanoparticles on the cell survival of and oxidative stress suppression were assessed under hypoxia.

2. Materials and Methods

2.1. Materials

Catalase (3500 units/mg) was obtained from Meryer (Shanghai, China). Poly (ethylene glycol) (PEG, Mw = 200), tetraethyl orthosilicate (TEOS), calcium chloride ($CaCl_2$), ammonia, and hydrogen peroxide solution (30% *w*/*w*) were purchased from Aladdin (Shanghai, China). Potassium permanganate ($KMnO_4$), sodium oxalate, and sulfuric acid were obtained from Jiangtian Chemical Technology Co., Ltd., Tianjin, China. All other chemicals were analytic grade and used without further purification.

2.2. Preparation of CPO@SiO_2 Nanoparticles

Calcium peroxide (CPO) nanoparticles were prepared in the presence of PEG according to Equation (1):

$$CaCl_2 + H_2O_2 + 2NH_3{\cdot}H_2O \rightarrow CaO_2 + 2NH_4Cl + 2H_2O \quad (1)$$

Briefly, 3 g of calcium chloride ($CaCl_2$) was dissolved in 30 mL of distilled water, followed by the addition of 15 mL of ammonia (30 wt%) under stirring. Subsequently, 120 mL of PEG was added as a dispersant. Then, 15 mL of hydrogen peroxide (30% *w*/*w*) was added dropwise at a controlled rate of 0.25 mL/min with continuous stirring for 1 h. This reaction was maintained for 2 h, after which the pH was adjusted to 11.5 using 0.1 M NaOH solution. After sedimentation, the supernatant was carefully removed, and the precipitate was washed three times with 0.01 M NaOH solution. Finally, CPO nanoparticles were obtained through freeze-drying under optimized conditions, with the cold trap temperature maintained at −60 °C and a system vacuum below 10 Pa.

To fabricate the SiO_2 layer on the nano CPO surface, 50 mg of CPO was dispersed in 12 mL of ammonia solution (28% *w*/*w*) and 100 mL of ethanol. After 30 min of ultrasonication, tetraethyl orthosilicate (TEOS) was added dropwise at the rate of one drop per 6 min under stirring. The reaction was carried out at room temperature for 6 h. After aging overnight, the solid precipitate was washed three times with 0.4% acetic acid solution. The final core–shell-structured CPO@SiO_2 nanoparticles were collected via centrifugation and freeze-drying. The thickness of the SiO_2 layer was modulated by varying the TEOS volume (2, 1, 0.5, 0.3, 0.2, and 0.1 mL), and the corresponding samples were designated as CPO@SiO_2-2.0T, CPO@SiO_2-1.0T, CPO@SiO_2-0.5T, CPO@SiO_2-0.3T, CPO@SiO_2-0.2T, and CPO@SiO_2-0.1T, respectively.

2.3. Characterizations

The chemical structures of the prepared CPO and CPO@SiO_2-1.0T samples were investigated using Fourier transform infrared spectroscopy (FTIR) (Nicolet 6700, Thermo Fisher, Waltham, MA, USA) in the range of 4000–500 cm^{-1}. The X-ray diffraction (XRD) patterns of CPO and CPO@SiO_2-1.0T were collected over a 2θ range of 10° to 80° with a scan rate of 8.0°/min at 36 kV and 200 mA.

The hydrodynamic sizes of the CPO@SiO_2 were measured using a Zetasizer Nano ZS, (Malvern Instruments, Malvern, UK) at 25 °C with a detection angle of 173°. Briefly, the prepared CPO and CPO@SiO_2 nanoparticles were dispersed in ethanol at a concentration of 1 mg/mL using ultrasonication (100 W, 15 min). The particle size distribution was measured using a Malvern Nano Zeta laser particle size analyzer (Malvern Instruments, Malvern, UK).

The morphology of CPO@SiO_2 was evaluated using scanning electron microscopy (SEM) at an accelerating voltage of 5.0 kV and transmission electron microscope (TEM) at an accelerating voltage of 200 kV, with the elemental mapping of Si, O, and Ca performed using TEM-EDS.

2.4. Determination of CPO Contents

The CaO_2 contents in the CPO@SiO_2 nanoparticles were determined via potassium permanganate titration. The procedure was as follows: first, a 0.02 M potassium permanganate ($KMnO_4$) aqueous solution was prepared. The $KMnO_4$ solution was standardized by titration against sodium oxalate (0.15–0.2 g, pre-dried at 120 °C for ≥1 h) dissolved in 1 M sulfuric acid, with the reaction carried out in a water bath maintained at 80–90 °C. For the sample analysis, the same titration procedure was performed using CPO@SiO_2-2.0T, CPO@SiO_2-1.0T, CPO@SiO_2-0.5T, or CPO@SiO_2-0.3T nanoparticles instead of sodium oxalate. The CaO_2 content in each sample was calculated based on the titration results. The relevant reaction equation is provided below:

$$2MnO_4^{-} + 5C_2O_4^{2-} + 16H^{+} = 2Mn^{2+} + 5CO_2\uparrow + 8H_2O \quad (2)$$

$$CaO_2\ (s) + 2H_2O \rightarrow Ca(OH)_2\ (s) + H_2O_2 \quad (3)$$

$$2MnO_4^{-} + 5H_2O_2 + 6H^{+} = 2Mn^{2+} + 5O_2\uparrow + 8H_2O \quad (4)$$

2.5. The Oxygen Generation Assay

First, 50 mL of catalase solution (100 U/mL) was added to a high-pressure airtight container, and dissolved oxygen was purged using nitrogen. Then, predetermined amounts of CPO or CPO@SiO_2 with varying shell thicknesses and concentrations were added. Oxygen content was detected using a PreSens Headspace Residual Oxygen Analyzer (Microx 4 Trace, PreSens, Regensburg, Germany), with dissolved oxygen (DO) (mg/L) measurements taken at designated time points. In order to explore the effect of different shell thicknesses while maintaining consistent total oxygen release across samples, we normalized the nanoparticle concentrations based on their CaO_2 content (Figure 1C), This normalization ensured that all experimental groups (except the control and CPO@SiO_2-1.0T groups) were adjusted to concentrations that would generate oxygen equivalent to 0.1 mg/mL of CPO@SiO_2-1.0T for all experimental groups. Furthermore, equivalent concentrations (0.02, 0.06, and 0.1 mg/mL) of CPO@SiO_2-1.0T were tested to evaluate the CPO@SiO_2 oxygen-generation profiles.

2.6. In Vitro Cytotoxicity Assay

L929 cells, widely used as a model for evaluating biomaterial cytotoxicity and cell–biomaterial interactions, were employed to assess the CPO@SiO_2's biocompatibility. The experimental procedure was conducted as follows: first, L929 cells were seeded into 96-well plates at a density of 1×10^4 cells/well and cultured for 6 h in RPMI-1640 medium containing 10 wt% fetal bovine serum (FBS) and a 1% mixture of penicillium and streptomycin. Then, the culture medium was replaced with 200 μL of fresh culture medium containing different concentrations of CPO@SiO_2, followed by 24 h incubation. After incubation, 20 μL of Cell Counting Kit-8 (CCK-8) solution was added to each well and incubated for another 2 h at 37 °C. The absorbance of the supernatant was measured at 450 nm using a microplate reader (PerkinElmer, Phoenix, AZ, USA), with the cell culture medium as the blank and untreated normal cells as the control. The cell viability was calculated using Equation (5):

$$\text{Cell viability } (\%) = \frac{(OD_{sample} - OD_{blank})}{(OD_{control} - OD_{blank})} \times 100\% \quad (5)$$

The cellular activity and morphology of L929 cells treated with CPO@SiO_2 were also investigated using acridine orange/propidium iodide (AO/PI) staining. Fluorescent images were acquired using an IX-51 fluorescence microscope (Olympus, Tokyo, Japan).

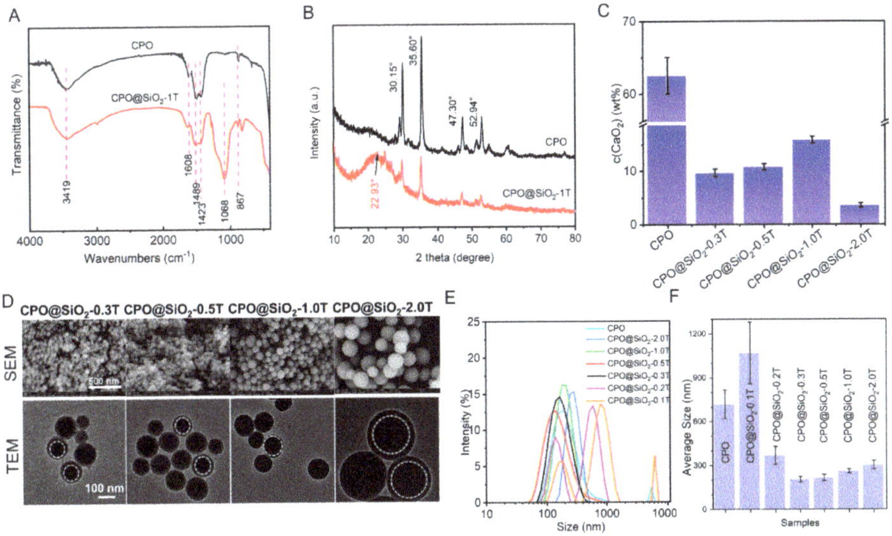

Figure 1. Characteristics and morphology of CPO@SiO_2 nanoparticles. (**A**) FT-IR spectra and (**B**) XRD patterns of CPO and CPO@ SiO_2-1.0T. (**C**) Calcium peroxide content in CPO and CPO@SiO_2. (**D**) SEM and TEM images of CPO@SiO_2-0.3T, CPO@SiO_2-0.5T, CPO@SiO_2-1.0T, and CPO@SiO_2-2.0T. (**E**) Size curves and (**F**) average sizes of different CPO@ SiO_2 nanoparticles measured using DLS.

2.7. Detection of Reactive Oxygen Species in Cells

L929 cells were seeded in 96-well plates at a density of 5×10^3 cells/well and cultured at 37 °C under normoxic conditions (21% O_2 and 5% CO_2). After 6 h, the culture medium was replaced with 200 μL of fresh medium containing CPO@SiO_2, and cells were transferred to a hypoxia incubator (3% O_2, 5% CO_2) at 37 °C. The cell viability was investigated at 6 h, 12 h, 24 h, and 48 h using CCK-8 assay. According to the results of the CCK-8 assay, superoxide dismutase (SOD) activity was detected using a Total Superoxide Dismutase Assay Kit with WST-8 (Beyotime, Shanghai, China), while the level of the lipid oxidation product malondialdehyde (MDA) was investigated using a Lipid Peroxidation MDA Assay Kit (Beyotime, China).

2.8. Statistical Analysis

All data are expressed as mean ± standard deviation (SD). The data from the five or six groups were compared. Statistical analyses were performed using Origin 2017 software. * $p < 0.01$, ** $p < 0.05$, and *** $p < 0.001$ were considered statistically significant.

3. Results

3.1. Formation of Nano CPO@SiO_2 with Core–Shell Structure

Currently, the method for preparing CPO particles is well established, but their size is usually at the micron level. Additionally, CPO microspheres do not easily disperse homogeneously in water. To overcome these issues, we prepared CPO nanoparticles using PEG as a dispersant, following Khodaveisi's report [43]. The selection of PEG as the solvent is a key factor in ensuring the nanoscale size of the prepared CPO particles. First, PEG has a higher viscosity than water, which promotes the dispersion of the prepared CPO nanoparticles and prevents sedimentation during prolonged preparation. Second, PEG prevents the possible dissolution and recrystallization of smaller CPO particles onto larger ones due to the use of water as a solvent (a phenomenon known as the Ostwald ripening effect), which would otherwise lead to an increase in particle size. Based on these advantages, the core–shell structure CPO@SiO_2 was prepared by coating the SiO_2 onto the CPO surface. In this system, the CPO core serves as the oxygen source and the SiO_2 shell layer reduces the contact of CPO with water and further modulates the oxygen generation rate.

The chemical structures of the prepared CPO@SiO_2 were investigated via FTIR and XRD. The adsorption band of the O-O peak at 867 cm^{-1} in CPO suggested that the CPO had been successfully synthesized [42,44,45]. The stretching vibration of the Si-O-Si appeared at 1068 cm^{-1} in CPO@SiO_2-1.0T (Figure 1A) [17,46–48]. These results suggested the formation of CPO@SiO_2 nanoparticles. In addition, the asymmetrical stretching vibration and symmetric stretching vibration absorption bands of O-Ca-O appeared at 1423 and 1489 cm^{-1} in both CPO and CPO@SiO_2 [45,49], which may come from CaO_2 and the calcium carbonate. The small peak at 1608 cm^{-1} is from the CO_3^{2-} of calcium carbonate [49,50], produced by the reaction of CPO with carbon dioxide during the preparation and preservation process. Similarly, as shown in Figure 1B, obvious characteristic diffraction peaks of CPO were observed at 30.15°, 35.60°, 47.30°, and 52.94° [43,44,51,52]. Owing to the formation of the SiO_2 layer, the intensity of these diffraction peaks obviously decreased in the CPO@SiO_2-1.0T sample. Moreover, the typical dispersion peak of amorphous silica at 22.93° could be observed in CPO@SiO_2-1.0T [53–55]. As shown in Figure 1C, the content of CaO_2 in CPO was 60–70%. As expected, the CaO_2 content obviously decreased to 1.8–16.4 wt% in CPO@SiO_2, owing to the introduction of the SiO_2 shell layer. However, the content of CaO_2 in CPO@SiO_2-0.3T to CPO@SiO_2-1.0T slightly increased. As shown in Figure S1, a clear spherical distribution of Si and O elements was observed from the mapping image of single nanoparticles in CPO@SiO2-0.5T and CPO@SiO2-1.0T. Ca, however, is too close to Si in atomic number and cannot be observed because it was wrapped in the SiO_2 shell.

The TEM analysis revealed that bare CPO exhibited an irregular morphology with partial agglomeration, displaying individual particles ranging from 20 to 30 nm in size (Figure S2). In comparison, CPO@SiO_2-0.1T and CPO@SiO2-0.2T showed significant agglomeration, to the extent that the distinct morphology of individual particles became indiscernible in the TEM images. Notably, with a progressive increase in SiO_2 shell thickness, the nanoparticles developed well-defined spherical morphologies (Figure 1D). Moreover, the core–shell structure became clearly visible CPO@SiO2-0.3T/0.5T/1.0T/2.0T nanoparticles, as evidenced by the distinct contrast between the inner CPO core and outer SiO_2 shell in the TEM images. Furthermore, the size of the CPO@SiO_2 nanoparticles increased with the increases in the thickness of the SiO_2 layer. The SEM and TEM images showed that the sizes of CPO@SiO_2-0.3T and CPO@SiO_2-0.5T are about 60–70 nm. As tetraethyl orthosilicate is added in higher amounts, the sizes of the CPO@SiO_2-1.0T and CPO@SiO_2-2.0T nanoparticles increase to 100–120 nm and 220–240 nm, respectively.

The DLS experiments yielded similar results, as shown in Figure 1E,F. The CPO@SiO_2-0.3T, CPO@SiO_2-0.5T, CPO@SiO_2-1.0T, and CPO@SiO_2-2.0T nanoparticles exhibited a concentration-dependent size progression, exhibiting hydrodynamic diameters of 205.1 $\pm$ 2.3 nm, 215.3 $\pm$ 3.1 nm, 260.3 $\pm$ 4.7 nm, and 302.5 $\pm$ 5.6 nm, respectively. These systems maintain monodisperse characteristics, as evidenced by polydispersity indices (PDIs) below 0.15. In contrast, three systems showed divergent behavior: (1) CPO displayed a bimodal distribution (190.1 nm and 5560 nm populations), indicative of hierarchical aggregation, as is consistent with the TEM results; (2) CPO@SiO_2-0.2T exhibited distinct subpopulations at 122.4 nm and 531.2 nm; and (3) CPO@SiO_2-0.1T showed a trimodal distribution (141.8 nm, 712.4 nm, and 59 nm), with the largest aggregate size matching the CPO control group. The observed size discrepancies between the DLS and TEM measurements arise from the measurement method: DLS measures the hydrodynamic diameter in solution, which includes solvation layers and is inherently larger than TEM's dry-state measurements. Based on their superior monodispersity, the CPO@SiO_2-0.3T, CPO@SiO_2-0.5T, CPO@SiO_2-1.0T, and CPO@SiO_2-2.0T formulations were selected for subsequent oxygen generation and biological studies.

3.2. The Oxygen-Generating Behaviors of CPO@SiO_2

The oxygen-generating behavior of the prepared CPO@SiO_2 nanoparticles occurs through hydrolysis as follows:

$$CaO_2 + 2H_2O \rightarrow Ca(OH)_2 + H_2O_2 \quad (6)$$

$$2H_2O_2 \rightarrow 2H_2O + O_2 \quad (7)$$

In this study, the SiO_2 shell layer was formed on the surface of CPO to control the production of oxygen. The oxygen-generating profiles of CPO@ SiO_2 were investigated in a high-pressure airtight container. As shown in Figure 2A, CPO@SiO_2-0.3T and CPO@SiO2-0.5T exhibited oxygen-generating curves similar to bare CPO. This similarity may stem from two plausible mechanisms: (a) an intact silica encapsulation with low defect density, or (b) subcritical shell thickness insufficient to retard the reaction between CaO_2 and water. In contrast, the oxygen-generating rate of CPO@SiO_2-1.0T and CPO@SiO_2-2.0T obviously decreased during the first 50 min, owing to the SiO_2 shell layer reducing the reaction rate of water with the CPO core (Figure 2B). While CPO reached its maximum value at 200 min, CPO@SiO_2-0.3T and CPO@SiO_2-0.5T delayed this peak at about 250 min. After 300 min, oxygen was released completely in CPO@SiO_2-1.0T and CPO@SiO_2-2.0T. Importantly, the SiO_2 shell layer did not affect the total yield, which remained consistent regardless of the shell thickness. These results suggested that the small amount of SiO_2 shell layer did not influence the production of oxygen. The total oxygen output could be controlled by adjusting the CPO@SiO_2 concentration, yielding final dissolved oxygen levels of 1.07 $\pm$ 0.5, 3.12 $\pm$ 0.3, and 5.34 $\pm$ 0.9 mg/L when the concentration of CPO@SiO_2-1.0T was 0.02 mg/mL, 0.06 mg/mL, and 0.10 mg/mL, respectively (Figure 2C). Therefore, the oxygen-generating behaviors of the CPO@SiO_2 could be controlled by adjusting the thickness of the SiO_2 shell layer and the nanoparticle concentration. This phenomenon is consistent with the work of Suvarnapathaki et al., who developed CaO_2-polycaprolactone

(PCL) scaffolds showing concentration-dependent oxygen release [56]. In addition, existing oxygen-releasing systems showed release durations of 6 to 48 h [57,58]. Our CPO@SiO_2 also achieved extended-duration oxygen generation (>5 h) in aqueous environments when the SiO_2 thickness and nanoparticle concentration were adjusted (Figure 2A). Notably, Mohammed et al. demonstrated superior sustainability using polylactic acid (PLA)/CPO showing three-day continuous oxygen release [59]. Naturally, the oxygen production duration across studies cannot be precisely compared, owing to variations in measurement techniques and experimental conditions. However, these findings highlight the potential for the future optimization of CPO-based formulations through hydrogel/scaffold integration to meet specific therapeutic oxygen delivery requirements.

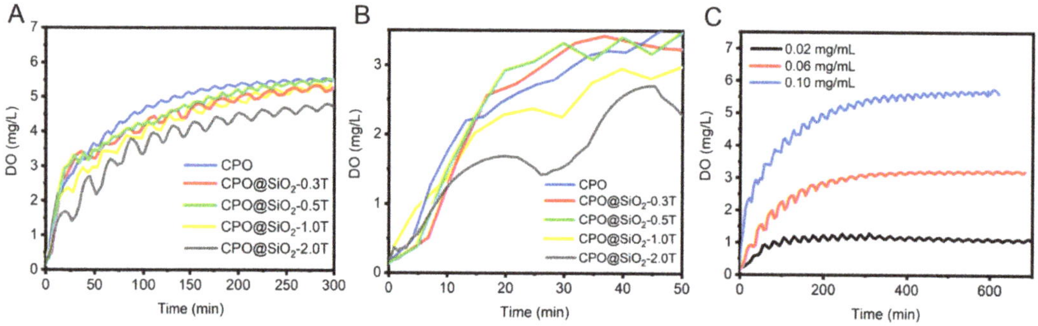

Figure 2. Oxygen-generating behaviors of CPO@SiO_2. (**A**,**B**) Oxygen-generating curves of CPO and CPO@ SiO_2 nanoparticles with different shell thicknesses of SiO_2 shell at 0.10 mg/mL. (**C**) Oxygen-generating curves of CPO@ SiO_2-1.0T with different concentrations.

3.3. *Cytocompatibility of CPO@SiO_2 Nanoparticles*

The cytotoxicity of the prepared CPO@SiO_2 nanoparticles was evaluated using L929 cells via a CCK-8 assay. As shown in Figure 3A, cell viability gradually increased with the increase in the CPO@SiO_2-1.0T concentration up to 0.1 mg/mL, then decreased at 0.15 mg/mL. These results indicated that the SiO_2 shell layer significantly enhanced biocompatibility. The maximum cell viability was achieved at a CPO@SiO_2-1.0T concentration of 0.1 mg/mL. This phenomenon can be explained as follows: when the nanoparticle concentration was below 0.1 mg/mL, CPO@SiO_2 continuously generated oxygen during cell culture. The oxygen production increased with the nanoparticle concentration, providing essential nutrients for cell proliferation. However, excessive nanoparticle concentrations adversely affected cellular growth. At 0.2 mg/mL, the nanoparticles exhibited cytotoxicity, resulting in cell viability lower than the control group. The AO/PI staining results were consistent with these findings. As shown in Figure 3C, the number of live cells gradually increased with the concentration of CPO@SiO_2-1.0T up to 0.1 mg/mL (Figure 3C), while some dead cells could be observed at 0.15 mg/mL due to the production of excess oxygen.

The thickness of the SiO_2 shell layer was the main factor influencing cytotoxicity. As shown in Figure 3B, the cell viability in the CPO group was obviously lower than that of the control, which may have been caused by the rapid release of oxygen. Although the cell viability in the CPO@SiO2-0.5T group slightly increased compared to that in the bare CPO group, it still showed certain cytotoxicity because of the small SiO_2 layer. As expected, the CPO@SiO_2-1.0T and CPO@SiO_2-2.0T showed excellent biocompatibility, with cell viability significantly exceeding the control group. This enhancement resulted from the suitable oxygen stimulation of cell proliferation. Under normoxic conditions, the CPO@SiO_2-1.0T released more oxygen, thereby promoting cell reproduction and enhancing cell activity. In contrast, the CPO@SiO_2-2.0T group released a relatively low amount of oxygen, which had a weaker effect on cell promotion than the CPO@SiO_2-1.0T group, so the cell activity was more consistent with the control group.

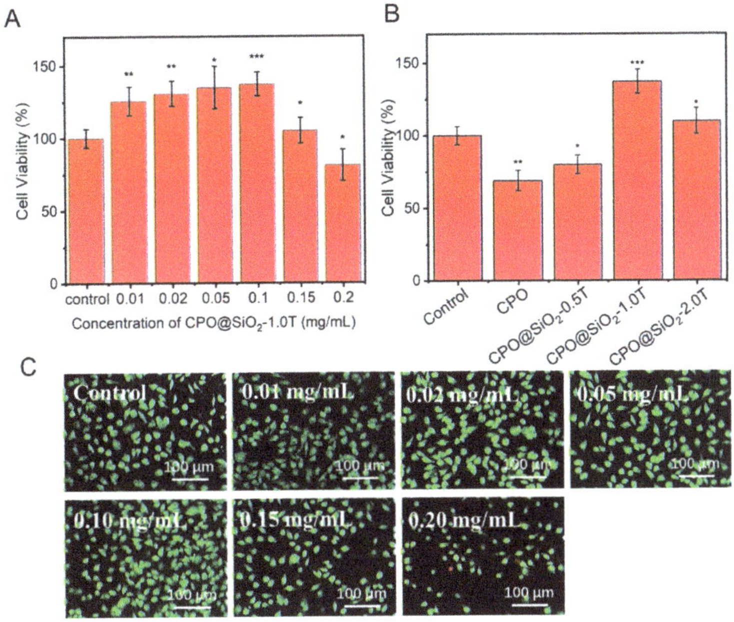

Figure 3. The cell viability of L929 cells cultured in the presence of (**A**) CPO@ SiO_2-1.0T with different concentrations and (**B**) 0.1 mg/mL of CPO@SiO_2 nanoparticles with different shell thicknesses (* $p < 0.05$, ** $p < 0.01$, *** $p < 0.001$). (**C**) AO/PI staining images of L929 cells cultured in the presence of CPO@SiO_2-1.0T with different concentrations.

3.4. Effect of CPO@SiO_2 on Oxidant Stress Damages in Cells

Hypoxia is typically the main microenvironment at the lesion site following tissue damage and within large-scale engineered tissue, a state that could induce oxidative stress damage to both host and implanted cells. In this study, L929 cells were cultured in a hypoxic incubator at 3% O_2 to establish an oxidative stress model. The protective effects of CPO@SiO_2 on cell survival were investigated using the CCK-8 assay.

As shown in Figure 4A, the cell viability gradually increased with the increase in CPO@ SiO_2-1.0T concentrations up to 0.1 mg/mL; then, it gradually decreased while remaining higher than the control group. This occurred because the generated oxygen content increased with the CPO@ SiO_2-1.0T concentration, which progressively alleviated the hypoxia of cells. It should be noted that the cell viability at 0.01 mg/mL was lower than that of the control, likely due to residual catalase activity causing cell damage at a low concentration of CPO@SiO_2-1.0T nanoparticles. With the increase in CPO@ SiO_2-1.0T concentrations, most of the catalase was used to decompose the H_2O_2 produced by CPO to generate O_2 [23]. For all tested concentrations of CPO@SiO_2-1.0T, cell viability exhibited distinct time-dependent characteristics under hypoxic conditions. The maximum viability was achieved at 12 h, followed by a gradual decline. This temporal pattern can be attributed to the sustained oxygen generation by CPO@SiO_2-1.0T during the initial 12 h period, which effectively mitigated hypoxia-induced cellular damage. Beyond this time point, the diminishing oxygen generation capacity correlated with reduced cell survival rates, consistent with our oxygen generation profile measurements.

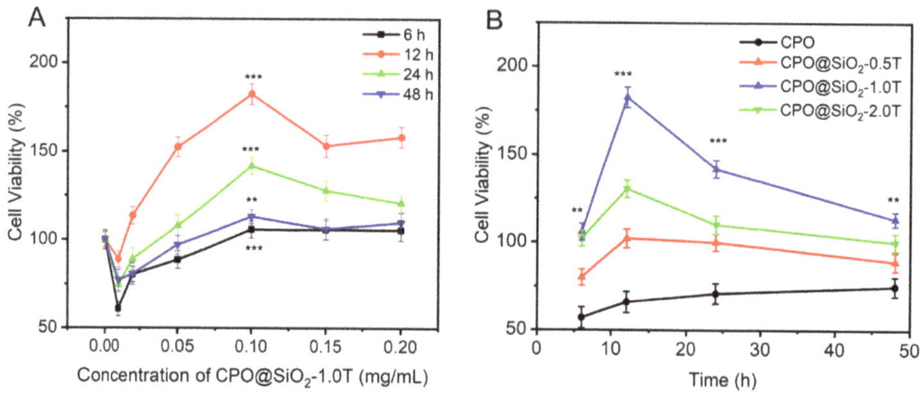

Figure 4. The viability of cells cultured under hypoxia for 48 h in the presence of (**A**) $CPO@SiO_2$-1.0T nanoparticles with different concentrations and (**B**) $CPO@SiO_2$ nanoparticles with different shell thicknesses at 0.1 mg/mL (** $p < 0.05$, *** $p < 0.001$).

Meanwhile, the cell viability also depended on the thickness of the SiO_2 shell (Figure 4B). Cells inevitably contacted CPO directly due to the lack of a protective SiO_2 layer; therefore, the cell viability in the CPO group was lower than that in the control. The $CPO@SiO_2$-0.5T group exhibited similar viability to bare CPO. Under hypoxic conditions, the slower oxygen release rate in the $CPO@SiO_2$-2.0T group would result in inadequate oxygen supply during the initial hypoxic phase, preventing effective protection against hypoxia-induced damage. Although the $CPO@SiO_2$-2.0T group had a longer oxygen release duration, the subsequent oxygen release was insufficient to fully repair the damaged cells. Consequently, the $CPO@SiO_2$-2.0T group displayed lower cell activity compared to the $CPO@SiO_2$-1.0T group. This highlights that protecting hypoxic cells requires attention not only to the duration of oxygen supply but also to minimizing initial damage [60].

3.5. The Effect of $CPO@SiO_2$ on ROS Expression Under Hypoxia

It is generally believed that oxidative stress under hypoxia produces a large number of ROS, which impairs the normal physiological functions of cells. SOD is an essential antioxidant enzyme that protects cells from ROS [61]. The hypoxic microenvironment led to increased SOD expression levels. The addition of $CPO@SiO_2$-1.0T can obviously reduce SOD activity. As shown in Figure 5A, the SOD activity of the cells in the control group was higher than that of all the $CPO@SiO_2$-1.0T groups at different concentrations across the four time points. With the increase in $CPO@SiO_2$-1.0T concentration, the SOD activity showed a decreasing trend, reaching the lowest value at 0.1 mg/mL. Comparing the four curves at different times, we observed that the SOD activity decreased over time. This may have been due to the continuous release of oxygen, which alleviated hypoxia. In addition, SOD activity in the CPO group was also lower than that in the control group, indicating that oxygen production in the CPO core could relieve hypoxia and thus reduce SOD activity. With the increase in shell thickness, SOD activity gradually decreased until $CPO@SiO_2$-1.0T, reaching the lowest value, and then it increased slightly in $CPO@SiO_2$-2.0T (Figure 5B). This also verified the protective effect of the SiO_2 shell on cells. As for $CPO@SiO_2$-2.0T, its

SOD activity was slightly higher than the CPO@SiO2-1.0T group, which may be due to the slow oxygen production rate, failing to alleviate hypoxia in the early stage.

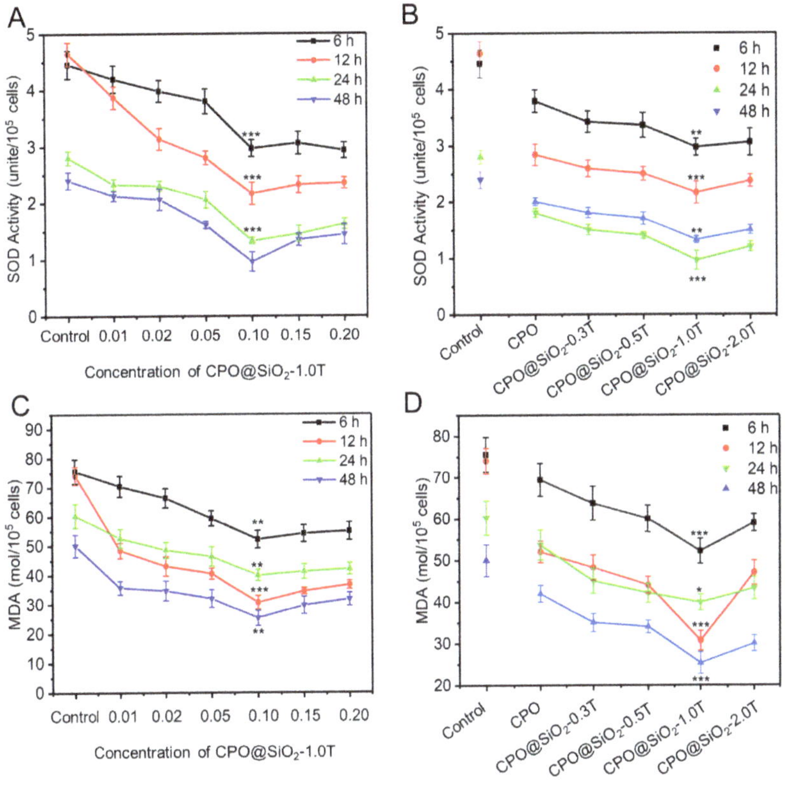

Figure 5. The SOD activity of L929 cells cultured within 48 h in the presence of (**A**) CPO@SiO_2 nanoparticles with different concentrations and (**B**) CPO@SiO_2-1.0T nanoparticles with different shell thicknesses at 0.1 mg/mL. The MDA level of L929 cells cultured within 48 h in the presence of (**C**) CPO@SiO_2 nanoparticles with different concentrations and (**D**) CPO@SiO_2-1.0T nanoparticles with different shell thicknesses at 0.1 mg/mL. (* $p < 0.01$, ** $p < 0.05$, *** $p < 0.001$).

In biological organisms, free radicals induce lipid peroxidation, with MDA being the final oxidation product. Therefore, the expression levels of MDA can reflect the degree of lipid peroxidation in the cells and indirectly reflect the degree of cell damage [62]. As shown in Figure 5C, cells in CPO@SiO_2-1.0T with a concentration of 0.1 mg/mL showed the lowest MDA levels, indicating reduced lipid peroxidation and minimal cell damage. Furthermore, at the same concentration, the cells in CPO@SiO_2-1.0T consistently yielded the lowest MDA levels (Figure 5D). Unlike the trend observed in SOD activity, MDA levels did not exhibit a gradual decline over time.

4. Conclusions

In this study, a series of core–shell-structured CPO@SiO_2 nanoparticles were prepared to alleviate the hypoxic microenvironment. The CPO core served as an oxygen-generating source, and the SiO_2 shell layer functioned to isolate the CPO core from water and regulate oxygen-generating behaviors. The results suggest that the SiO_2 shell thickness could be modulated by adjusting the amount of TEOS. The CPO@SiO_2 nanoparticles exhibited sizes ranging from 205 to 302 nm with narrow PDIs. The SiO_2 shell layer significantly decreased the particle agglomeration and effectively modulated oxygen generation by adjusting the SiO_2 shell thickness. In addition, the prepared CPO@ SiO_2 nanoparticles showed good biocompatibility. Furthermore, they were able to alleviate the hypoxic microenvironment, suppress intracellular ROS, reduce oxidative stress damage, and improve cell survival under hypoxic conditions.

Supplementary Materials: The following supporting information can be downloaded at: https://www.mdpi.com/article/10.3390/ma18112568/s1, Figure S1. TEM bright field and mapping of CPO@ SiO_2-0.5T and CPO@SiO_2-1.0T nanoparticles. Figure S2. TEM images of CPO, CPO@SiO_2-0.1T, and CPO@SiO_2-0.2T nanoparticles.

Author Contributions: Conceptualization, X.X. and W.L.; methodology, X.S.; software, X.X. and X.S.; validation, X.X. and X.Y.; formal analysis, X.X. and R.W.; investigation, X.S. and R.W.; resources, W.L.; data curation, X.S.; writing—original draft preparation, X.S.; writing—review and editing, X.X. and W.L.; visualization, X.X.; supervision, W.L.; project administration, W.L.; funding acquisition, X.X. All authors have read and agreed to the published version of the manuscript.

Funding: This study was funded by the National Natural Science Foundation of China (52371231) and the Startup Fund for Talent Introduction of Shanxi Electronic Science and Technology Institute (2023RKJ004).

Institutional Review Board Statement: Not applicable.

Informed Consent Statement: Not applicable.

Data Availability Statement: The original contributions presented in the study are included in the article and Supplementary Materials. Further inquiries can be directed to the corresponding authors.

Conflicts of Interest: The authors declare no conflicts of interest.

References

1. Willemen, N.G.A.; Hassan, S.; Gurian, M.; Li, J.; Allijn, I.E.; Leijten, J. Oxygen-Releasing Biomaterials: Current Challenges and Future Applications. *Trends Biotechnol.* **2021**, *3*, 1144–1159. [CrossRef] [PubMed]
2. Hao, T.; Li, J.; Yao, F.; Dong, D.; Wang, Y.; Yang, B.; Wang, C. Injectable Fullerenol/Alginate Hydrogel for Suppression of Oxidative Stress Damage in Brown Adipose-Derived Stem Cells and Cardiac Repair. *ACS Nano* **2017**, *11*, 5474–5488. [CrossRef] [PubMed]
3. Bialik, S.; Geenen, D.; Sasson, I.; Cheng, R.; Horner, J.W.; Evans, S.M.; Lord, E.; Koch, C.J.; Kitsis, R.N. Myocyte Apoptosis During Acute Myocardial Infarction in The Mouse Localizes to Hypoxic Regions but Occurs Independently of p53. *J. Clin. Investig.* **1997**, *100*, 1363–1372. [CrossRef] [PubMed]
4. Juul, S. Erythropoietin in the Central Nervous System, and Its Use to Prevent Hypoxic-Ischemic Brain Damage. *Acta Paediatr.* **2002**, *91*, 36–42. [CrossRef]
5. Ashok, B.S.; Ajith, A.; Sivanesan, S. Hypoxia-Inducible Factors as Neuroprotective Agent in Alzheimer's Disease. *Clin. Exp. Pharmacol. Physiol.* **2017**, *44*, 327–334. [CrossRef]
6. Chung, H.M.; Won, H.; Sung, J. Responses of Adipose-Derived Stem Cells During Hypoxia: Enhanced Skin-Regenerative Potential. *Expert Opin. Biol. Ther.* **2009**, *9*, 1499–1508. [CrossRef]
7. Nys, K.; Maes, H.; Dudek, A.M.; Agostinis, P. Uncovering the Role of Hypoxia Inducible Factor-1α In Skin Carcinogenesis. *Biochim. Biophys. Acta* **2011**, *1816*, 1–12. [CrossRef]
8. Zhang, Z.; Rong, Z.; Wu, G.; Wang, Y.; Tan, Z.; Zheng, J.; Jin, Y.; Liang, Z.; Liu, C.; Guo, J.; et al. Gelatin-CaO_2/SAP/PLGA Composite Scaffold Enhances the Reparation of Critical-Sized Cranial Defects by Promoting Seed Cell Survival. *Appl. Mater. Today* **2021**, *22*, 100960. [CrossRef]
9. Pedraza, E.; Coronel, M.; Fraker, C.K.; Ricordi, C.; Stabler, C. Preventing Hypoxia-Induced Cell Death in Beta Cells and Islets via Hydrolytically Activated, Oxygen-Generating Biomaterials. *Proc. Natl. Acad. Sci. USA* **2012**, *109*, 4245–4250. [CrossRef]
10. Farris, A.; Rindone, A.; Grayson, W. Oxygen Delivering Biomaterials for Tissue Engineering. *J. Mater. Chem. B* **2016**, *4*, 3422–3432. [CrossRef]

11. Li, Z.; Guo, X.; Guan, J. An Oxygen Release System to Augment Cardiac Progenitor Cell Survival and Differentiation Under Hypoxic Condition. *Biomaterials* **2012**, *33*, 5914–5923. [CrossRef] [PubMed]
12. Abdi, S.; Choi, J.; Lau, H.; Lim, J. Controlled Release of Oxygen From PLGA-Alginate Layered Matrix and Its In Vitro Characterization on the Viability of Muscle Cells Under Hypoxic Environment. *Tissue Eng. Regen. Med.* **2013**, *10*, 131–138. [CrossRef]
13. Niu, H.; Li, C.; Guan, Y.; Dang, Y.; Li, X.; Fan, Z.; Shen, J.; Ma, L.; Guan, J. High Oxygen Preservation Hydrogels to Augment Cell Survival Under Hypoxic Condition. *Acta Biomater.* **2020**, *105*, 56–67. [CrossRef]
14. Lin, L.; Huang, T.; Song, J.; Ou, X.; Wang, Z.; Deng, H.; Tian, R.; Liu, Y.; Wang, J.; Liu, Y.; et al. Synthesis of Copper Peroxide Nanodots for H_2O_2 Self-Supplying Chemodynamic Therapy. *J. Am. Chem. Soc.* **2019**, *141*, 9937–9945. [CrossRef]
15. He, J.; Fu, J.; Qi, C.; Lin, J.; Huang, P. Metal peroxides For Cancer Treatment. *Bioact. Mater.* **2021**, *6*, 2698–2710. [CrossRef] [PubMed]
16. Cassell, O.; Hofer, S.; Morrison, W.; Knight, K. Vascularisation of Tissue-Engineered Grafts: The Regulation of Angiogenesis In Reconstructive Surgery and In Disease States. *Br. J. Plast. Surg.* **2002**, *55*, 603–610. [CrossRef] [PubMed]
17. Rastinfard, A.; Nazarpak, M.; Moztarzadeh, F. Controlled Chemical Synthesis of CaO_2 Particles Coated with Polyethylene Glycol: Characterization of Crystallite Size and Oxygen Release Kinetics. *RSC Adv.* **2018**, *8*, 91–101. [CrossRef]
18. Yeh, C.; Wang, R.; Chang, W.; Shih, Y. Synthesis and Characterization of Stabilized Oxygen-Releasing CaO_2 Nanoparticles for Bioremediation. *J. Environ. Manag.* **2018**, *212*, 17–22. [CrossRef]
19. Lee, E.; Jung, J.; Alam, Z.; Yi, H.; Kim, H.; Choi, J.; Hurh, S.; Kim, Y.; Jeong, J.; Yang, J.; et al. Effect of An Oxygen-Generating Scaffold on the Viability and Insulin Secretion Function of Porcine Neonatal Pancreatic Cell Clusters. *Xenotransplantation* **2018**, *25*, e12378. [CrossRef]
20. Mizukami, Y.; Takahashi, Y.; Shimizu, K.; Konishi, S.; Takakura, Y.; Nishikawa, M. Calcium Peroxide-Containing Polydimethylsiloxane-Based Microwells for Inhibiting Cell Death in Spheroids through Improved Oxygen Supply. *Biol. Pharm. Bull.* **2021**, *44*, 1458–1464. [CrossRef]
21. McQuilling, J.; Sittadjody, S.; Pendergraft, S.; Farney, A.; Opara, E. Applications of Particulate Oxygen-Generating Substances (POGS) in the Bioartificial Pancreas. *Biomater. Sci.* **2017**, *5*, 2437–2447. [CrossRef]
22. Lu, Z.; Jiang, X.; Chen, M.; Feng, L.; Kang, Y.Z. An oxygen-Releasing Device to Improve the Survival of Mesenchymal Stem Cells in Tissue Engineering. *Biofabrication* **2019**, *11*, 045012. [CrossRef] [PubMed]
23. Kang, J.I.; Park, K.M.; Park, K.D. Oxygen-Generating alginate Hydrogels as a Bioactive Acellular Matrix for Facilitating Wound Healing. *J. Ind. Eng. Chem.* **2019**, *69*, 397–404. [CrossRef]
24. Motealleh, M.; Kehr, N.S. Injectable Oxygen-Generating Nanocomposite Hydrogels with Prolonged Oxygen Delivery for Enhanced Cell Proliferation Under Hypoxic and Normoxic Conditions. *J. Mater. Chem. B* **2020**, *8*, 4195–4201. [CrossRef]
25. Montesdeoca, C.Y.C.; Afewerki, S.; Stocco, T.D.; Corat, M.A.F.; Paula, M.M.M.; Marciano, F.R.; Lobo, O. Oxygen-Generating Smart Hydrogels Supporting Chondrocytes Survival in Oxygen-Free Environments. *Colloids Surf. B Biointerfaces* **2020**, *194*, 111192. [CrossRef]
26. Alemdar, N.; Leijten, J.; Camci-Unal, G.; Hjortnaes, J.; Ribas, J.; Paul, A.; Mostafalu, P.; Gaharwar, A.K.; Qiu, Y.; Sonkusale, S.; et al. Oxygen-Generating Photo-Cross-Linkable Hydrogels Support Cardiac Progenitor Cell Survival by Reducing Hypoxia-Induced Necrosis. *ACS Biomater. Sci. Eng.* **2017**, *3*, 1964–1971. [CrossRef] [PubMed]
27. Park, S.; Park, K.M. Hyperbaric oxygen-generating hydrogels. *Biomaterials* **2018**, *182*, 234–244. [CrossRef] [PubMed]
28. Tamaddon, M.; Gilja, H.; Wang, L.; Romandini, I.; Filippis, I.R.D.; Zaffagnini, S.; Filardo, J. Osteochondral Scaffolds for Early Treatment of Cartilage Defects in Osteoarthritic Joints: From Bench to Clinic. *Biomater. Transl.* **2020**, *1*, 3–17.
29. Xu, X.; Song, J. Segmental Long Bone Regeneration Guided by Degradable Synthetic Polymeric Scaffolds. *Biomater. Transl.* **2020**, *1*, 33–45.
30. Ai, C.; Lee, Y.H.D.; Tan, X.H.; Tan, S.H.S.; Hui, J.H.P.; Goh, J.C.-H. Osteochondral tissue engineering: Perspectives for Clinical Application and Preclinical Development. *J. Orthop. Transl.* **2021**, *30*, 93–102. [CrossRef]
31. Li, L.; Yu, F.; Zheng, L.; Wang, R.; Yan, W.; Wang, Z.; Xu, J.; Wu, J.; Shi, D.; Zhu, L.; et al. Natural Hydrogels for Cartilage Regeneration: Modification, Preparation and Application. *J. Orthop. Transl.* **2019**, *17*, 26–41. [CrossRef]
32. Hsieh, T.E.; Lin, S.J.; Chen, L.C. Optimizing an Injectable Composite Oxygen-Generating System for Relieving Tissue Hypoxia. Front. *Bioeng. Biotechnol.* **2020**, *8*, 511.
33. Abudula, T.; Gauthaman, K.; Hammad, A.H.; Chen, C.C.; Lai, P.L.; Huang, C.C. Oxygen-Releasing Antibacterial Nanofibrous Scaffolds for Tissue Engineering Applications. *Polymers* **2020**, *12*, 1233. [CrossRef]
34. Oh, S.H.; Ward, C.L.; Atala, A.; Yoo, J.J.; Harrison, B.S. Oxygen Generating Scaffolds for Enhancing Engineered Tissue Survival. *Biomaterials* **2009**, *30*, 757–762. [CrossRef] [PubMed]

35. Lv, X.G.; Li, Z.; Chen, S.Y.; Xie, M.K.; Huang, J.W.; Peng, X.F.; Yang, R.X.; Wang, H.P.; Xu, Y.M.; Feng, C. Structural and Functional Evaluation of Oxygenating Keratin/Silk Fibroin Scaffold and Initial Assessment of Their Potential for Urethral Tissue Engineering. *Biomaterials* **2016**, *84*, 99–110. [CrossRef] [PubMed]
36. Erdem, A.; Darabi, M.A.; Nasiri, R.; Sangabathuni, S.; Ertas, Y.N.; Alem, H.; Hosseini, V.; Shamloo, A.; Nasr, A.S.; Ahadian, S.; et al. 3D Bioprinting: 3D Bioprinting of Oxygenated Cell-Laden Gelatin Methacryloyl Constructs. *Adv. Healthcare Mater.* **2020**, *9*, 2070047. [CrossRef]
37. Ortega, J.A.; Sirois, C.L.; Memi, F.; Glidden, N.; Zecevic, N. Oxygen Levels Regulate the Development of Human Cortical Radial Glia Cells. *Cerebral. Cortex.* **2017**, *27*, 3736–3751. [CrossRef]
38. Wang, R.; Sun, Y.F.; Wang, H.; Liu, T.; Shavandi, A.; Nie, L.; Yunusov, K.E.; Jiang, G.H. Core-Shell Structured Microneedles with Programmed Drug Release Functions for Prolonged Hyperuricemia Management. *J. Mater. Chem. B* **2024**, *12*, 1064–1076. [CrossRef]
39. Wu, X.; Han, X.; Guo, Y.; Liu, Q.; Sun, R.; Wen, Z.; Dai, C. Application Prospect of Calcium Peroxide Nanoparticles in Biomedical Field. *Rev. Adv. Mater. Sci.* **2023**, *62*, 20220308. [CrossRef]
40. Hwang, U.; Kim, J.; Kim, N.K.; Choi, K.; Chung, J.Y.; Kim, T.; Suhr, J.; Nam, J.D. Surface Charge Control of Hierarchical Ceria/Silica Hybrid Shells for Enhanced Dispersion Stability. *Appl. Surf. Sci.* **2022**, *571*, 151173. [CrossRef]
41. Kim, H.S.; Lee, J.H.; Mandakhbayar, N.; Jin, G.Z.; Kim, S.J.; Yoon, J.Y.; Jo, S.B.; Park, J.H.; Singh, R.K.; Jang, J.H.; et al. Therapeutic Tissue Regenerative Nanohybrids Self-Assembled from Bioactive Inorganic Core / chitosan Shell Nanounits. *Biomaterials* **2021**, *274*, 120857. [CrossRef] [PubMed]
42. Wu, D.; Zhu, Z.-Q.; Tang, H.-X.; Shi, Z.-E.; Kang, J.; Liu, Q.; Qi, J. Efficacy-shaping Nanomedicine by loading Calcium Peroxide into Tumor Microenvironment-Responsive Nanoparticles for the Antitumor Therapy of Prostate Cancer. *Theranostics* **2020**, *10*, 9808–9829. [CrossRef]
43. Khodaveisi, J.; Banejad, H.; Afkhami, A.; Olyaie, E.; Lashgari, S.; Dashti, R. Synthesis of Calcium Peroxide Nanoparticles as an Innovative Reagent for in Situ Chemical Oxidation. *J. Hazard. Mater.* **2011**, *192*, 1437–1440. [CrossRef] [PubMed]
44. Madan, S.S.; Wasewar, K.L.; Kumar, C.R. Adsorption Kinetics, Thermodynamics, and Equilibrium of α-toluic Acid onto Calcium Peroxide Nanoparticles. *Adv. Powder Technol.* **2016**, *27*, 2112–2120. [CrossRef]
45. Balci, B.; Aksoy, N.; Erkurt, F.E.; Budak, F.; Basibuyuk, M.; Zaimoglu, Z.; Turan, E.S.; Yilmaz, S. Removal of a Reactive Dye from Simulated Textile Wastewater by Environmentally Friendly Oxidant Calcium Peroxide. *Int. J. Chem. React. Eng.* **2021**, *19*, 1231–1243. [CrossRef]
46. Mendes, G.; Faria, M.; Carvalho, A.; Gonçalves, M.C.; de Pinho, M.N. Structure of Water in Hybrid Cellulose Acetate-silica Ultrafiltration Membranes and Permeation Properties. *Carbohydr. Polym.* **2018**, *189*, 342–351. [CrossRef] [PubMed]
47. Tai, X.; Ma, J.; Du, Z.; Wang, W.; Wu, J. A Simple Method for Synthesis of Thermal Responsive Silica Nanoparticle/PNIPAAm Hybrids. *Powder Technol.* **2013**, *233*, 47–51. [CrossRef]
48. Du Mont, J.W.; Marquardt, A.E.; Cano, A.M.; George, S.M. Thermal Atomic Layer Etching of SiO_2 by a "Conversion-Etch" Mechanism Using Sequential Reactions of Trimethylaluminum and Hydrogen Fluoride. *ACS Appl. Mater. Interfaces* **2017**, *9*, 10296–10307. [CrossRef]
49. Ali, M.; Farooq, U.; Lyu, S.; Sun, Y.; Li, M.; Ahmad, A.; Shan, A.; Abbas, Z. Synthesis of Controlled Release Calcium Peroxide Nanoparticles (CR-nCPs): Characterizations, H_2O_2 Liberate Performances and Pollutant Degradation Efficiency. *Sep. Purif. Technol.* **2020**, *241*, 116729. [CrossRef]
50. Xue, P.; Hou, M.; Sun, L.; Li, Q.; Zhang, L.; Xu, Z.; Kang, Y. Calcium-carbonate Packaging Magnetic Polydopamine Nanoparticles Loaded with Indocyanine Green for Near-infrared Induced Photothermal/photodynamic Therapy. *Acta Biomater.* **2018**, *81*, 242–255. [CrossRef]
51. Takada, S.; Hata, N.; Seino, Y.; Fujii, N.; Kikkawa, T. Skeletal Silica Characterization in Porous-silica Low-dielectric-constant Films by Infrared Spectroscopic Ellipsometry. *J. Appl. Phys.* **2005**, *97*, 113504. [CrossRef]
52. Wu, B.; Su, L.; Dai, X.; Chai, X. Development of Montmorillonite-supported Nano CaO_2 for Enhanced Dewatering of Waste-activated Sludge by Synergistic Effects of Filtration Aid and Peroxidation. *Chem. Eng. J.* **2017**, *307*, 418–426. [CrossRef]
53. Wang, J.; Lu, C.; Xiong, J. Self-Cleaning and Depollution of Fiber Reinforced Cement Materials Modified by Neutral TiO_2/SiO_2 Hydrosol Photoactive Coatings. *Appl. Surf. Sci.* **2014**, *298*, 19–25. [CrossRef]
54. Musić, S.; Filipović-Vinceković, N.; Sekovanić, L. Precipitation of Amorphous SiO_2 Particles and Their Properties. *Braz. J. Chem. Eng.* **2011**, *2*, 89–94. [CrossRef]
55. Dhanalekshmi, K.I.; Meena, K.S. Comparison of Antibacterial Activities of Ag@TiO_2 and Ag@SiO_2 Core-Shell Nanoparticles. Spectrochim. *Acta A Mol. Biomol. Spectrosc.* **2014**, *128*, 887–890. [CrossRef]
56. Suvarnapathaki, S.; Nguyen, M.A.; Goulopoulos, A.A.; Lantigua, D.; Camci-Unal, G. Engineering Calcium Peroxide Based Oxygen Generating Scaffolds for Tissue Survival. *Biomater. Sci.* **2021**, *9*, 2519–2532. [CrossRef] [PubMed]

57. Seekell, R.P.; Lock, A.T.; Peng, Y.; Cole, A.R.; Perry, D.A.; Kheir, J.N.; Polizzotti, B.D. Oxygen Delivery Using Engineered Microparticles. *Proc. Natl. Acad. Sci. USA* **2016**, *113*, 12380–12385. [CrossRef]
58. Cook, C.A.; Hahn, K.C.; Morrissette-McAlmon, J.B.; Grayson, W.L. Oxygen Delivery from Hyperbarically Loaded Microtanks Extends Cell Viability in Anoxic Environments. *Biomaterials* **2015**, *52*, 376–384. [CrossRef] [PubMed]
59. Mohammed, A.; Saeed, A.; Elshaer, A.; Melaibari, A.A.; Memi'c, A.; Hassanin, H.; Essa, K. Fabrication and Characterization of Oxygen-Generating Polylactic Acid/Calcium Peroxide Composite Filaments for Bone Scaffolds. *Pharmaceuticals* **2023**, *16*, 627. [CrossRef]
60. Pastwińska, J.; Walczak-Drzewiecka, A.; Kozłowska, E.; Harunari, E.; Ratajewski, M.; Dastych, J. Hypoxia Modulates Human Mast Cell Adhesion to Hyaluronic *Acid. Immunol. Res.* **2022**, *70*, 152–160. [CrossRef]
61. Azadmanesh, J.; Borgstahl, G.E.O. A Review of the Catalytic Mechanism of Human Manganese Superoxide Dismutase. *Antioxidants* **2018**, *7*, 25. [CrossRef] [PubMed]
62. Rodríguez-García, A.; García-Vicente, R.; Morales, M.L.; Ortiz-Ruiz, A.; Martínez-López, J.; Linares, M. Protein Carbonylation and Lipid Peroxidation in Hematological Malignancies. *Antioxidants* **2020**, *9*, 1212. [CrossRef] [PubMed]

materials

MDPI

Review

Advances in Polypyrrole Nanofiber Composites: Design, Synthesis, and Performance in Tissue Engineering

Lu Hao [1,2,*], Demei Yu [3], Xinyu Hou [1,2] and Yixuan Zhao [1]

1 Department of Materials Engineering, Shaanxi Polytechnic Institute, Xianyang 712000, China; 18691006725@163.com (X.H.); 13098159434@163.com (Y.Z.)
2 Key Laboratory of Shaanxi Higher Education Institutions for Ultra-Flexible Forming Technology of Electronic Glass, Shaanxi Polytechnic Institute, Xianyang 712000, China
3 State Key Laboratory of Electrical Insulation and Power Equipments, MOE Key Laboratory for Non-Equilibrium Synthesis and Modulation of Condensed Matter, School of Chemistry, Xi'an Jiaotong University, Xi'an 710049, China; dmyu@xjtu.edu.cn
* Correspondence: haoluatssd@163.com

Abstract: This review is different from previous studies focusing on polypyrrole (PPy) in universal fields such as sensors and supercapacitors. It is the first TO systematically review the specific applications of PPy-based electrospun nanofiber composites in the biomedical field, focusing on its biocompatibility regulation mechanism and tissue repair function. Although PPy exhibits exceptional electrical conductivity, redox activity, and biocompatibility, its clinical translation is hindered by processing challenges and poor degradability. These limitations can be significantly mitigated through composite strategies with degradable nanomaterials, enhancing both process compatibility and biofunctionality. Leveraging the morphological similarity between electrospun nanofibers and the natural extracellular matrix (ECM), this work comprehensively analyzes the topological characteristics of three composite fiber architectures—randomly distributed, aligned, and core–shell structures—and elucidates their application mechanisms in nerve regeneration, skin repair, bone mineralization, and myocardial tissue reconstruction (e.g., facilitating oriented cell migration and regulating differentiation through specific signaling pathway activation). The study further highlights critical challenges in the field, including PPy's poor solubility, limited spinnability, insufficient mechanical strength, and scalability limitations. Future efforts should prioritize the development of multifunctional gradient composites, intelligent dynamic-responsive scaffolds, and standardized biosafety evaluation systems to accelerate the substantive translation of these materials into clinical applications.

Keywords: PPy; electrospun nanofibers; composite; tissue engineering

1. Introduction

In the past few decades, one-dimensional nanostructures with nanoscale and molecular properties have made significant progress and are constantly meeting people's needs, such as carbon nanotubes, inorganic semiconductors and metal nanotubes/wires, conjugated polymer nanofibers/tubes, etc. [1–4]. These nanostructures have a profound impact on the basic research and potential applications of nanoelectronics or molecular electronics, nanodevices, nanocomposites, biological nanotechnology, and medicine [1–4].

Conductive polymers (CPs) are a special type of conjugated polymers with inherent conductivity. Due to their unique polymer properties, excellent electrical and mechanical properties, stability, and biocompatibility, they have broad application prospects in fields such as energy storage [5], flexible electronics [6], and bioelectronics [7]. Polypyrrole (PPy) is a kind of biocompatible conductive polymer. Due to its excellent conductivity, excellent redox properties, biocompatibility, easy synthesis, and environmental stability, it has potential applications in the biomedical field [8,9]. In the field of drug delivery, PPy nanofiber composites enable precise controlled release through electrochemical stimulation or environmental responsiveness (e.g., pH/temperature changes). The high specific surface area of their nanofibrous structure significantly enhances drug loading efficiency while reducing systemic toxicity. In biosensing applications, these materials allow the construction of highly sensitive detection systems by immobilizing glucose oxidase or modifying specific aptamers, facilitating the development of portable diagnostic devices and laboratory-on-a-chip technologies. Their photothermal antibacterial properties, combined with sustained antimicrobial agent release, provide effective anti-biofilm solutions for surgical sutures and implant surfaces. Additionally, in medical imaging, these composites serve as multifunctional contrast agents, enhancing MRI contrast via superparamagnetic particle loading or enabling deep-tissue visualization through photoacoustic effects. In flexible wearable devices, self-powered electronic skins leveraging their high tensile strain sensitivity have been developed to monitor real-time electrocardiogram (ECG), electromyogram (EMG) signals, and sweat metabolites.

In recent years, the research on PPy in the field of tissue engineering has attracted particular attention. PPy can regulate cell activities, including electrical stimulation (ES) to synthesize deoxyribonucleic acid (DNA), migration, and proliferation of cells in the biological environment [10]. It was found that conductive PPy-based scaffolds could support cell adhesion and proliferation regardless of whether electrical stimulation was applied, but electrical stimulation significantly enhanced cell activity. In addition, PPy-based scaffolds with electrical stimulation significantly increased the expression and secretion of nerve growth factor and brain-derived neurotrophic factor [11]. However, PPy has an amorphous structure and is insoluble, making it difficult to process, and has poor degradability. Composites with degradable nanomaterials can solve this problem. Due to the similarity between nanofibers and natural extracellular matrix (ECM), the preparation of nanofiber scaffolds by combining PPy and other materials is the main research topic in tissue engineering.

Electrospun nanofibers are an important class of materials in the current research of tissue engineering scaffold materials [12,13], mainly due to their following characteristics: (1) Electrospun nanofiber films have a good porous structure, high surface volume ratio, and space volume. These properties facilitate the loading and release of bioactive molecules, such as proteins, nanomedicines, nucleic acids, etc. Moreover, a higher surface area is conducive to the contact and adhesion between cells and scaffold materials; (2) The electrospinning device can be used to produce fibers with diameters of tens of microns or even several nanometers. Nanoscale fibers are similar to the ordered multi-layer fiber structure of natural extracellular matrix. As a scaffold material, it can simulate ECM well, which is conducive to cell adhesion and proliferation on the surface. A single polymer or a variety of polymer composites can be fabricated into nanofiber materials by electrospinning technology. These polymers include a large number of natural polymers, synthetic polymers, and their composites to meet the needs of different tissue engineering materials.

The doping process of PPy regulates its conductivity, mechanical properties, and biological activity by introducing counter ions (such as Cl^-, p-toluenesulfonate) or functional molecules (drugs, growth factors). The dopants are divided into inorganic materials (metal ions, carbon-based nanoparticles to enhance conductivity and mechanical strength), biomolecules (heparin, drugs to achieve anticoagulant or controlled release), and composite systems (such as Ag/dexamethasone synergistic antibacterial and anti-inflammatory), and are integrated into the scaffold through oxidative polymerization (oxidant selection, reaction medium optimization) and electrospinning process (in situ/post-treatment doping). Doping not only enhances interfacial adhesion (such as a high degree of deacetylation of chitosan to enhance PPy adhesion) but also achieves cell response regulation (such as neural differentiation) or stimulation-triggered drug release through electroactivity. However, it is necessary to balance the stability and biocompatibility of dopants and explore multi-mode response strategies to promote clinical application.

This paper utilizes the Web of Science database and employs "PPy" and "tissue engineering" as keywords to screen over 100 relevant SCI papers with successful cases published between 2010 and 2024. The number of published papers per year from 2010 to 2024 (the abscissa is the year, and the ordinate is the number of papers) is also provided in Figure 1. It can be seen from the figure that since 2010, the number of papers related to research has generally increased year by year.

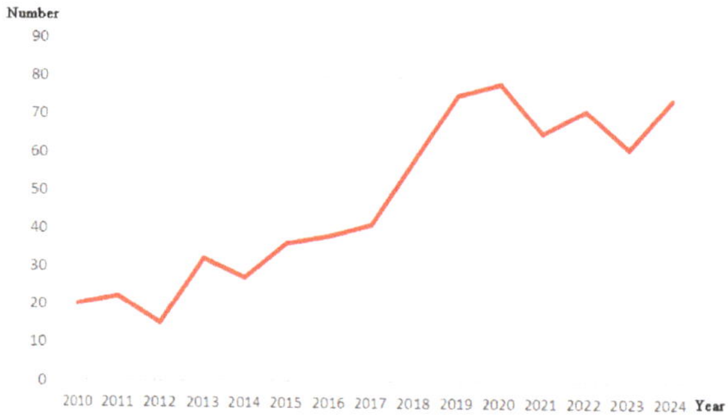

Figure 1. Number of papers published each year on the application of polypyrrole nanocomposites in tissue engineering.

This paper mainly summarizes the research progress of PPy-based electrospun nanofibers as scaffold materials in the field of tissue engineering in recent years. PPy-based electrospun nanofiber composites mainly include three types of structures: randomly distributed PPy-based nanofibers, aligned PPy-based nanofibers, and core–shell structure PPy-based nanofibers. Then, the application of PPy-based electrospun nanofiber composites as scaffold materials in the fields of nerve, skin, bone, skeletal muscle, and myocardium is summarized.

2. Structural Classification of PPy-Based Electrospun Nanofiber Composites

Although PPy has great potential in tissue engineering applications, PPy still has some limitations related to its inherent properties. Although the main benefit of adding PPy to the scaffold is the injection of electroactive properties, for in vivo implants, PPy-based scaffolds must also have mechanical stability, biocompatibility, and biological activity. The concept of using pure PPy and ES in tissue engineering has been reported to facilitate cell maturation. However, they are limited in the form of thin films, and there are no data on their mechanical properties and biocompatibility; therefore, pure PPy is not suitable for in vivo implants [14]. In addition, it is known that CPs are very brittle; thus, it is difficult to use high-concentration CPs to fabricate conductive scaffolds [15]. A common strategy to overcome its inherent brittleness is to mix a small but sufficient amount of CPs with non-conductive polymers (such as PLA, PCL, chitosan, etc.) to simulate the conductivity of natural tissues. These polymers have less brittleness than CPs as a matrix, thus forming composite materials. In turn, the introduction of CPs helps to improve the moderate mechanical strength and Young's modulus of non-conductive polymers, which are usually too low to be used as biological scaffolds. In this sense, CPs can be regarded as filler particles (i.e., load-bearing media) in composites under the premise that there is sufficient interaction between CPs and polymer matrix interfaces, which will help to strengthen the mechanical properties of non-conductive polymer matrix (i.e., load-bearing media). The discussion in this review focuses on the preparation of PPy polymer composites using electrospinning technology, because it is the most common type of PPy-based electroactive scaffolds so far.

2.1. Randomly Distributed PPy-Based Nanofibers

Randomly distributed PPy-based nanofibers are the most common type of PPy composite nanomaterials prepared by electrospinning [16–21]. Roca et al. [20] prepared PPy-coated polylactic acid (PLA) conductive nanofiber membranes and studied the effects of different reaction parameters on their physicochemical and dielectric properties. It is found that the conduction mechanism is based on the transformation law of PPy content: for the samples without PPy, there is a temperature-dependent relaxation process; for samples with high PPy content, a temperature-dependent conduction process; and for samples with low PPy content, a combination of these two processes. A uniform and continuous coating was obtained from 23 wt% PPy, and the percolation effect near 27 wt% PPy was observed. A higher percentage of PPy allows for higher conductivity (above 0.20 S cm^{-1}), but aggregates appear in 34 wt% of PPy. Sun et al. [21] coated PPy on the surface of electrospun poly (l-lactic-ε-caprolactone)/silk fibroin (PLCL/SF) to prepare a PPy-coated nerve-guided catheter (NGC). The preparation process is shown in Figure 2.

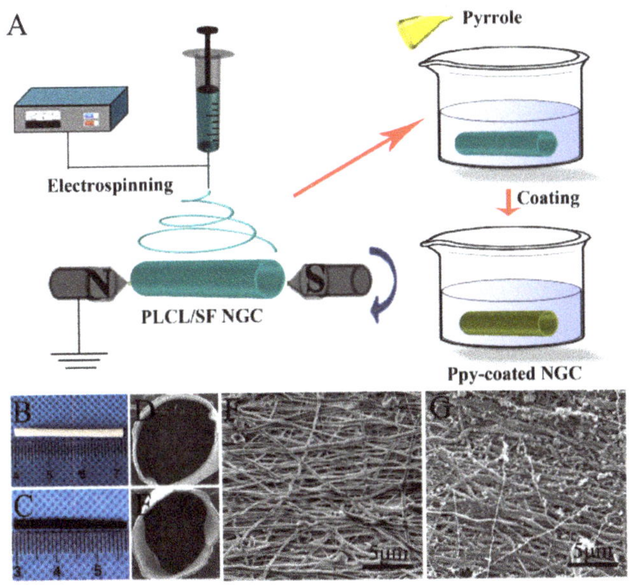

Figure 2. (**A**) Schematic illustration of Ppy-coated PLCL/SF NGC by electrospinning and in situ polymerization. Digital photos of (**B**) PLCL/SF NGC and (**C**) Ppy-coated NGC. Scanning electron microscope images of (**D**) PLCL/SF NGC cross-section, (**E**) Ppy-coated NGC cross-section, (**F**) the nanofiber surface of PLCL/SF NGC, and (**G**) the nanofiber surface of Ppy-coated NGC [21]. Copyright 2019 Materials Science and Engineering: C.

2.2. *Aligned PPy-Based Nanofibers*

The application of aligned nanofibers in tissue engineering is mainly to control the growth direction of cells. The aligned nanofiber scaffolds can be used for tissue engineering research, such as artificial blood vessels and ligaments [22–29]. Mancino et al. [22] proposed an electrospinning structure of a new type of conductive composite fiber with oriented distribution combining gelatin, polylactic acid-glycolic acid, and PPy, which can be used as a heart patch. Three different types of cells (neonatal rat ventricular myocytes (NRVM), human lung fibroblasts (MRC-5) and induced pluripotent stem cells (iPSCs)) were used to study the cell-biomaterial interaction. According to its phenotype, all cell types show good viability and a unique structure on the surface of the construct. The growth of NRVMs on aligned and randomly distributed nanofiber scaffolds is shown in Figure 3.

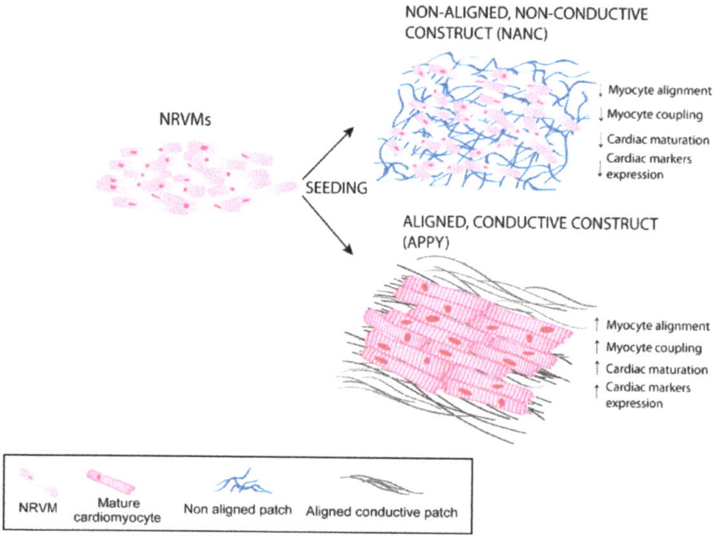

Figure 3. The growth of NRVMs on aligned and randomly distributed nanofiber scaffolds [22]. Copyright 2022 Nanomedicine: Nanotechnology, Biology and Medicine.

Studies have shown that the proliferation and differentiation rate of cells on aligned PPy-based nanofiber scaffolds is faster than that of randomly distributed nanofiber scaffolds [30–33]. Shrestha et al. [30] designed a self-electrically stimulated double-layer nerve guidance catheter (NGC), which was realized by chitosan-grafted polyurethane and well-dispersed functionalized multi-walled carbon nanotubes (fMWCNTs) nanofiber mats after uniform coating of PPy. The NGC is assembled by an electrospinning mat, with a directional arrangement of the inner layer, covering a randomly oriented outer layer. The interconnected NGC structural framework exhibits a cell-biomaterial interface and improves physical and chemical properties, including electrical conductivity, mechanical strength, and cell compatibility. As a natural host matrix for natural extracellular matrix (ECM), it plays an important role in neural tissue engineering. During in vitro cell culture, the regeneration, proliferation, and migration of Schwann cells (S42) and the differentiation of rat pheochromocytoma cells (PC12) were greatly accelerated on the mats with aligned nanofibers compared with the mats with randomly oriented nanofibers. The morphology and phenotype of the spontaneous growth of axon bundles are preferentially guided along the axis of aligned nanofibers, which maintains strong adaptability in axon regeneration.

The equipment and SEM images for the preparation of aligned polypyrrole nanocomposite fibers are shown in Figure 4.

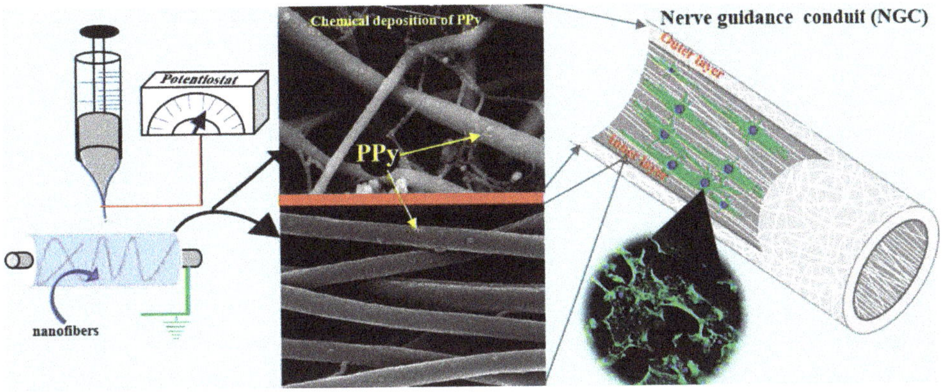

Figure 4. The equipment and SEM images for the preparation of aligned polypyrrole nanocomposite fibers [30]. Copyright 2018 Carbon.

In the field of tissue engineering, many studies focus on the surface morphology of conductive materials. However, there is little research on how electrical stimulation of this material affects nerve regeneration. Elashnikov et al. [34] studied the effect of electrical stimulation on randomly arranged and uniaxially arranged Polypyrrole-coated cellulose acetate butyrate (PPy/CAB) nanofibers. Research has shown that uniaxially arranged nanocomposite fibers have a strong impact on cell morphology and adhesion, with or without electrical stimulation. In addition, fluorescence microscopy showed that cells electrically stimulated on PPy/CAB exhibited longer axonal growth.

2.3. Core–Shell Structure PPy-Based Nanofibers

Functionalized core–shell-structured composite nanofibers can be used to control the release of growth factors or drugs, as well as to develop high-sensitivity sensors and tissue engineering scaffold composites [35–38]. Xiong et al. [35] uniformly coated polydopamine (PDA) and PPy on the surface of poly (L-lactide) (PLLA) nanofibers by in situ polymerization to form a new PPy/PDA/PLLA three-layer core–shell structure. The schematics of the structural and mechanical properties of the new material are shown in Figure 5. The uniformly coated PPy and PDA layers can significantly improve the hydrophilicity, conductivity (Resistance was around 291 Ω), near-infrared photothermal antibacterial properties, wound hemostasis speed, antioxidant capacity, and active oxygen scavenging ability, respectively. In addition, PPy/PDA/PLLA nanofibers showed good biocompatibility. The rat wound healing model confirmed that PPy/PDA/PLLA nanofibers could significantly accelerate wound repair in vivo.

Coaxial electrospinning represents an advanced adaptation of conventional electrospinning technology, distinguished primarily by its implementation of a multi-lumen coaxial spinneret during the fabrication process. This innovative configuration allows two distinct material components to be simultaneously delivered through separate concentric channels, ultimately self-assembling into core–shell fibrous composites. The technique's principal advantages lie in its capacity to achieve precise nanoscale control over sheath thickness, tunable mechanical integrity, and engineered degradation profiles—all while maintaining the inherent biocompatibility of the constituent materials. This unique combination of features enables customized fiber characteristics without compromising biological functionality, making it particularly valuable for biomedical applications requiring tailored material performance. Functionalized core–shell-structured composite nanofibers can be used to control the release of growth factors or drugs, as well as to develop high sensitivity. Zhang et al. [39] prepared lysine-doped PPy/regenerated spider silk protein (RSSP)/poly (L-lactic acid) (PLLA)/nerve growth factor (NGF) (L-PRPN) composite scaffolds by coaxial electrospray and electrospinning. This L-PRPN composite scaffold has a microfiber structure with a core–shell structure as the backbone and nanofibers as branches, and has good biocompatibility, cell adhesion, and relatively stable conductivity (the average conductivity was 16.70 ± 0.07 s cm^{-1}).

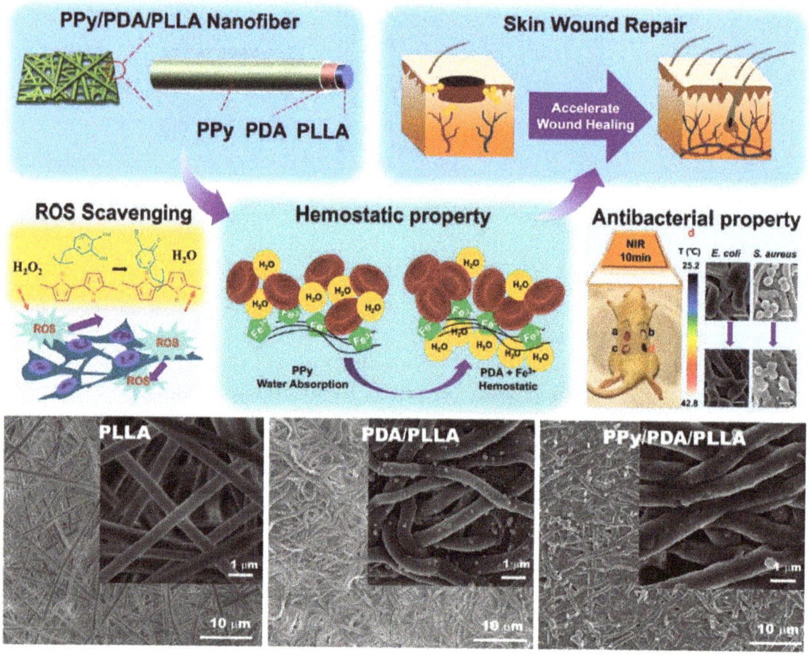

Figure 5. Mechanism of core–shell PPy/PDA/PLLA composites in the biological field [35]. Copyright 2022 International Journal of Biological Macromolecules.

In addition, to address the regulatory challenges arising from the complex interactions between processing parameters and material properties in electrospinning of polymeric tissue engineering scaffolds, Golbabaei et al. [40] proposed an innovative integration of SMILES molecular encoding and machine learning algorithms to construct a material-process-property prediction model. It achieves the first intelligent prediction of scaffold fiber diameter and conductivity and validates the model's reliability through experimental verification. This strategy significantly narrows the optimization scope of process parameters and provides an efficient data-driven solution for optimizing electrospinning processes of diverse biopolymers.

3. Application of Electrospun PPy-Based Composites in Tissue Engineering

The electrical signals in the human body regulate various types of tissues, such as neural communication activity, heartbeat activity, bone regeneration, muscle contraction, and wound healing [41]. Due to the direct current and stable potential, the important role of bioelectric field leads to tissue repair and cell migration to the wound site [42]. PPy allows fine-tuning of chemical, electrical, and physical properties to meet the needs of the tissue parts using PPy [43]. In order to achieve a greater impact of electroactive scaffolds, their implementation should not only meet the general characteristics of biomaterials, but also meet the specific needs of each tissue that may change with other tissues due to differences in extracellular matrix and function. In addition to electrical conductivity, certain mechanical properties unique to each tissue should also be considered, such as the high compressive strength and modulus of bone tissue and the typical elasticity of skin tissue. Therefore, this section focuses on (i) the electrical properties of PPy-based electroactive scaffolds suitable for the specified tissue and its electrical stimulation phenomenon, and (ii) some other properties (such as mechanical properties) of PPy-based electroactive scaffolds related to the target tissue.

3.1. Bone Tissue Engineering

The electrical conductivities of cancellous bone and cortical bone were 1.6–2.0 × 10^{-3} S/cm and 5.8–6.3 × 10^{-4} S/cm, respectively. One of the common strategies to simulate the electrical properties of bone tissue is to add conductive fillers, such as PPy, to increase the conductivity of the scaffold [29,44–51]. Chitosan's electroactive scaffold properties are tunable through molecular weight (MW) and degree of deacetylation (DD). A high DD (≥80%) boosts cationic charge density, enhancing electrostatic/coordination interactions with PPy or additives, while the controlled MW tailors porosity and degradation for optimized drug release. When combined with PPy via electrospinning, chitosan's bioactivity and film-forming ability synergize with PPy's conductivity. Zarei et al. [45] demonstrated this by fabricating PPy/chitosan/collagen scaffolds (0–25 wt% PPy). Conductivity increased with PPy content due to enhanced conductive particle contacts, yet all scaffolds exhibited low cytotoxicity from bioactive chitosan/collagen. Notably, the 10 wt% PPy scaffold achieved optimal performance, balancing electrical functionality and biocompatibility for cell proliferation. Liu et al. [49] self-polymerized dopamine (DA) on the surface of polylactic acid (PLLA)/hydroxyapatite (HA) nanowire composite fibers using electrochemical deposition to form an adhesive polydopamine (PDA) film, and constructed a stable silver nanoparticle (Ag-NPs) coating on it by electrochemically driven Ag^+ coordination and PPy-mediated chelation to achieve a stable and slow release of Ag-NPs. The synthesis mechanism of composite fibers is shown in Figure 6. The PLLA/HA/PDA/PPy/Ag composite fiber promoted the nucleation and growth of apatite on the surface, and had good cell compatibility with osteoblasts, indicating the ability to induce osteogenic differentiation. In addition, the combination with silver ions serves to stabilize and control the slow release of Ag^+, which contributes to the long-term antibacterial performance of the material.

In addition to increasing the concentration of PPy, the use of different PPy morphologies is also an attractive strategy to improve the electrical properties of electroactive scaffolds [52,53].

PPy is usually very fragile, with a Young's modulus of 180 MPa, which is very low compared to the Young's modulus of bone (especially cortical bone). This characteristic is contrary to the requirements of bone scaffolds, which require strength and ductility to avoid brittle fracture. In order to overcome this problem, the mechanical properties of PPy can be optimized by doping or combining PPy with metals, ceramics, or other polymers with higher mechanical properties than PPy. Maharjan et al. [54] prepared PCL/PPy composite nanofibers by electrospinning for bone tissue engineering scaffolds.

The mechanical strength test of the PCL/PPy scaffold showed that the Young's modulus (YM) of the composite was two to four times higher than that of the single PPy material, and the tensile strength (TS) was increased by three to four times. With the increase in PPy concentration in the composite scaffold, YM and TS also gradually increased.

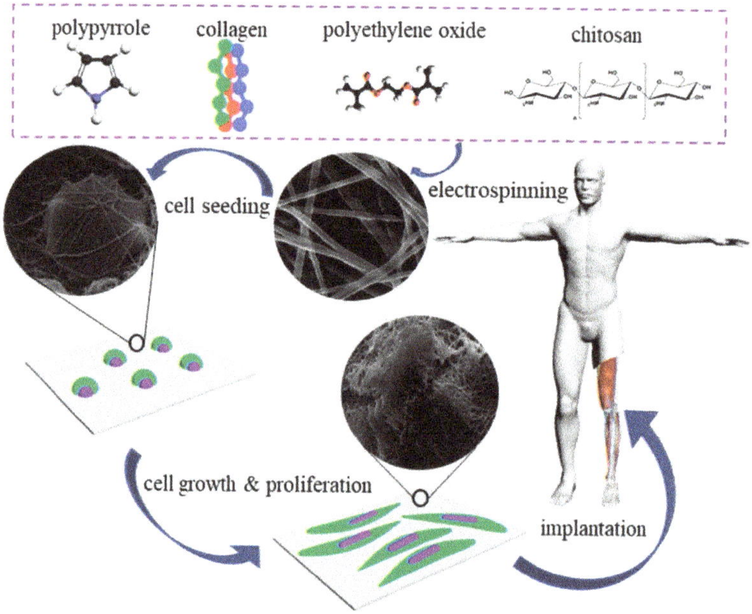

Figure 6. A conductive electrospun nanofiber scaffold containing conductive PPy polymer was fabricated to accelerate the healing of damaged tissues [45]. Copyright 2021 International Journal of Biological Macromolecules.

Tissue scaffolds that allow the control of the behavior of their cells are of particular interest in tissue engineering. Hardy et al. [55] described the preparation of conductive fiber-based bone tissue scaffolds (electrospun polycaprolactone nonwoven mats with interpenetrating networks of PPy and polystyrene sulfonate) that can electrically stimulate human bone marrow mesenchymal stem cells and enhance ALP activity and calcium deposition. The application of electrospun PPy nanofibers in bone engineering is shown in Table 1.

Table 1. Application of electrospun PPy composite nanofibers in bone engineering.

Fabrication Method	Composition of the Composite	Arrangement Patterns of Nanofiber Scaffolds	Cell Type	In Vitro Performance	Ref.
Electrospinning	PAN/PPy	Aligned and random	Mouse bone marrow mesenchymal stem cells	Alignment process increased the tensile strength of nanofibers 3.9-fold, while the tensile strain of nanofibers decreased by 78%	[29]
Electrospinning	PPy/chitosan/ collagen/PEO	Randomly arranged	Fibroblast cells	Enhanced the conductivity up to 164.274×10^{-3} s/m	[45]

Table 1. *Cont.*

Fabrication Method	Composition of the Composite	Arrangement Patterns of Nanofiber Scaffolds	Cell Type	In Vitro Performance	Ref.
Electrospinning and plasma polymerization	PLA/HA/Pyrrole/ Iodine matrices	Randomly arranged and porous	Mesenchymal stem cells	Presenting a variety of apparent pore sizes to allow for the passage of nutrients to bone cells. The increased cell proliferation and significantly improved cell viability	[46]
Electrospinning and electrodepositing	Nanohydroxyapatite/ PBAT/PPy	Randomly arranged	MG-63 cell	More hydrophilic with improved cell differentiation	[47]
Electrospinning and electrodepositing	PBAT/PPy/nHAp	Randomly arranged	Osteoblasts	Long-term antibacterial property, bioactivity, and osteoinductivity	[49]
Electrospinning	PLLA/PPy	An oriented direction and bead-free morphology	Bone mesenchymal stromal cells	Accelerated the osteogenic differentiation of the seeded cells	[50]
Electrospinning and in situ polymerization	PPy/Fe_3O_4/PLGA	Randomly arranged	MC3T3-E1 pre-osteoblasts	Good biocompatibility, hydrophilicity, and thermal stability	[51]
Electrospinning and in situ polymerization	PCL/PPy	Randomly arranged	MC3T3-E1 cell	Exhibited enhanced cell adhesion, proliferation, and differentiation in electrical stimulation conditions	[54]
Electrospinning and in situ polymerization	PCL/PPy/PSS	Randomly arranged	Human mesenchymal stem cells	Enhance the differentiation toward osteogenic outcomes	[55]

3.2. Neural Tissue Engineering

Neural networks in the human body play a unique and important role in all physiological processes, including cell recognition, sensory, and motor functions. The application of electroactive scaffolds is attractive and promising for further promoting the growth and differentiation of neurons and the formation of neural networks [21,30,33,34,38,56–67]. The constructed PPy/SF conductive NGC, together with ES, can effectively promote axonal regeneration and myelin formation in vivo. In addition, it was found that the MAPKs signal transduction pathway was activated by ES at the conduction conduit. The mechanical properties of scaffolds for nerve tissue engineering should imitate the mechanical properties of ECM to promote the neural differentiation of cells. The use of conductive polymers in electroactive scaffolds can reduce or improve the mechanical properties of the scaffolds. Therefore, it is necessary to determine the optimal composition of the conductive polymer in the electroactive scaffold to obtain improved mechanical properties [68–71]. Imani et al. [68] chemically synthesized a gelatin-PPy conductive copolymer by grafting pyrrole with different weight ratios on the gelatin chain, and then grafted it onto electrospun PLA nanofibers to improve its conductivity and biological interaction. The preparation process of gelatin/PPy/PLA nanofiber composite scaffolds is shown in Figure 7. The research revealed that increasing the PPy content enhances the composite's electrical conductivity, with the highest value (11.95 ± 1.10 mS/m) achieved at 20 wt% PPy content. Pheochromocytoma (PC12) cells were used to study the composite scaffold in vitro, including MTT and cell adhesion studies. The results show that the scaffolds containing 15% and 20% PPy can provide acceptable conditions for the adhesion and growth of nerve cells and can be introduced as a promising scaffold in nerve tissue engineering applications. Talebi et al. [70] prepared two kinds of graphene-containing PCL/gelatin/PPy and PCL/polyglycerol sebacate/PPy composite scaffold materials with intrinsic conductive properties using the electrospinning process and studied the effect of graphene on the properties of PCL/gelatin/PPy and PCL/polyglycerol sebacate/PPy. The results showed that the addition of graphene (3 wt%) significantly adjusted the physical and mechanical properties of the scaffold, increasing the conductivity from 0.1 to 3.9 ± 0.3 S/m, and the toughness was 76 MPa (PCL/gelatin/PPy) and 143.4 MPa (PCL/PGS/PPy). In addition, in the PCL/gelatin/PPy system, the elastic modulus of the scaffolds with 0,1 and 2 wt% graphene were reported to be 210, 300, and 340 kPa, and 72, 85, and 92 kPa in the PCL/PGS/PPy system.

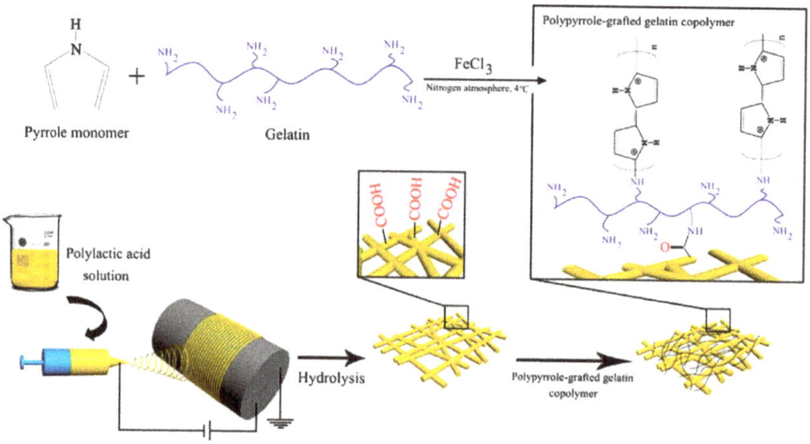

Figure 7. Schematic diagram for the preparation of gelatin/PPy/PLA nanofiber composite scaffolds [68]. Copyright 2021 Polymer.

Increased hydrophilicity contributes to the attachment of cells to the scaffold, thereby enabling better cell proliferation and differentiation [72–74]. Sadeghi et al. [72] developed a ternary PCL/chitosan/PPy composite scaffold using electrospun nanofibers, combining chitosan's hydrophilicity and PPy's electroactivity. The scaffold exhibited a 66% reduction in water contact angle due to chitosan, while PC12 cell studies revealed a 356% increase in proliferation and neurite extension compared to pure PCL. The synergy between high-DD chitosan-enhancing PPy interfacial adhesion to mitigate CP brittleness and PPy's electroactivity (enabling real-time monitoring or drug release) underpinned the scaffold's multifunctionality. By optimizing chitosan's DD, MW, and PPy concentration, such composites balance electroactivity, mechanical resilience, and biocompatibility, advancing implants for tissue regeneration and therapeutic applications like electrically controlled drug delivery.

In order to further improve the processability, solubility, and hydrophilicity of PPy and enhance its further application in tissue engineering, our research group [75,76] prepared a series of PPy derivatives as composite scaffolds for electroactive materials, and applied them in the field of tissue engineering to study the effects of the surface morphology and electrical properties of the scaffolds on neural cell behavior.

In addition, some researchers have used electrospun PPy-based composites for photoelectric biomaterials, which is a potential treatment option for neurodegenerative diseases (such as macular degeneration). Yuan et al. [77] developed a simple antioxidant, biocompatibility, and fibrous membrane heterojunction, which is composed of photosensitive polymer poly (3-hexylthiophene) (P3HT), PCL, and PPy. Compared with the traditional PCL-P3HT, the material not only has a lower impedance but also enhances the neurogenesis of PC-12 cells under light stimulation. Its mechanism is shown in Figure 8. This work demonstrates the importance of maintaining the PV-CP heterojunction while providing an optoelectronic platform for neural and optical tissue engineering. The application of electrospun PPy nanofibers in tissue engineering and regenerative medicine is shown in Table 2.

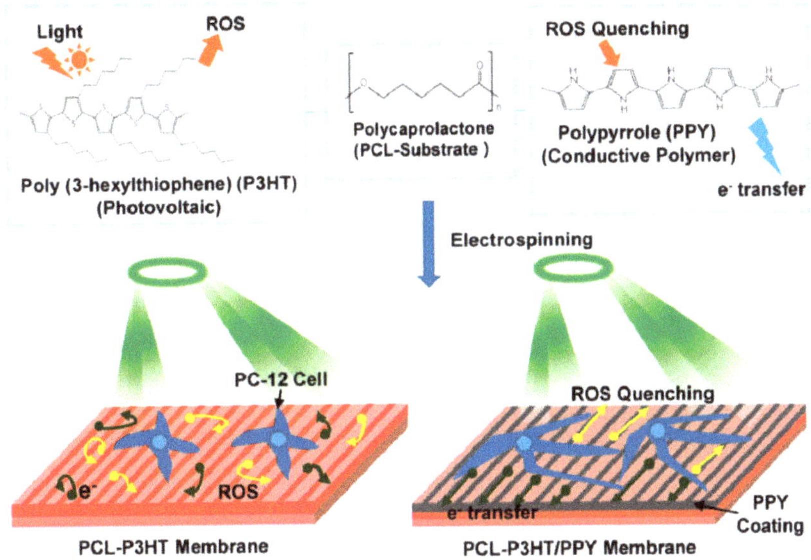

Figure 8. Preparation and photoelectric mechanism of P3HT/PCL/PPy heterojunction [77]. Copyright 2022 Acta Biomaterialia.

Table 2. Application of electrospun PPy composite nanofibers in tissue engineering and regenerative medicine.

Fabrication Method	Composition of the Composite	Cell Type	In Vitro Performance	Ref.
Electrospinning	PPy coating (MWCNT/chitosan-g-polycarbamate)	S42 and PC12	Supports better viability, growth, and axon formation. Aligned fiber arrangement positively regulates cell response.	[30]
Electrospinning and electrodeposition	Chitosan/PPy coating poly-L-lactone/poly (ε-caprolactone)	PC12	Supports the viability, differentiation, and axon growth of cultured cells. Electrical stimulation improves genes (NF-L and Trk A).	[33]
Electrospinning	PPy-coated cellulose acetate butyrate	SH-SY5Y	The cell adhesion of the coating substrate is reduced. Aligned fibers and electrical stimulation positively regulate neurite outgrowth.	[34]

Table 2. *Cont.*

Fabrication Method	Composition of the Composite	Cell Type	In Vitro Performance	Ref.
3D Printing, electrodeposition and electrospinning	PPy-coated silk fibroin	Rat Schwann cells	Supports good cell attachment, growth, and axon growth. Printing parameters seriously affect cell response.	[57]
Electrospinning	PPy-coated PLCL/silk fibroin	PC12	Showed better proliferation. Electrical stimulation improved cell response in terms of proliferation and differentiation phenotypes.	[61]
Electrospinning	PPy/PBAT	Neuro2A	Supports cell adhesion and axon growth.	[64]
Electrospinning	PPy-coated polyacrylonitrile	Rat neurons and glial cells	Forms cluster morphology and axon formation. Electrical stimulation induces cell proliferation and maturation.	[65]
Electrospinning	PPy-coated poly (ε-caprolactone)	Mouse Schwann cells	Supports cell growth.	[67]
Electrospinning	Polyornithine-coated polylactic acid/PPy	PC12	Supports good cell viability and growth. Electrical stimulation and fiber arrangement significantly improved the differentiated cell phenotype.	[74]

3.3. Skin Tissue Engineering

Skin trauma is one of the most common injuries caused by burns, diabetes, trauma, surgery, wound beds, and aging problems. Scaffold applications can expand ECM, provide potential opportunities for cell adhesion, proliferation, and migration, and, ultimately, enable new skin tissue regeneration (such as keratinocytes and fibroblasts). In the management of skin trauma, the development of electroactive scaffolds has a beneficial effect on enhancing the repair process of congenital trauma (such as local inflammation, cell infiltration, and neovascularization) [78]. In order to improve the electrical properties of skin scaffolds to meet the requirements of skin conductivity, researchers have conducted some research [79,80]. Massoumi et al. [79] prepared uniform, conductive, and biocompatible nanofiber scaffolds using an electrospinning PCL solution and synthesized PEGs-b-$(PPy)_4$ copolymer (the conductivity was 0.24–0.31 S/cm). The properties of the prepared nanofibers as tissue engineering scaffolds were tested in terms of biological (biocompatibility and biodegradability) and physicochemical (electroactivity, electrical conductivity, mechanical properties, and morphology) characteristics. As a result, it was found that the prepared electrospun nanofibers were suitable scaffolds for tissue engineering applications requiring electroactivity. Roman-Doval et al. [80] prepared PPy/iodine-coated biopolymer scaffolds for skin tissue engineering using electrospinning technology and developed and manufactured artificial skin tissues for dermal substitutes (Figure 9). Studies have shown that the polypyrrole/iodine complex is a scaffold with the best coating for the growth of HaCaT cells, improving cell viability, adhesion, and cell healing.

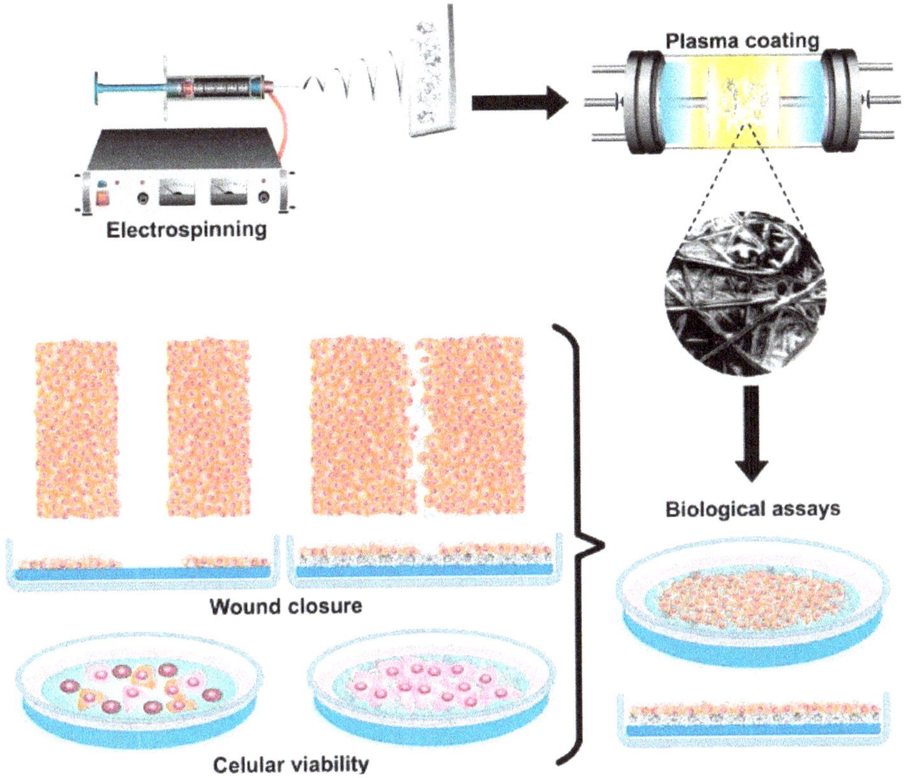

Figure 9. Electrospun the polypyrrole/iodine scaffolds of biopolymers used in skin tissue engineering to develop and create artificial skin tissue for replacement dermal substitutes [80]. Copyright 2019 Journal of Materials Science.

In addition, Ma et al. [81] fabricated a conductive polypyrrole@keratin nanofiber membrane (PPy@KNM) with excellent biocompatibility and non-cytotoxicity via a combination of electrospinning and in situ chemical polymerization. This conductive membrane effectively accelerated cell growth and proliferation under electrical stimulation, demonstrating its suitability as a functional biomedical material for promoting wound healing.

3.4. Skeletal Muscle Tissue Engineering

Skeletal muscle tissue accounts for about 45% of the total weight of the human body and is responsible for generating the power of various biological motor functions. Muscles can be regarded as electromechanical actuators that convert electrical energy transmitted from the nervous system into mechanical energy. Skeletal muscle has the ability to regenerate from mild injury. However, severe injuries caused by major trauma or medical reasons (such as myopathy or long-term denervation) usually lead to irreversible loss of muscle function. Because of its ability to simulate the function of muscle tissue as an electromechanical actuator, PPy has gained attention in the development of electroactive muscle scaffolds. These actuators even exceed the performance of natural muscle tissue in terms of working density, which makes them replicate many muscle-like actions in vitro and in vivo [82,83]. Harjo et al. [82] made glucose–gelatin nanofiber scaffolds have electrical conductivity and electrical activity via the chemical and electrochemical deposition of PPy (doped with trifluoromethanesulfonic acid). Since it may be a potential electromechanically activated cell growth matrix, the linear driving properties of this material in cell culture medium (CCM) were studied. Based on the flux of the main Na^+ ions from CCM, the material shows a relatively stable cation drive in CCM regardless of the deposition conditions.

The progress in tissue engineering and biomaterial development may provide scalable solutions to the problem of large skeletal muscle defects. Previous research on the development of skeletal muscle regeneration scaffolds has focused on strategies to increase electrical conductivity, which improves satellite cell attachment and differentiation. However, these strategies usually increase scaffold stiffness, and some studies have shown that this may be harmful to the development of myoblasts. Browe et al. [84] synthesized a PPy-PCL copolymer, which was designed to increase the conductivity of the scaffold without significantly affecting the stiffness. Different scaffolds were prepared using electrospinning, and their applicability to the proliferation and differentiation of myoblasts was evaluated. These scaffolds include aligned and randomly aligned pure PCL, 10% PPy-PCL, 20% PPy-PCL, and 40% PPy-PCL iterations. Only the scaffold containing 40% PPy-PCL has measurable electrical conductivity, and the addition of PPy-PCL has no significant effect on the stiffness of the scaffold. Compared with a pure PCL scaffold, the PPy-PCL copolymer significantly increased the adhesion of C2C12 myoblasts; however, the change in PPy-PCL concentration did not significantly change cell adhesion. In addition, by measuring the fusion index and the number of nuclei in each myotube, the scaffold with PPy-PCL promoted myoblast differentiation more than the scaffold made of PCL. In almost all measurements, the scaffolds composed of aligned nanofibers were superior to the scaffolds composed of randomly distributed nanofibers. These results suggest that electrical conductivity may not be a key factor in improving skeletal muscle scaffolds. On the contrary, the guidance cues of cell attachment and orientation may have a greater impact on myoblast differentiation.

3.5. Myocardial Tissue Engineering

The myocardium is a type of conductive tissue; therefore, conductive materials are used to simulate its inherent properties to repair damaged tissue. However, the inhibition of electrical conductivity may occur during the formation of fibrotic tissue or the remodeling of cardiac myocytes that damage cardiac function. Electroactive scaffolds are used as a strategy to help repair and improve the conductivity of the network to promote electrical connections between cells in the scaffold [85–90] Gelmi et al. [86] introduced an electroactive scaffold that provides electrical, mechanical, and topographic guidance for induced human pluripotent stem cells (iPS). The polylactic acid-glycolic acid fiber scaffolds coated with conductive PPy can provide direct electrical stimulation and mechanical stimulation to iPS. The electroactive scaffold has no cytotoxic effect on iPS, and the expression of cardiac biomarkers increases for both stimulation and non-stimulation regimens. This study is the first application of PPy as a supporting electroactive material for iPS and the first development of a fiber scaffold capable of dynamic mechanical drive.

The patch design for the treatment of heart disease should match the mechanical behavior of the myocardium. Olvera et al. [91] fabricated auxetic patches by melt electrospinning writing (MEW), which can overcome the limited elastic range of the traditional square patch design. Nutritional plaques can adapt to the strain and stress of the human myocardium during relaxation and contraction. The results show that the geometry of the auxetic patch can be fine-tuned to reflect the anisotropic mechanical properties. The effective stiffness anisotropy ratio (E-1/E-2) of the acrylic patch is consistent with the directionally dependent mechanics of the heart. In addition, the in situ polymerization of doped PPy on the auxetic patch imparts electrical conductivity (2.73 ± 0.3 S/m) similar to that of human myocardium.

The ability to sense tissue function online, provide stimuli to control contractility, and effectively release drugs within the engineered tissue microenvironment can enhance tissue assembly and improve the therapeutic outcomes for implanted engineered tissues. In order to endow the cardiac patch with this ability, Feiner et al. [92] developed an elastic and biodegradable electronic scaffold. The preparation process and mechanism are shown in Figure 10. The scaffold is composed of electrospun albumin fibers, which are used as both the substrate of the evaporated gold electrode and the passivation layer. The myocardial cell tissue inoculated on the electronic scaffold becomes a functional heart tissue, and its function is recorded online. In addition, electronic scaffolds can activate engineered tissues to control their function and trigger drug release. After implantation, these electronic scaffolds degrade, leading to the dissociation of inorganic materials from the scaffold. The establishment of this technology can be used for various degradable devices for tissue engineering, as well as original cell-free devices for electronic components used in the short term in vivo. The integration with gold was achieved through direct deposition of gold electrodes onto electrospun fiber scaffolds. This configuration enables the recording of extracellular signals from engineered tissues while simultaneously controlling the contraction and release of the anti-inflammatory drug dexamethasone.

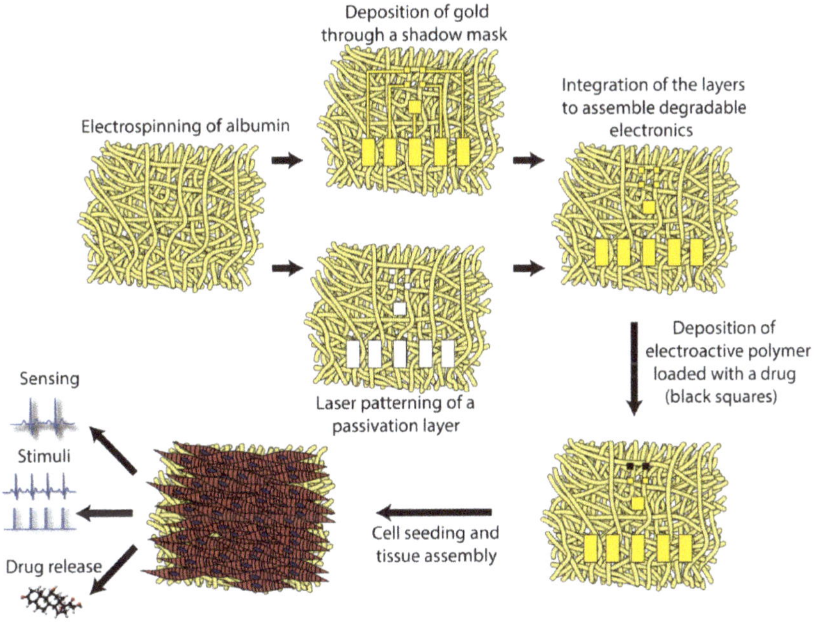

Figure 10. Schematic diagram of preparation of cardiac tissue engineering scaffold [92]. Copyright 2018 Journal of Controlled Release.

4. Conclusions and Outlook

This paper briefly reviews the main types of electrospun PPy nanocomposite fiber materials and their related applications in the field of tissue engineering. Discussions were conducted on preparation, physical, chemical, mechanical, and biological properties, as well as biomedical applications.

Electrospinning is a widely used technology and an ideal platform for preparing continuous ultrafine fibers. It has many advantages, and its performance can be adjusted by modification, which may solve the problem of the applicability of PPy and bring new application fields. In the past decade, the research on the preparation of PPy ultrafine fibers using electrospinning has developed rapidly. Although the direct electrospinning of pure PPy is difficult due to its low solubility and inherent brittleness, several strategies have been developed for the manufacture of PPy-based fibers, including the co-electrospinning of PPy with other spinnable carrier polymers and the synthesis of electrospun fiber templates, which are developed to solve this problem and improve the electrical, mechanical, thermal, and biological properties of electrospun PPy fibers. However, in order to further improve the performance of devices based on electrospun PPy materials and achieve the aforementioned interesting applications, many major challenges still need to be solved.

Firstly, due to the inherent properties of traditional CPs with low solubility and rigid frameworks, it is still a challenge to improve their solubility or spinnability by modifying or synthesizing new PPy derivatives, or finding/synthesizing better organic solvents or functional dopants. For example, the functional dopant sodium bis (2-ethylhexyl) sulfosuccinate (NaDEHS) can improve the solubility of PPy and form soluble PPy in organic solvents for electrospinning [93]. Ionic liquids are very stable and can be used as dopants or solvents. The improvement of solubility and processability of PPy in common organic solvents (including water and ionic liquids) is critical for more practical applications.

Secondly, in order to further improve the physical/chemical properties (e.g., thermal stability) and equipment performance (e.g., signal flow and stability) of electrospun PPy fibers, one approach is to fabricate polymers or nanofillers containing multiple functions, such as carbon nanotubes, graphene nanoplates, and metal or semiconductor nanoparticles/quantum dots, because polymer-based composites have shown improved performance and better applicability. In addition to adjusting the material composition or introducing functional materials during electrospinning, another important method is surface or chemical modification of PPy-based fibers through the adsorption of bioactive molecules, interception of small/large molecules, or covalent chemical coupling. This is an effective method to change conductivity, hydrophilicity/hydrophobicity, and biological activity, which is critical for neuroprobe and biosensor applications (e.g., surface modification of Au electrodes coated with PPy composite nanofibers by adsorbing ssDNA [94]).

Thirdly, electrospinning provides a multifunctional platform for the controllability of the morphology and structure of PPy fibers, which may bring new application fields. For example, core–shell nanofibers and hollow nanotubes synthesized using coaxial electrospinning or electrospun fiber templates can be used for drug delivery devices; spiral/wave or patterned fibers can be used for stretchable and flexible devices; the fiber array arranged using near-field electrospinning or nanofiber bundles can improve the performance of FET or bioactuator; 3D conductive fiber scaffold with controllable morphology, porosity, and pore size is suitable for tissue engineering applications. In addition, the reported electrospun PPy fibers/tubes usually exhibit a relatively large diameter from 100 nm to a few microns, and electrospun PPy nanofibers/tubes with an outer diameter of less than 100 nm are rarely reported. Therefore, precise control of the diameter, morphology, structure, and assembly of electrospun CP fibers remains a challenge in many cases of PPy.

Fourthly, the large-scale production of PPy-based fibers using electrospinning remains a challenge for commercial applications. At present, out of academic interest, most research on the manufacture and application of electrospun PPy fibers uses single-needle electrospinning in the laboratory. More attention should be paid to the engineering understanding and technical optimization of PPy fiber electrospinning process using various multi-needle/multi-nozzle electrospinning and nozzleless electrospinning technologies [95–99] to increase its application range. In addition, the poor mechanical properties and biocompatibility of PPy are additional challenges for regenerative medicine, especially for devices implanted in the human body. PPy composites containing other more biocompatible components are likely to exhibit better performance in biomedical applications. It should be noted that the biocompatibility of PPy composites depends on the type of additives and the synthesis method. For example, compared with metal nanoparticles prepared by traditional chemical methods, metal nanoparticles synthesized using green methods have much less cytotoxicity.

Finally, although there is compelling evidence that electrospun biopolymer fiber scaffolds have a significant impact, the manufacture of CPs has begun as biosensors, biological actuators, tissue engineering scaffolds, drug delivery devices, and neural interfaces. In order to support these important biological applications, two key issues must be addressed. The first problem is that the organic solvent is often used in solution electrospinning. The use of volatile and toxic solvents shows that cells or tissues cannot directly contact the electrospun structure without completely removing the solvent, which raises concerns about the cytotoxicity caused by residual solvents and solvent accumulation. The second problem is that the traditional method of electrospinning has challenges in the precise production and replication of structured 3D fiber scaffolds [100], which may affect the application of electrospun nanofiber scaffolds in engineering soft tissues. Melt-electrospinning (or melt-electrospinning writing) [101] is considered a promising candidate to address these two challenges, because melt-electrospinning nanofibers can be directly attached to cells without affecting their survival rate [102], and 3D (or even tubular) scaffolds with regular structures and high porosity (>85%) can be prepared by melt-electrospinning writing [101,103–105]. However, the challenge is that CPs are difficult to melt into electrospun fibers. Therefore, a combination of electrospinning and other manufacturing methods or post-processing (e.g., melt-electrospinning fiber template synthesis) may be a viable option. Due to the combination of the electroactivity of CPs, the biocompatibility of biopolymers, and good mechanical properties, electrospun CPs/biopolymer nanofibers are ideal candidates for tissue templates, artificial organ implants, and drug delivery that require electrical stimulation.

Tissue engineering relies heavily on the ability to expose cells that migrate to damaged tissue to precisely defined biophysical, biochemical, and biomechanical cues. By incorporating conductive PPy into the scaffold, a more cell-friendly tissue engineering construct can be developed. Although several studies have demonstrated the superiority of PPy in animal models, it has not yet been studied in humans. The main problem to be solved in the future is the huge gap between in vitro and in vivo outcomes. The research progress in this field may help researchers in this field to better predict the biological fate of PPy and reduce the gap between in vitro and in vivo results. A better transformation relationship between researchers, biotech companies, and clinicians is needed to elevate PPy-based conductive materials to a higher level and achieve clinical trials of these materials. Although it must be acknowledged that clinical trials are very expensive, researchers may consider using humanized animal models as an alternative test platform. The application of PPy-based nanofibers in surgical implants and tissue engineering remains challenging, including a mechanical-electroactive trade-off due to PPy's inherent brittleness and low conductivity (<1 S/cm) from insufficient doping, coupled with poorly understood interfacial dynamic behaviors (e.g., fatigue under cyclic loading). Additionally, biocompatibility risks from dopant residues (e.g., oxidative byproducts, acidic ions) and the lack of long-term metabolic clarity hinder clinical translation, while existing systems' reliance on single-mode electrical stimulation limits spatiotemporal regulation (e.g., growth factor gradients) or multimodal synergy (e.g., photo-electro-magnetic coupling). To overcome these limitations, future efforts should focus on bioinspired interface engineering (e.g., dynamic covalent bonds for toughness) and gradient/core–shell designs to harmonize mechanical-electroactive properties, alongside bio-derived dopants (natural polyelectrolytes, DNA) and metabolic tracking to reduce immunogenicity. Integrating multimodal smart-responsive systems (e.g., light-electro triggered drug release) and 4D-printed dynamic reconfiguration technologies could address tissue regeneration's temporal demands. Concurrently, advancing scalable manufacturing (high-throughput spinning-polymerization) and clinical-grade validation are critical to resolving challenges in long-term electroactivity, degradation-regeneration synchronization, and vascularization, ultimately accelerating translation into precision applications like neural interfaces and dynamic cardiac patches. Furthermore, the application of PPy and its composites in stimuli-responsive systems may have considerable development, such as in the manufacture of electro-reactive microneedles. In addition, PPy can be incorporated into pH-sensitive materials to develop pH-responsive platforms. The device can release its cargo in response to pH changes in the external environment or within the cell. Research in this field may lead to the emergence of new 'smart materials' with better processability, improved mechanical properties, biocompatibility, and excellent tissue function.

Author Contributions: Conceptualization, L.H. and D.Y.; Funding acquisition, L.H. and X.H.; Investigation, L.H., D.Y. and Y.Z.; Project administration, L.H. and D.Y.; Resources, X.H.; Supervision, D.Y.; Validation, X.H.; Visualization, Y.Z.; Writing—original draft, L.H. and X.H.; Writing—review & editing, Y.Z. All authors have read and agreed to the published version of the manuscript.

Funding: The research was financially supported by Xianyang Key Laboratory Project (L2023-CXNL-CXPT-ZDSYS-011); Research Start-up Funds for Advanced Talents of Project of Shaanxi Polytechnic Institute (203000536).

Institutional Review Board Statement: Not applicable.

Informed Consent Statement: Not applicable.

Data Availability Statement: No new data were created or analyzed in this study. Data sharing is not applicable to this article.

Conflicts of Interest: The authors declare no conflicts of interest.

References

1. Secchi, E.; Marbach, S.; Niguès, A.; Stein, D.; Siria, A.; Bocquet, L. Massive radius-dependent flow slippage in carbon nanotubes. *Nature* **2016**, *537*, 210–213. [CrossRef] [PubMed]
2. Tian, J.; Zhao, Z.; Kumar, A.; Boughton, R.I.; Liu, H. Recent progress in design, synthesis, and applications of one-dimensional TiO_2 nanostructured surface heterostructures: A review. *Chem. Soc. Rev.* **2014**, *43*, 6920–6937. [CrossRef] [PubMed]
3. Fang, Y.; Luan, D.; Gao, S.; Lou, X.W. Rational Design and Engineering of One-Dimensional Hollow Nanostructures for Efficient Electrochemical Energy Storage. *Angew. Chem. Int. Ed. Engl.* **2021**, *60*, 20102–20118. [CrossRef]
4. Ghosh, S.; Kouamé, N.A.; Ramos, L.; Remita, S.; Dazzi, A.; Deniset-Besseau, A.; Beaunier, P.; Goubard, F.; Aubert, P.-H.; Remita, H. Conducting polymer nanostructures for photocatalysis under visible light. *Nat. Mater.* **2015**, *14*, 505–511. [CrossRef] [PubMed]
5. Wang, K.; Wu, H.; Meng, Y.; Wei, Z. Conducting Polymer Nanowire Arrays for High Performance Supercapacitors. *Small* **2013**, *10*, 14–31. [CrossRef]
6. Xu, S.; Shi, X.-L.; Dargusch, M.; Di, C.; Zou, J.; Chen, Z.-G. Conducting polymer-based flexible thermoelectric materials and devices: From mechanisms to applications. *Prog. Mater. Sci.* **2021**, *121*, 100840. [CrossRef]
7. Nezakati, T.; Seifalian, A.; Tan, A.; Seifalian, A.M. Conductive Polymers: Opportunities and Challenges in Biomedical Applications. *Chem. Rev.* **2018**, *118*, 6766–6843. [CrossRef]

8. Ramanavicius, S.; Ramanavicius, A. Conducting Polymers in the Design of Biosensors and Biofuel Cells. *Polymers* **2021**, *13*, 49. [CrossRef]
9. Mao, J.; Zhang, Z. Polypyrrole as Electrically Conductive Biomaterials: Synthesis, Biofunctionalization, Potential Applications and Challenges. In *Cutting-Edge Enabling Technologies for Regenerative Medicine*; Springer: Singapore, 2018; Volume 1078, pp. 347–370.
10. Kim, S.; Jang, L.K.; Jang, M.; Lee, S.; Hardy, J.G.; Lee, J.Y. Electrically Conductive Polydopamine–Polypyrrole as High Performance Biomaterials for Cell Stimulation In Vitro and Electrical Signal Recording In Vivo. *ACS Appl. Mater. Interfaces* **2018**, *10*, 33032–33042. [CrossRef] [PubMed]
11. Huang, J.; Hu, X.; Lu, L.; Ye, Z.; Zhang, Q.; Luo, Z. Electrical regulation of Schwann cells using conductive polypyrrole/chitosan polymers. *J. Biomed. Mater. Res. Part A* **2010**, *93A*, 164–174. [CrossRef]
12. Jun, I.; Han, H.S.; Edwards, J.R.; Jeon, H. Electrospun Fibrous Scaffolds for Tissue Engineering: Viewpoints on Architecture and Fabrication. *Int. J. Mol. Sci.* **2018**, *17*, 745. [CrossRef] [PubMed]
13. Rahmati, M.; Mills, D.K.; Urbanska, A.M.; Saeb, M.R.; Venugopal, J.R.; Ramakrishna, S.; Mozafari, M. Electrospinning for tissue engineering applications. *Prog. Mater. Sci.* **2021**, *117*, 20102–20118.
14. Stewart, E.; Kobayashi, N.R.; Higgins, M.J.; Quigley, A.F.; Jamali, S.; Moulton, S.E.; Kapsa, R.M.; Wallace, G.G.; Crook, J.M. Electrical stimulation using conductive polymer polypyrrole promotes differentiation of human neural stem cells: A biocompatible platform for translational neural tissue engineering. *Tissue Eng. Part C Methods* **2015**, *21*, 385–393. [CrossRef]
15. Hermenegildo, B.; Ribeiro, C.; Peřinka, N.; Martins, P.; Trchová, M.; Hajná, M.; Stejskal, J.; Lanceros-Méndez, S. Electroactive poly(vinylidene fluoride) electrospun fiber mats coated with polyaniline and polypyrrole for tissue regeneration applications. *React. Funct. Polym.* **2022**, *170*, 105118. [CrossRef]
16. Guo, B.; Ma, P.X. Conducting polymers for tissue engineering. *Biomacromolecules* **2018**, *19*, 1764–1782. [CrossRef]
17. Sánchez-Pech, J.C.; Rosales-Ibáñes, R.; Cauich-Rodriguez, J.V.; Carrillo-Escalante, H.J.; Rodríguez-Navarrete, A.; Avila-Ortega, A.; Hernández-Sánchez, F. Design, synthesis, characterization, and cytotoxicity of PCL/PLGA scaffolds through plasma treatment in the presence of pyrrole for possible use in urethral tissue engineering. *J. Biomater. Appl.* **2019**, *34*, 840–850. [CrossRef]
18. Hatamzadeh, M.; Sarvari, R.; Massoumi, B.; Agbolaghi, S.; Samadian, F. Liver tissue engineering via hyperbranched polypyrrole scaffolds. *Int. J. Polym. Mater. Polym. Biomater.* **2020**, *69*, 1112–1122. [CrossRef]
19. Harjo, M.; Kesküla, A.; Leemets, K.; Khorram, M.S.; Saar, R.; Järvekülg, M.; Tamm, T.; Kiefer, R. Polypyrrole coatings on gelatin fiber scaffolds: Material and electrochemical characterizations in organic and aqueous electrolyte. *Synth. Met.* **2017**, *232*, 25–30. [CrossRef]
20. Roca, F.G.; García-Bernabé, A.; Moreno, V.C.; Martínez-Ramos, C.; Pradas, M.M. Solid Polymer Electrolytes Based on Polylactic Acid Nanofiber Mats Coated with Polypyrrole. *Macromol. Mater. Eng.* **2020**, *306*, 2000584. [CrossRef]
21. Sun, B.; Zhou, Z.; Li, D.; Wu, T.; Zheng, H.; Liu, J.; Wang, G.; Yu, Y.; Mo, X. Polypyrrole-coated poly(l-lactic acid-co-ε-caprolactone)/silk fibroin nanofibrous nerve guidance conduit induced nerve regeneration in rat. *Mater. Sci. Eng. C* **2019**, *94*, 190–199. [CrossRef]
22. Mancino, C.; Hendrickson, T.; Whitney, L.V.; Paradiso, F.; Abasi, S.; Tasciotti, E.; Taraballi, F.; Guiseppi-Elie, A. Electrospun electroconductive constructs of aligned fibers for cardiac tissue engineering. *Nanomed. Nanotechnol. Biol. Med.* **2022**, *44*, 102567. [CrossRef]
23. Guo, H.; Qiao, T.; Jiang, S.; Li, T.; Song, P.; Zhang, B.; Song, X. Aligned poly (glycolide-lactide) fiber membranes with conducting polypyrrole. *Polym. Adv. Technol.* **2016**, *28*, 484–490. [CrossRef]
24. Roca, F.G.; Estellés, J.M.; Pradas, M.M.; Martínez-Ramos, C. Axonal extension from dorsal root ganglia on fibrillar and highly aligned poly(lactic acid)-polypyrrole substrates obtained by two different techniques: Electrospun nanofibres and extruded microfibres. *Int. J. Biol. Macromol.* **2020**, *163*, 1959–1969. [CrossRef] [PubMed]
25. Nekouian, S.; Sojoodi, M.; Nadri, S. Fabrication of conductive fibrous scaffold for photoreceptor differentiation of mesen-chymal stem cell. *J. Cell Physiol.* **2019**, *234*, 15800–15808. [CrossRef] [PubMed]
26. Yang, A.; Huang, Z.; Yin, G.; Pu, X. Fabrication of aligned, porous and conductive fibers and their effects on cell adhesion and guidance. *Colloids Surf. B Biointerfaces* **2015**, *134*, 469–474. [CrossRef]
27. Cortés-Ortiz, E.; Olayo-Valles, R.; Rodríguez-Talavera, R.; González-Torres, M.; Vargas-Muñoz, S.; Olayo, R.; Godinez-Fernández, R.; Juárez, O.E.U.; Morales-Corona, J. Plasma Functionalized Scaffolds of Polyhydroxybutyrate Electrospun Fibers for Pancreatic Beta Cell Cultures. *Front. Mater.* **2021**, *8*, 600738. [CrossRef]
28. Sengupta, P.; Ghosh, A.; Bose, N.; Mukherjee, S.; Roy Chowdhury, A.; Datta, P. A comparative assessment of poly(vinylidene fluoride)/conducting polymer electrospun nanofiber membranes for biomedical applications. *J. Appl. Polym. Sci.* **2020**, *137*, 49115. [CrossRef]
29. Yardimci, A.I.; Baskan, O.; Yilmaz, S.; Mese, G.; Ozcivici, E.; Selamet, Y. Osteogenic differentiation of mesenchymal stem cells on random and aligned PAN/PPy nanofibrous scaffolds. *J. Biomater. Appl.* **2019**, *34*, 640–650. [CrossRef] [PubMed]

30. Shrestha, S.; Shrestha, B.K.; Kim, J.I.; Ko, S.W.; Park, C.H.; Kim, C.S. Electrodeless coating polypyrrole on chitosan grafted polyurethane with functionalized multiwall carbon nanotubes electrospun scaffold for nerve tissue engineering. *Carbon* **2018**, *136*, 430–443. [CrossRef]
31. Osorio-Londoño, D.M.; Godínez-Fernández, J.R.; Acosta-García, M.C.; Morales-Corona, J.; Olayo-González, R.; Morales-Guadarrama, A. Pyrrole Plasma Polymer-Coated Electrospun Scaffolds for Neural Tissue Engineering. *Polymers* **2021**, *13*, 3876. [CrossRef]
32. Patel, M.; Min, J.H.; Hong, M.H.; Lee, H.J.; Kang, S.; Yi, S.; Koh, W.G. Culture of neural stem cells on conductive and microgrooved polymeric scaffolds fabricated via electrospun fiber-template lithography. *Biomed. Mater.* **2020**, *15*, 045007. [CrossRef] [PubMed]
33. Xu, Y.; Huang, Z.; Pu, X.; Yin, G.; Zhang, J. Fabrication of Chitosan/Polypyrrole-coated poly(L-lactic acid)/Polycaprolactone aligned fibre films for enhancement of neural cell compatibility and neurite growth. *Cell Prolif.* **2019**, *52*, e12588. [CrossRef] [PubMed]
34. Elashnikov, R.; Rimpelová, S.; Děkanovský, L.; Švorčík, V.; Lyutakov, O. Polypyrrole-coated cellulose nanofibers: Influence of orientation, coverage and electrical stimulation on SH-SY5Y behavior. *J. Mater. Chem. B* **2019**, *7*, 6500–6507. [CrossRef]
35. Xiong, F.; Wei, S.; Sheng, H.; Wu, S.; Liu, Z.; Cui, W.; Sun, Y.; Wu, Y.; Li, B.; Xuan, H.; et al. Three-layer core-shell structure of polypyrrole/polydopamine/poly(l-lactide) nanofibers for wound healing application. *Int. J. Biol. Macromol.* **2022**, *222*, 1948–1962. [CrossRef]
36. Dai, W.; Shi, J.; Liu, S.; Xu, Z.; Shi, Y.; Zhao, Y.; Yang, Y. Preparation and properties of fiber-based conductive composite scaffolds for pe-ripheral nerve regeneration. *Chin. J. Reparative Reconstr. Surg.* **2019**, *33*, 356–362.
37. Guo, W.; Yang, Z.; Qin, X.; Wei, Y.; Li, C.; Huang, R.; Zhou, C.; Wang, H.; Jin, L.; Wang, H. Fabrication and Characterization of the Core-Shell Structure of Poly(3-Hydroxybutyrate-4-Hydroxybutyrate) Nanofiber Scaffolds. *BioMed Res. Int.* **2021**, *2021*, 8868431. [CrossRef]
38. Jing, J.; Ao, Q.; Wang, L.; Huang, Z.; Cai, Q.; Chen, G.; Yang, X.; Zhong, W. Constructing conductive conduit with conductive fibrous infilling for peripheral nerve regeneration. *Chem. Eng. J.* **2018**, *345*, 566–577. [CrossRef]
39. Zhang, H.; Wang, K.; Xing, Y.; Yu, Q. Lysine-doped polypyrrole/spider silk protein/poly(l-lactic) acid containing nerve growth factor composite fibers for neural application. *Mater. Sci. Eng. C* **2015**, *56*, 564–573. [CrossRef]
40. Golbabaei, M.H.; Varnoosfaderani, M.S.; Hemmati, F.; Barati, M.R.; Pishbin, F.; Ebrahimi, S.A.S. Machine learning-guided morphological property prediction of 2D electrospun scaffolds: The effect of polymer chemical composition and processing parameters. *RSC Adv.* **2024**, *14*, 15178–15199. [CrossRef]
41. Zhang, X.; Li, L.; Ouyang, J.; Zhang, L.; Xue, J.; Zhang, H.; Tao, W. Electroactive electrospun nanofibers for tissue engineering. *Nano Today* **2021**, *39*, 101196. [CrossRef]
42. Ning, C.; Zhou, Z.; Tan, G.; Zhu, Y.; Mao, C. Electroactive polymers for tissue regeneration: Developments and perspectives. *Prog. Polym. Sci.* **2018**, *81*, 144–162. [CrossRef] [PubMed]
43. Ohki, T.; Yamato, M.; Ota, M.; Takagi, R.; Kondo, M.; Kanai, N.; Okano, T.; Yamamoto, M. Application of regenerative medical technology using tissue-engineered cell sheets for endoscopic submucosal dissection of esophageal neoplasms. *Dig. Endosc.* **2015**, *27*, 182–188. [CrossRef] [PubMed]
44. Jadoun, S.; Riaz, U.; Budhiraja, V. Biodegradable conducting polymeric materials for biomedical applications: A review. *Med. Devices Sens.* **2021**, *4*, e10141. [CrossRef]
45. Zarei, M.; Samimi, A.; Khorram, M.; Abdi, M.M.; Golestaneh, S.I. Fabrication and characterization of conductive polypyrrole/chitosan/collagen electrospun nanofiber scaffold for tissue engineering application. *Int. J. Biol. Macromol.* **2021**, *168*, 175–186. [CrossRef] [PubMed]
46. Flores-Sánchez, M.G.; Islas-Arteaga, N.C.; Raya-Rivera, A.M.; Esquiliano-Rendon, D.R.; Morales-Corona, J.; Uribe-Juarez, O.E.; Vivar-Velázquez, F.I.; Ortiz-Vázquez, G.P.; Olayo, R. Effect of a plasma synthesized polypyrrole coverage on polylactic acid/hydroxyapatite scaffolds for bone tissue engineering. *J. Biomed. Mater. Res. Part A* **2021**, *109*, 2199–2211. [CrossRef]
47. de Castro, J.G.; Rodrigues, B.V.M.; Ricci, R.; Costa, M.M.; Ribeiro, A.F.C.; Marciano, F.R.; Lobo, A.O. Designing a novel nanocomposite for bone tissue engineering using electrospun conductive PBAT/polypyrrole as a scaffold to direct nanohydroxyapatite electrodeposition. *RSC Adv.* **2016**, *6*, 32615–32623. [CrossRef]
48. Elias, C.d.M.V.; Filho, A.L.M.M.; da Silva, L.R.; do Amaral, F.P.d.M.; Webster, T.J.; Marciano, F.R.; Lobo, A.O. In Vivo Evaluation of the Genotoxic Effects of Poly (Butylene adipate-co-terephthalate)/Polypyrrole with Nanohydroxyapatite Scaffolds for Bone Regeneration. *Materials* **2019**, *12*, 1330. [CrossRef]
49. Liu, F.; Wang, X.; Chen, T.; Zhang, N.; Wei, Q.; Tian, J.; Wang, Y.; Ma, C.; Lu, Y. Hydroxyapatite/silver electrospun fibers for anti-infection and osteoinduction. *J. Adv. Res.* **2020**, *21*, 91–102. [CrossRef]
50. Jing, W.; Huang, Y.; Wei, P.; Cai, Q.; Yang, X.; Zhong, W. Roles of electrical stimulation in promoting osteogenic differentiation of BMSCs on conductive fibers. *J. Biomed. Mater. Res. Part A* **2019**, *107*, 1443–1454. [CrossRef]

51. Li, K.; Zhang, S.; Wang, S.; Zhu, F.; Liu, M.; Gu, X.; Li, P.; Fan, Y. Positive effect of magnetic-conductive bifunctional fibrous scaffolds on guiding double electrical and magnetic stimulations to pre-osteoblasts. *J. Biomed. Nanotechnol.* **2019**, *15*, 477–486. [CrossRef]
52. Jin, Z.; Wei, S.; Jin, W.; Lu, B.; Xu, Y. Preparation and structure–property relationship study of piezoelectric–conductive composite polymer nanofiber materials for bone tissue engineering. *Polymers* **2024**, *16*, 1952. [CrossRef] [PubMed]
53. Peña, B.; Laughter, M.; Jett, S.; Rowland, T.J.; Taylor, M.R.G.; Mestroni, L.; Park, D. Injectable hydrogels for cardiac tissue engineering. *Macromol. Biosci.* **2018**, *18*, e1800079. [CrossRef] [PubMed]
54. Maharjan, B.; Kaliannagounder, V.K.; Jang, S.R.; Awasthi, G.P.; Bhattarai, D.P.; Choukrani, G.; Park, C.H.; Kim, C.S. In-situ polymerized polypyrrole nanoparticles immobilized poly(ε-caprolactone) electrospun conductive scaffolds for bone tissue engineering. *Mater. Sci. Eng. C* **2020**, *114*, 111056. [CrossRef]
55. Hardy, J.G.; Villancio-Wolter, M.K.; Sukhavasi, R.C.; Mouser, D.J.; Aguilar, D.; Geissler, S.A.; Kaplan, D.L.; Schmidt, C.E. Electrical Stimulation of Human Mesenchymal Stem Cells on Conductive Nanofibers Enhances their Differentiation toward Osteogenic Outcomes. *Macromol. Rapid Commun.* **2015**, *36*, 1884–1890. [CrossRef] [PubMed]
56. Zhao, Y.; Liang, Y.; Ding, S.; Zhang, K.; Mao, H.-Q.; Yang, Y. Application of conductive PPy/SF composite scaffold and electrical stimulation for neural tissue engineering. *Biomaterials* **2020**, *255*, 120164. [CrossRef]
57. Zhao, Y.-H.; Niu, C.-M.; Shi, J.-Q.; Wang, Y.-Y.; Yang, Y.-M.; Wang, H.-B. Novel conductive polypyrrole/silk fibroin scaffold for neural tissue repair. *Neural Regen. Res.* **2018**, *13*, 1455–1464. [CrossRef]
58. Alhosseini, S.N.; Moztarzadeh, F.; Karkhaneh, A.; Dodel, M.; Khalili, M.; Arshaghi, T.E.; Elahirad, E.; Mozafari, M. Improved cellular response on functionalized polypyrrole interfaces. *J. Cell Physiol.* **2019**, *234*, 15279–15287. [CrossRef]
59. Thunberg, J.; Kalogeropoulos, T.; Kuzmenko, V.; Hägg, D.; Johannesson, S.; Westman, G.; Gatenholm, P. In situ synthesis of conductive polypyrrole on electrospun cellulose nanofibers: Scaffold for neural tissue engineering. *Cellulose* **2015**, *22*, 1459–1467. [CrossRef]
60. Gopalakrishnan-Prema, V.; Mohanan, A.; Shivaram, S.B.; Madhusudanan, P.; Raju, G.; Menon, D.; A Shankarappa, S. Electrical stimulation of co-woven nerve conduit for peripheral neurite differentiation. *Biomed. Mater.* **2020**, *15*, 065015. [CrossRef]
61. Sun, B.; Wu, T.; Wang, J.; Li, D.; Wang, J.; Gao, Q.; Bhutto, M.A.; El-Hamshary, H.; Al-Deyab, S.S.; Mo, X. Polypyrrole-coated poly(L-lactic acid-co-ε-caprolactone)/silk fibroin nanofibrous membranes promoting neural cell proliferation and differentiation with electrical stimulation. *J. Mater. Chem. B* **2016**, *4*, 6670–6679. [CrossRef]
62. Zhou, Z.-F.; Zhang, F.; Wang, J.-G.; Chen, Q.-C.; Yang, W.-Z.; He, N.; Jiang, Y.-Y.; Chen, F.; Liu, J.-J. Electrospinning of PELA/PPY Fibrous Conduits: Promoting Peripheral Nerve Regeneration in Rats by Self-Originated Electrical Stimulation. *ACS Biomater. Sci. Eng.* **2016**, *2*, 1572–1581. [CrossRef]
63. Namhongsa, M.; Daranarong, D.; Sriyai, M.; Molloy, R.; Ross, S.; Ross, G.M.; Tuantranont, A.; Tocharus, J.; Sivasinprasasn, S.; Topham, P.D.; et al. Surface-Modified Polypyrrole-Coated PLCL and PLGA Nerve Guide Conduits Fabricated by 3D Printing and Electrospinning. *Biomacromolecules* **2022**, *23*, 4532–4546. [CrossRef]
64. Granato, A.E.; Ribeiro, A.C.; Marciano, F.R.; Rodrigues, B.V.; Lobo, A.O.; Porcionatto, M. Polypyrrole increases branching and neurite extension by Neuro2A cells on PBAT ultrathin fibers. *Nanomed. Nanotechnol. Biol. Med.* **2018**, *14*, 1753–1763. [CrossRef]
65. Xu, Q.; Jin, L.; Li, C.; Kuddannayai, S.; Zhang, Y. The effect of electrical stimulation on cortical cells in 3D nanofibrous scaffolds. *RSC Adv.* **2018**, *8*, 11027–11035. [CrossRef]
66. Zhou, X.; Yang, A.; Huang, Z.; Yin, G.; Pu, X.; Jin, J. Enhancement of Neurite Adhesion, Alignment and Elongation on Conductive Polypyr-role-Poly(lactide acid) Fibers with Cell-Derived Extracellular Matrix. *Colloids Surf. B Biointerfaces* **2017**, *149*, 217–225. [CrossRef]
67. Pan, X.; Sun, B.; Mo, X. Electrospun polypyrrole-coated polycaprolactone nanoyarn nerve guidance conduits for nerve tissue engineering. *Front. Mater. Sci.* **2018**, *12*, 438–446. [CrossRef]
68. Imani, F.; Karimi-Soflou, R.; Shabani, I.; Karkhaneh, A. PLA electrospun nanofibers modified with polypyrrole-grafted gelatin as bioactive electroconductive scaffold. *Polymer* **2021**, *218*, 123487. [CrossRef]
69. Zhou, J.-F.; Wang, Y.-G.; Cheng, L.; Wu, Z.; Sun, X.-D.; Peng, J. Preparation of polypyrrole-embedded electrospun poly(lactic acid) nanofibrous scaffolds for nerve tissue engineering. *Neural Regen. Res.* **2016**, *11*, 1644–1652.
70. Talebi, A.; Labbaf, S.; Atari, M.; Parhizkar, M. Polymeric Nanocomposite Structures Based on Functionalized Graphene with Tunable Properties for Nervous Tissue Replacement. *ACS Biomater. Sci. Eng.* **2021**, *7*, 4591–4601. [CrossRef]
71. Chen, X.; Liu, C.; Huang, Z.; Pu, X.; Shang, L.; Yin, G.; Xue, C. Preparation of carboxylic graphene oxide-composited polypyrrole conduits and their effect on sciatic nerve repair under electrical stimulation. *J. Biomed. Mater. Res. Part A* **2019**, *107*, 2784–2795. [CrossRef] [PubMed]
72. Sadeghi, A.; Moztarzadeh, F.; Mohandesi, J.A. Investigating the effect of chitosan on hydrophilicity and bioactivity of conductive electrospun composite scaffold for neural tissue engineering. *Int. J. Biol. Macromol.* **2019**, *121*, 625–632. [CrossRef]

73. Shafei, S.; Foroughi, J.; Chen, Z.; Wong, C.S.; Naebe, M. Short Oxygen Plasma Treatment Leading to Long-Term Hydrophilicity of Conductive PCL-PPy Nanofiber Scaffolds. *Polymers* **2017**, *9*, 614. [CrossRef] [PubMed]
74. Tian, L.; Prabhakaran, M.P.; Hu, J.; Chen, M.; Besenbacher, F.; Ramakrishna, S. Synergistic effect of topography, surface chemistry and conductivity of the electrospun nanofibrous scaffold on cellular response of PC12 cells. *Colloids Surf. B Biointerfaces* **2016**, *145*, 420–429. [CrossRef] [PubMed]
75. Zha, F.; Chen, W.; Lv, G.; Wu, C.; Hao, L.; Meng, L.; Zhang, L.; Yu, D. Effects of surface condition of conductive electrospun nanofiber mats on cell behavior for nerve tissue engineering. *Mater. Sci. Eng. C* **2021**, *120*, 111795. [CrossRef]
76. Zha, F.; Chen, W.; Hao, L.; Wu, C.; Lu, M.; Zhang, L.; Yu, D. Electrospun cellulose-based conductive polymer nanofibrous mats: Composite scaffolds and their influence on cell behavior with electrical stimulation for nerve tissue engineering. *Soft Matter* **2020**, *16*, 6591–6598. [CrossRef]
77. Yuan, B.; Aziz, M.R.F.; Li, S.; Wu, J.; Li, D.; Li, R.-K. An Electro-spun Tri-component Polymer Biomaterial with Optoelectronic Properties for Neuronal Differentiation. *Acta Biomater.* **2022**, *139*, 82–90. [CrossRef]
78. Debels, H.; Hamdi, M.; Abberton, K.; Morrison, W. Dermal matrices and bioengineered skin substitutes: A critical review of current options. *Plast. Reconstr. Surg. Glob. Open* **2015**, *3*, e284. [CrossRef]
79. Massoumi, B.; Hatamzadeh, M.; Firouzi, N.; Jaymand, M. Electrically conductive nanofibrous scaffold composed of poly(ethylene glycol)-modified polypyrrole and poly(ε-caprolactone) for tissue engineering applications. *Mater. Sci. Eng. C* **2019**, *98*, 300–310. [CrossRef]
80. Román-Doval, R.; Tellez-Cruz, M.M.; Rojas-Chávez, H.; Cruz-Martínez, H.; Carrasco-Torres, G.; Vásquez-Garzón, V.R. Enhancing electrospun scaffolds of PVP with polypyrrole/iodine for tissue engineering of skin regeneration by coating via a plasma process. *J. Mater. Sci.* **2019**, *54*, 3342–3353. [CrossRef]
81. Ma, Y.; Lei, T.; Yang, X.; Zhang, L.; Wang, Y.; Liu, J.; Li, X.; Yang, Q.; He, S.; Fan, J. Electrically Conductive Polypyrrole@keratin Nanofber Membranes with Potential for Tissue Restoration. *Fibers Polym.* **2024**, *25*, 751–763. [CrossRef]
82. Harjo, M.; Torop, J.; Järvekülg, M.; Tamm, T.; Kiefer, R. Electrochemomechanical Behavior of Polypyrrole-Coated Nanofiber Scaffolds in Cell Culture Medium. *Polymers* **2019**, *11*, 1043. [CrossRef] [PubMed]
83. Ebadi, S.V.; Fashandi, H.; Semnani, D.; Rezaei, B.; Fakhrali, A. Overcoming the potential drop in conducting polymer artificial muscles through metallization of electrospun nanofibers by electroplating process. *Smart Mater. Struct.* **2020**, *29*, 085036. [CrossRef]
84. Browe, D.; Freeman, J. Optimizing C2C12 myoblast differentiation using polycaprolactone–polypyrrole copolymer scaffolds. *J. Biomed. Mater. Res. Part A* **2019**, *107*, 220–231. [CrossRef] [PubMed]
85. Baei, P.; Hosseini, M.; Baharvand, H.; Pahlavan, S. Electrically conductive materials for in vitro cardiac microtissue engineering. *J. Biomed. Mater. Res. Part A* **2020**, *108*, 1203–1213. [CrossRef]
86. Gelmi, A.; Cieslar-Pobuda, A.; de Muinck, E.; Los, M.; Rafat, M.; Jager, E.W.H. Direct Mechanical Stimulation of Stem Cells: A Beating Electromechanically Active Scaffold for Cardiac Tissue Engineering. *Adv. Healtc. Mater.* **2016**, *5*, 1471–1480. [CrossRef]
87. Liang, Y.; Mitriashkin, A.; Lim, T.T.; Goh, J.C.-H. Conductive polypyrrole-encapsulated silk fibroin fibers for cardiac tissue engineering. *Biomaterials* **2021**, *276*, 121008. [CrossRef] [PubMed]
88. Fakhrali, A.; Semnani, D.; Salehi, H.; Ghane, M. Electro-conductive nanofibrous structure based on PGS/PCL coated with PPy by in situ chemical polymerization applicable as cardiac patch: Fabrication and optimization. *J. Appl. Polym. Sci.* **2022**, *139*, 52136. [CrossRef]
89. Aziz, M.R.F.; Wlodarek, L.; Alibhai, F.; Wu, J.; Li, S.; Sun, Y.; Santerre, J.P.; Li, R. A Polypyrrole-Polycarbonate Polyurethane Elastomer Alleviates Cardiac Arrhythmias via Improving Bio-Conductivity. *Adv. Health Mater.* **2023**, *12*, e2203168. [CrossRef]
90. Fakhrali, A.; Tamimi, M.; Semnani, D.; Salehi, H.; Ghodsi, A.; Rajabi, S.; Pezeshki-Modaress, M.; Ebadi, S.V.; Abdi, S.; Ghane, M. Electroconductive Nanofiber/Myocardium Gel Scaffolds Applicable for Myocardial Infarction Therapy. *ACS Appl. Polym. Mater.* **2024**, *6*, 5593–5607. [CrossRef]
91. Olvera, D.; Molina, M.S.; Hendy, G.; Monaghan, M.G. Electroconductive Melt Electrowritten Patches Matching the Mechanical Anisotropy of Human Myocardium. *Adv. Funct. Mater.* **2020**, *30*, 1909880. [CrossRef]
92. Feiner, R.; Fleischer, S.; Shapira, A.; Kalish, O.; Dvir, T. Multifunctional degradable electronic scaffolds for cardiac tissue engineering. *J. Control. Release* **2018**, *281*, 189–195. [CrossRef] [PubMed]
93. Chronakis, I.S.; Grapenson, S.; Jakob, A. Conductive polypyrrole nanofibers via electrospinning: Electrical and morphological properties. *Polymer* **2006**, *47*, 1597–1603. [CrossRef]
94. Wang, X.; Wang, X.; Wang, X.; Chen, F.; Zhu, K.; Xu, Q.; Tang, M. Novel electrochemical biosensor based on functional composite nanofibers for sensitive detection of p53 tumor suppressor gene. *Anal. Chim. Acta* **2013**, *765*, 63–69. [CrossRef]
95. Liu, S.; Liu, S.-L.; Long, Y.-Z.; Liu, L.-Z.; Zhang, H.-D.; Zhang, J.-C.; Han, W.-P.; Liu, Y.-C. Fabrication of p-type ZnO nanofibers by electrospinning for field-effect and rectifying devices. *Appl. Phys. Lett.* **2014**, *104*, 042105. [CrossRef]

96. He, J.-H.; Kong, H.-Y.; Yang, R.-R.; Dou, H.; Faraz, N.; Wang, L.; Feng, C. Review on fiber morphology obtained by bubble electrospinning and blown bubble spinning. *Therm. Sci.* **2012**, *16*, 1263–1279. [CrossRef]
97. Liu, Y.; Zhang, L.; Sun, X.-F.; Liu, J.; Fan, J.; Huang, D.-W. Multi-jet electrospinning via auxiliary electrode. *Mater. Lett.* **2015**, *141*, 153–156. [CrossRef]
98. Brettmann, B.K.; Tsang, S.; Forward, K.M.; Rutledge, G.C.; Myerson, A.S.; Trout, B.L. Free surface electrospinning of fibers containing microparticles. *Langmuir* **2012**, *28*, 9714–9721. [CrossRef]
99. Persano, L.; Camposeo, A.; Tekmen, C.; Pisignano, D. Industrial upscaling of electrospinning and applications of polymer nano-fibers: A review. *Macromol. Mater. Eng.* **2013**, *298*, 504–520. [CrossRef]
100. Sun, B.; Long, Y.; Zhang, H.; Li, M.; Duvail, J.; Jiang, X.; Yin, H. Advances in three-dimensional nanofibrous macrostructures via electrospinning. *Prog. Polym. Sci.* **2014**, *39*, 862–890. [CrossRef]
101. Dalton, P.D.; Vaquette, C.; Farrugia, B.L.; Dargaville, T.R.; Brown, T.D.; Hutmacher, D.W. Electrospinning and additive manufacturing: Converging technologies. *Biomater. Sci.* **2013**, *1*, 171–185. [CrossRef]
102. Dalton, P.D.; Klinkhammer, K.; Salber, J.; Klee, D.; Möller, M. Direct in vitro electrospinning with polymer melts. *Biomacromolecules* **2006**, *13*, 686–690. [CrossRef] [PubMed]
103. Brown, T.D.; Dalton, P.D.; Hutmacher, D.W. Direct writing by way of melt electrospinning. *Adv. Mater.* **2011**, *23*, 5651–5657. [CrossRef] [PubMed]
104. Brown, T.D.; Slotosch, A.; Thibaudeau, L.; Taubenberger, A.; Loessner, D.; Vaquette, C.; Dalton, P.D.; Hutmacher, D.W. Design and fabrication of tubular scaffolds via direct writing in a melt electrospinning mode. *Biointerphases* **2012**, *7*, 13–16. [CrossRef] [PubMed]
105. Brown, T.D.; Dalton, P.D.; Hutmacher, D.W. Melt electrospinning today: An opportune time for an emerging polymer process. *Prog. Polym. Sci.* **2016**, *56*, 116–166. [CrossRef]

materials

Article

Synthesis of N-Doped Graphene Quantum Dots from Cellulose and Construction of a Fluorescent Probe for 6-Mercaptopurin Quantitative Detection

Qiang Xu [1,2], Jiayi Dong [3], Guiqin Yan [3], Rongnan Yi [2,*] and Xiaojing Yang [4,*]

1 College of Environmental Science and Engineering, Shanxi University of Electronic Science and Technology, Linfen 041004, China; xuqiang@sxdzkj.edu.cn
2 Key Laboratory of Food & Environment & Drug Monitoring and Testing of Universities in Hunan Province, Hunan Police Academy, Changsha 410138, China
3 School of Life Science, Shanxi Normal University, Taiyuan 030006, China; cleevine@163.com (J.D.); gqyan2013@163.com (G.Y.)
4 School of Resources, Environment and Materials, Guangxi University, Nanning 530004, China
* Correspondence: yrn@hnu.edu.cn (R.Y.); kjcyxj@gxu.edu.cn (X.Y.)

Abstract: With cellulose as the precursor and ethylenediamine as the N source, N-doped graphene quantum dots (N-GQDs) were synthesized by a simple and feasible one-pot hydrothermal method. The whole process did noSchemet need a strong acid or strong base and avoided interference from inorganic salt residues. The whole process lasted only 3 h and avoided any complex postprocessing. Because of the outstanding optical properties of N-GQDs, a high-efficiency 6-mercaptopurine fluorescent probe based on the inner filter effect of fluorescence was established. The detection range was 0.2–60 μM and the detection limit was 0.05 μM. This method can preliminarily detect 6-mercaptopurine in human urine and avoids any sample preparation or extraction in advance and brings satisfactory results.

Keywords: cellulose; N-doped graphene quantum dots; inner filter effect of fluorescence; 6-mercaptopurine

1. Introduction

Cancers are one of the leading causes of pathopoiesis and death worldwide [1]. The decline in the total mortality rate of cancers and the increase in the total number of surviving patients reflect that significant achievements have been made in cancer prevention and cure. Nevertheless, cancers are still challenges against the persistent public health of the world. So far, chemotherapy is pivotal in therapy of various cancers. The technological development of chemotherapy has resulted in considerable revolution in cancer treatment and has greatly improved the survival rate. The total survival rate of acute lymphocyte leukemia in children is over 90% [1,2]. Such success largely depends on the daily use of 6-mercaptopurine (6-MP) for maintenance treatment. Moreover, 6-MP can be used to cure chorionic epithelioma, malignant hydatidiform mole, malignant lymphoma, and frequently occurring myeloma at the blastic phase. Hence, the demand for 6-MP is very large in clinical treatment, but too low a dose will decrease the anti-inflammatory and anti-immune effects, while too high a dose will cause nausea, vomiting, stomatitis, and diarrhea or even restrain marrows and damage livers [3]. Thus, rapid and accurate detection of 6-MP will guide clinical drug use. At present, the 6-MP detection methods include chemiluminescence analysis [4,5], electrochemical detection [6,7], high-performance liquid chromatography [8–11], mass spectrum [12], molecular imprinting [13],capillary electrophoresis [14], Raman analysis [15], and fluorescence [16–18]. Despite their high sensitivity and selectivity, these methods are limited by the use of expensive equipment, complex sample pretreatment, low repeatability, and tedious detection. Hence, a simple and fast 6-MP quantitative detection method with high sensitivity and selectivity is urgently needed [19].

With the development of fluorescent probes, quantum dots (QDs) are widely used in fluorescent probes thanks to their strong optical stability, long fluorescence lifetime, wide excitation peaks, narrow symmetrical emission peaks, and larger Stoke shifts. Especially, graphene quantum dots (GQDs), as a new type of fluorescent nanomaterials [20], are featured by chemical inertness, strong resistance against photobleaching, low cytotoxicity, and high biocompatibility, and thus are widely used in cell imaging, biological sensing, and drug delivery. The synthesis of GQDs has attracted the attention of many researchers. The existing synthetic methods can be divided into top–down methods and bottom–up methods. The top–down methods are mostly based on the Hummers approach [21] and adopt graphite or graphite-like carbon fibers, graphene films, or graphite powder as precursors. However, top–down methods rely on strong acids, strong bases and enhancers (e.g., potassium dichromate) for chemical oxidation stripping, and have a long preparation time, and the by-products will cause environmental pollution. The bottom–up methods are mostly based on aromatic structures (e.g., citric acid, acetylacetone, glucoses) as precursors [22] and synthesize GQDs from carbonization under microwave-, ultrasound-, or hydrothermal-assisted conditions. Despite the simple processes, special equipment is needed, and the removal of small molecules is very tedious. Moreover, the difficulty in surface modification limits the potential application of GQDs. The surface modification of GQDs by doping with other atoms not only reserves the intrinsic optical and chemical properties of GQDs, but also is favorable for wider application of GQDs. Thus, it is urgent to seek new precursors and develop a safe, simple, and low-cost synthetic strategy [20].

In this study based on the green chemical method [23], N-doped GQDs (N-GQDs) were synthesized by a one-step hydrothermal method with cellulose (which is the most abundant macromolecular compound in the nature) as the precursor and ethylenediamine (EDA) as the N source. The whole process did not need a strong acid or strong base and avoided interference from inorganic salt residues and heavy metal ions. The whole process lasted only 3 h and avoided any complex postprocessing. The N-GQDs as prepared have strong photostability and salt resistance. After that, a novel high-selectivity and high-sensitivity fluorescent probe for 6-MP detection was established on the basis of the inner filter effect (IFE) of fluorescence. Moreover, it was successfully used to detect 6-MP in human urine, which confirms this N-GQDs-based fluorescent probe has high application prospects (diagram shown in Scheme 1).

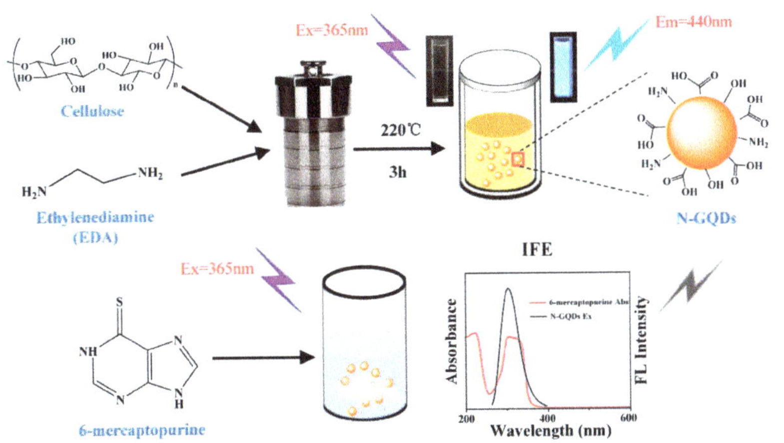

Scheme 1. Schematic diagram of N-GQDs synthesis and 6-MP detection process.

2. Experimental Section

2.1. Reagents and Instruments

Cellulose was provided by Aladdin Reagent Co., Ltd. (Shanghai, China), ethylenediamine, glucose and glycine were purchased from Tianjin Komeo Chemical Reagent Co., Ltd. (Tianjin, China), lactose, D-fructose and D-galactose were purchased from Shanghai Reagent Plant No. 2 (Shanghai, China), L-tyrosine was provided by Kelon Chemical Reagent Factory (Chengdu, China), and DL-methionine was provided by Beijing Chemical Reagent Co., Ltd. (Beijing, China). Alanine, L-proline, L-cysteine, arginine, glutamic acid, and lysine were all purchased from Tianjin Guangfu Fine Chemical Research Institute (Tianjin, China). All chemicals were of analytical grade and can be used without further purification. The ultrapure water (18.2 MΩcm) used in the experiment was prepared by the German microporous ultrapure water system.

The morphology and structure of N-GQDs were characterized by a Tecnai G2 F20 electron microscope (FEI, Hillsboro, OR, USA) at an accelerating voltage of 200 kV, X-ray photoelectron spectroscopy (Thermo Scientific Nexsa, Waltham, MA, USA), and UltimaIV-185 powder X-ray diffraction (XRD). The fluorescence spectrum and resonance light scattering spectrum (RLS) of N-GQDs were recorded by a Cary Eclipse fluorescence spectrophotometer (Varian, Melbourne, Australia). The UV–visible absorption spectrum was measured by TU-1901 double-beam UV-vis spectrophotometer. Fourier transform infrared spectroscopy (FT-IR) was measured by a Varian660-IR infrared spectrometer. The fluorescence lifetime of N-GQDs was measured by an FLS1000/FS5 steady-state transient fluorescence spectrometer (Edinburgh, Livingston, UK). The concentrated solution of N-GQDs was obtained by a rotary evaporation meter (RE-2000A), and the N-GQDs powder was obtained by a freeze dryer (ZX-LGJ-1A). An AEU-200 electronic balance was used to weigh drugs.

2.2. Preparation of N-GQDs

In brief, 1 g of cellulose and 16 mL of ethylenediamine were added into 64 mL of ultrapure water. Then, the mixture was transferred to a 100 mL stainless steel autoclave lined with polytetrafluorethylene and hydrothermally treated in an oven at 220 °C for 3 h. After cooling to room temperature, the solution was filtrated by a 0.22 μm filter membrane and then concentrated in a vacuum rotary evaporator, followed by freeze drying to form N-GQDs powder.

2.3. 6-MP Detection by the N-GQDs Fluorescent Probe

Into each of the eleven 5 mL colorimetric tubes, 500 μL of phosphate buffer solution (PBS) and 100 μL of 1 mg/mL N-GQDs solution were added. Then, the 13 tubes were added with 0 (blank control), 10, 20, 30, 40, 50, 75, 100, 125, 150, 200, 250, and 300 μL of the 1 mM 6-MP solution respectively. All tubes were diluted with ultrapure water to 5 mL and placed at room temperature for 5 min, followed by measurement of FL intensity F (excitation wavelength at 340 nm). The FL intensity of the blank control was set as F_0. Each experiment was repeated three times.

3. Results and Discussion

3.1. Characterization of GQDs

Transmission electron microscopy (TEM) shows that the N-GQDs have high dispersivity and narrow particle size distribution (Figure 1A), as the particle sizes are within 1.5–5.0 nm, with a mean of 3 nm (Figure 1B). High-resolution (HR)-TEM (inset in Figure 1A) shows a clear crystal structure, with the lattice distance of 0.213 nm, which is consistent with the (100) diffraction plane of graphene. The X-ray diffraction (XRD) patterns demonstrate a broad diffraction peak at $2\theta = 22.6°$ (Figure S1), which is attributed to the (002) diffraction of graphite, and the corresponding interlayer space is the same as that of massive graphite (0.34 nm). N-GQDs consist of a single layer or several layers of carbon atoms, which form hexagonal lattice structures through sp^2 hybridization of carbon atoms. At the scale of quantum dots, they primarily exhibit two-dimensional characteristics and may also form three-dimensional structures through stacking or encapsulation.

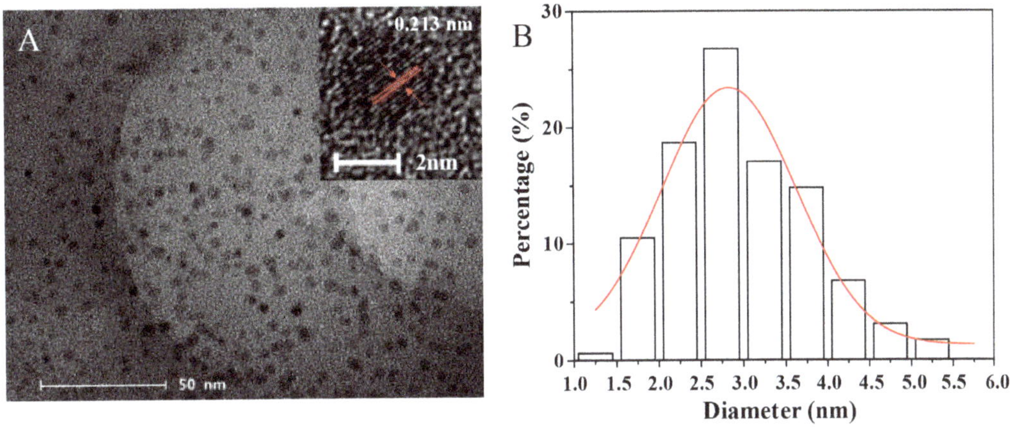

Figure 1. (**A**) TEM image of N-GQDs (inset: HRTEM image of one nanoparticle); (**B**) particle size distribution of N-GQDs.

Fourier transform infrared spectroscopy (FTIR) shows N-GQDs compared with cellulose demonstrate obviously stronger peaks at 2352 and 1400 cm^{-1}, which correspond to the stretching vibration of O-C-O and C-N [24] (Figure 2A). Moreover, the characteristic peaks at 3421, 2931, 1643, and 1047 cm^{-1} originate from the stretching vibration of -OH, C-H, C=O, and C-H, respectively. To further study the chemical environment around the carbon and nitrogen atoms in the GQDs [25], we conducted X-ray photoelectron spectroscopy (XPS) (Figure 2B). The peaks at 284.5, 399.7, and 531.6 eV are attributed to C1s, N1s, and O1s, respectively, and the C1s spectrum (Figure 2C) demonstrates five peaks at 284.5, 285.8, 285.2, 286.8, and 288.6 eV [26,27], which are ascribed to C-C/C=C, C-N and C=O, respectively [28]. High-resolution scanning at the N1s [28] region (Figure 2D) reveals two peaks at 398.5 and 399.7 eV, which indicate the presence of pyridine N and pyrrole N [29]. FTIR and XPS both confirm that hydrothermal reactions alter the structure of cellulose, and the nitrogen atom may replace the C atom in the fibrillar structure. This doping may occur at the center or edge positions. The surfaces of N-GQDs are functionalized with multiple carboxyl and amine groups, which increases their solubility in water.

3.2. Optical Properties of N-GQDs

The ultraviolet spectra of N-GQDs (Figure 3A) show a strong peak at 230 nm, ascribed to the π–π^* transition, and a weak shoulder peak at 300 nm, corresponding to the n–π^* transition [30]. The evident peak at 340 nm is the ultraviolet absorption of N-GQDs. Under excitation at 340 nm, there is an evident emission peak at 440 nm. Under natural irradiation, the QDs solution is colorless and transparent, but, in the dark, it emits obvious blue light under the fluorescence excitation at 365 nm (inset in Figure 3A). The photoluminescence (PL) spectrum of N-GQDs shows that when the excitation wavelength is prolonged from 310 to 430 nm, the fluorescence emission peak shifts from 400 to 480 nm (Figure 3B). The peak value first increases and then decreases and red shifts, which is consistent with the PL behaviors in other reports. As for biomedical applications, acidity/alkalinity is a key factor that must be considered. The FL intensity of N-GQDs changes significantly with pH [31] and is obviously quenched under acid conditions (pH > 7) and significantly enhanced under base conditions (pH > 7) (Figure 3C). The fluorescence intensity at pH = 11 is approximately 3.5 times that at pH = 2. Such an obvious response to pH is attributed to the protonation and deprotonation of the surface carboxyl group in N-GQDs. The fluorescence intensity of

N-GQDs does not change obviously within pH 4 and 9, which proves that N-GQDs can basically maintain a stable fluorescence in biological environments. Under excitation at 375 nm, the emission intensity of N-GQDs decays with time (Figure 3D) and the average fluorescent lifetime of N-GQDs is 4.19 ± 0.01 ns. With quinine sulphate as a reference, the fluorescent quantum yield of N-GQDs is 30.8% under the excitation at 340 nm [21].

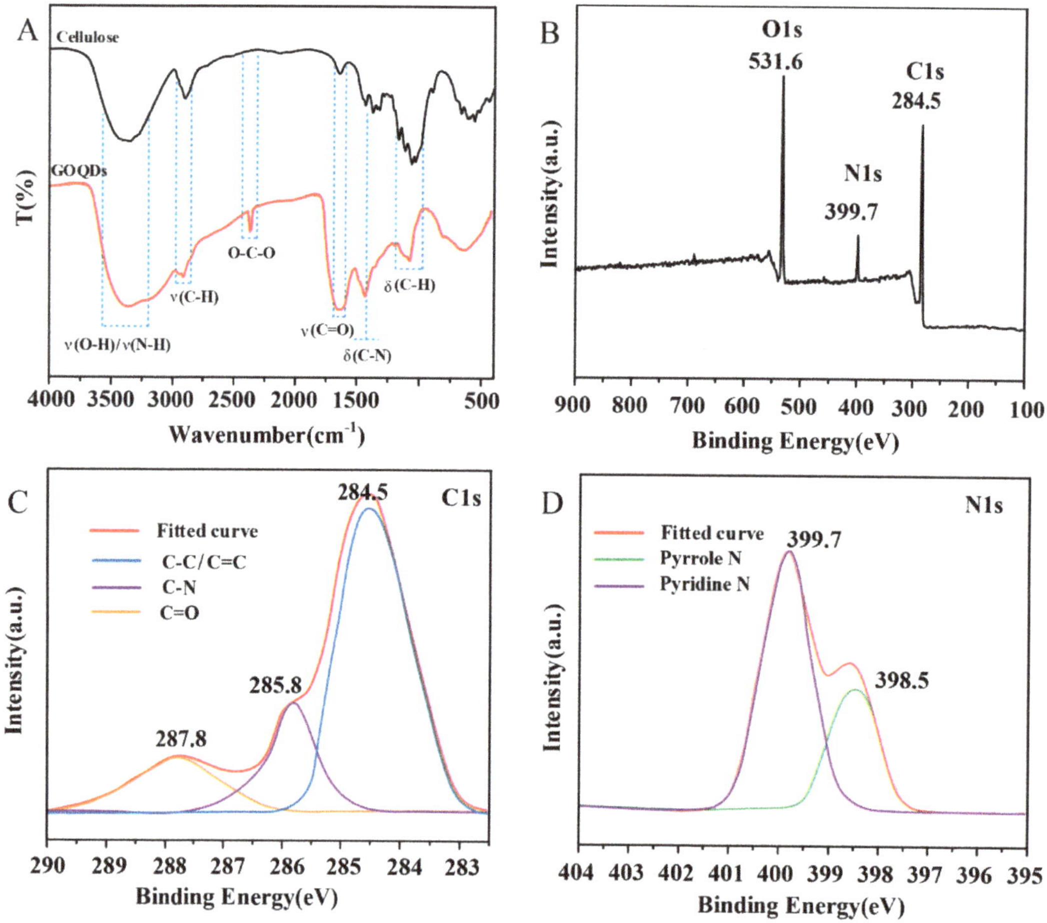

Figure 2. (**A**) FT-IR spectra of cellulose and N-GQDs; (**B**) XPS spectrum of N-GQDs; (**C**) C1s and (**D**) N1s high-resolution XPS spectra of N-GQDs.

3.3. Construction of a Fluorescent Probe for 6-MP Quantitative Detection

The feasibility of N-GQDs in 6-MP detection was evaluated first. Figure S2 shows evident quenching of fluorescence of N-GQDs when 30 μM 6-MP was dripped. Hence, such effective fluorescence quenching can be utilized into 6-MP quantitative detection. On this basis, the possible detection mechanism was studied. The 6-MP shows an evident absorption peak at 345 nm (Figure 4A), which is overlapped with the excitation and emission spectra of N-GQDs. Thus, the cause for the fluorescence quenching of N-GQDs may be attributed to the inner filter effect (IFE) or fluorescence resonance energy transfer (FRET). Since FRET belongs to the dynamic quenching mechanism, such energy transfer only occurs at the excited state and will significantly alter the lifespan of fluorophores [32]. To further uncover the mechanism of fluorescence quenching of N-GQDs [33], we measured the time-resolved fluorescence spectrum of N-GQDs with the presence or absence of 6-MP. Clearly, the presence of 6-MP on the fluorescent lifetime of N-GQDs can be ignored (Figure 4B). These results validate that the fluorescence quenching of N-GQDs is a static behavior and is mainly attributed to IFE.

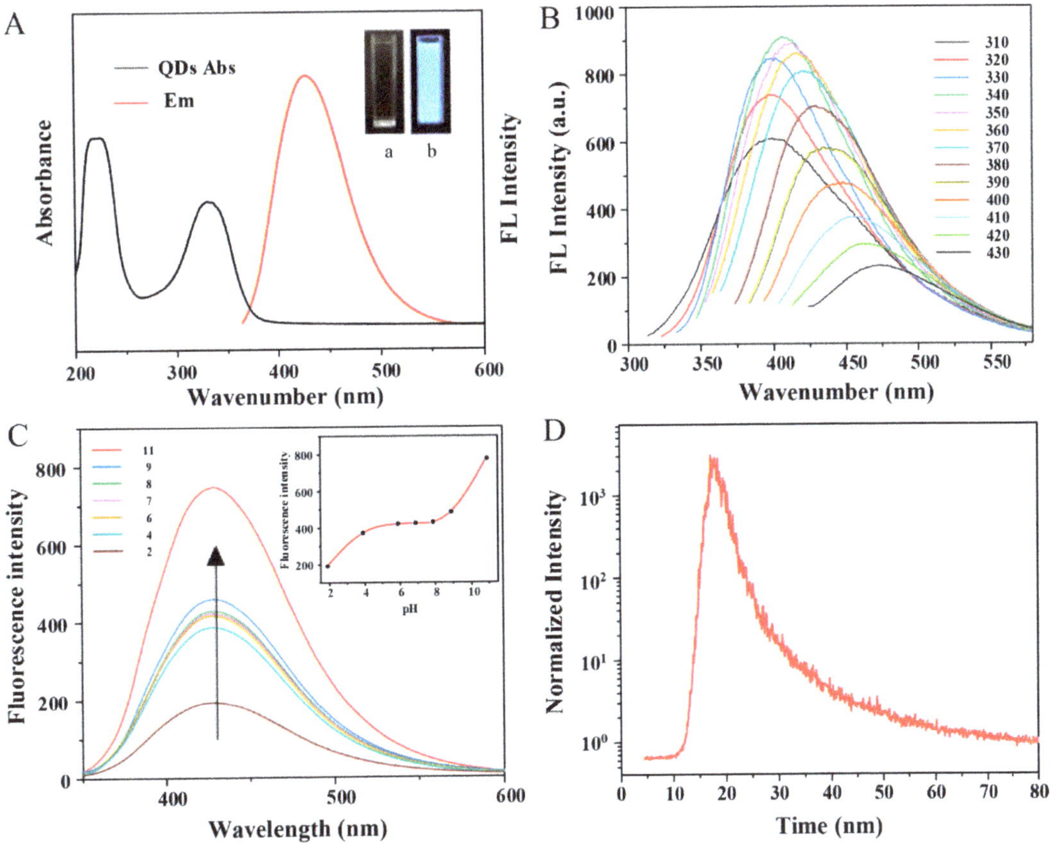

Figure 3. (**A**) UV-vis spectrum and fluorescence emission spectrum of N-GQDs (inset: samples under irradiation in natural light (a) and UV irradiation (365 nm, b)); (**B**) FL emission intensity of N-GQDs under excitation from 300 to 520 nm at an interval of 10 nm; (**C**) fluorescence spectra of N-GQDs dispersed in buffer solutions at different pH; (**D**) typical time-resolved fluorescence attenuation curves of N-GQDs at 440 nm (λ_{ex} = 375 nm).

For the most sensitive detection of 6-MP, we optimized the testing time and pH of the N-GQDs fluorescent probe. The optimal criterion for selecting fluorescence quenching efficiency is $(F_0 - F)/F_0$, where F_0 and F are the fluorescence intensity of N-GQDs without and with 6-MP, respectively. After the target substance is added, the testing time is a key factor of the detection system. When 6-MP (30 μM) was added into N-GQDs, the fluorescence was weakened and gradually stabilized after 5 min (Figure 4C). Hence, the reaction time between 6-MP and N-GQDs was selected to be 5 min. The effects of acidity/basicity on the fluorescent probe were studied by changing pH. The $(F_0 - F)/F_0$ was unchanged obviously within pH 5.0–9.0 (Figure 4D), indicating 6-MP can be detected within this wide range of acidity/basicity, which does not obviously interfere with the fluorescent probe.

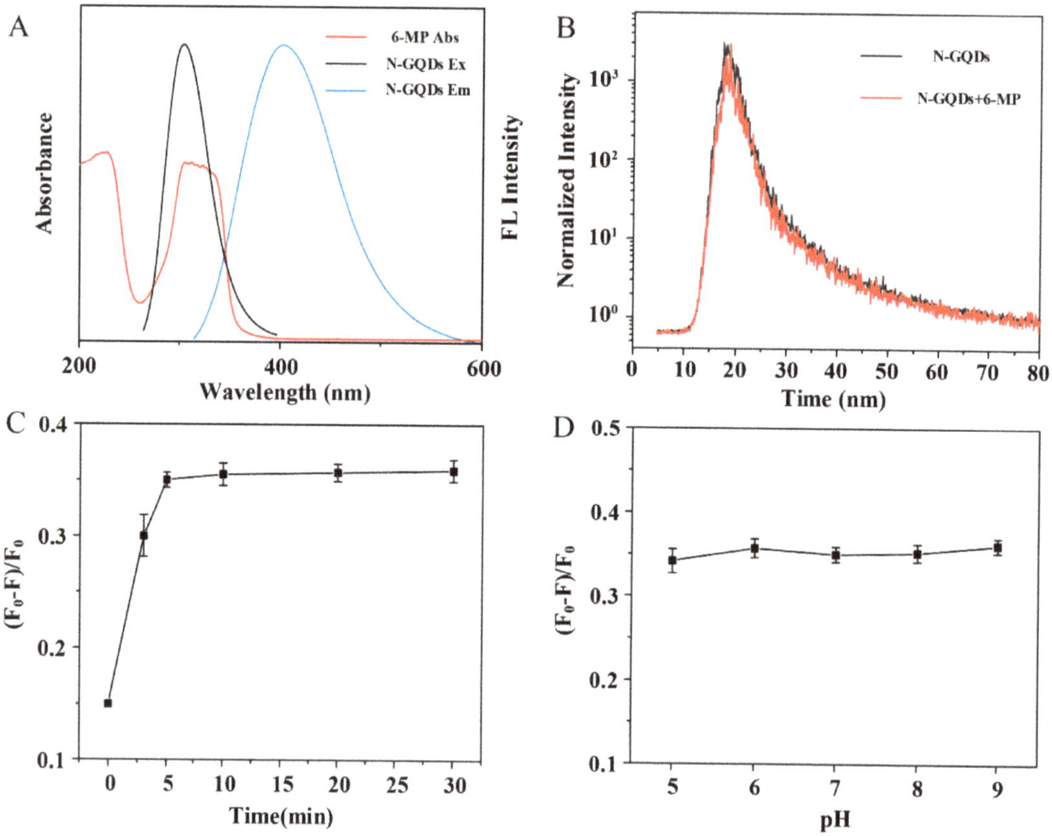

Figure 4. (**A**) UV-vis, excitation, and emission spectra of N-GQDs; (**B**) FL intensity decay curves of N-GQDs (λ_{ex} = 375 nm) with or without 6-MP; effects of (**C**) testing time and (**D**) acidity/basicity (pH 5–9) on the N-GQDs fluorescent probe, where F_0 and F are the FL intensity (λ_{ex} = 340 nm) of N-GQDs without and with 6-MP, respectively.

Then, the responses of the N-GQDs fluorescent probe to 6-MP were studied under the optimal conditions. As the 6-MP concentration rises, the FL intensity of N-GQDs is gradually weakened (Figure 5A). On this basis, the relation curve between fluorescence intensity and the target substance was established (Figure 5B). The linear range of N-GQDs in 6-MP detection is 0.2–60 μM, and the linear equations are $y = 0.0168x + 0.0048$ (R^2 = 0.9986) and $y = 0.0095x + 0.0925$ (R^2 = 0.996), and the detection limit ($3\sigma/n$) is 0.05 μM. The comparison of different 6-MP detection methods is illustrated in Table S1. Results indicate our new method is more sensitive in 6-MP detection than other methods. In comparison, our new N-GQDs fluorescent probe can be easily synthesized at low cost and without preprocessing. Moreover, the stable PL, low toxicity, and high biocompatibility of N-GQDs contribute to their huge potential in drug analysis.

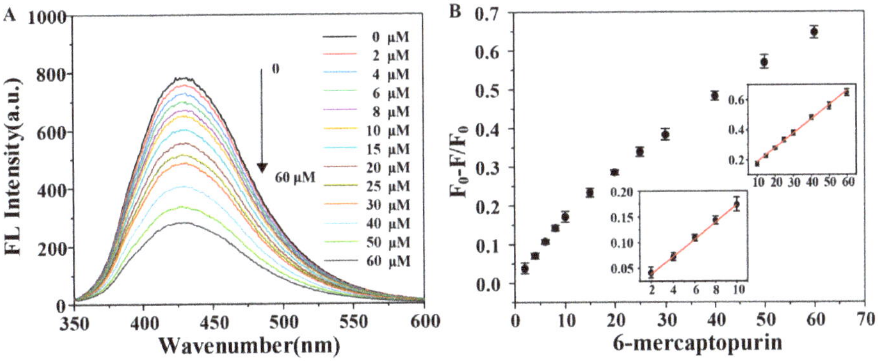

Figure 5. (**A**) Fluorescence emission spectra of N-GQDs (λ_{ex} = 340 nm) at different 6-MP concentrations (0, 2, 4, 6, 8, 10, 15, 20, 25, 30, 40, 50, 60 µM); (**B**) response relationship between 6-MP concentration and FL intensity of N-GQDs (inset: linear relationship between 6-MP concentration (2–10, 10–60 µM) and FL intensity of N-GQDs).

3.4. Interference from Foreign Substances

In addition to sensitivity, the selectivity and anti-interference ability of the new fluorescent probe over the target substance in a complex real environment are also very important. Hence, several common coexisting substances (Glu, Lac, Lys, Fru, Gala, Ala, Pro, Cys, Tyr, Arg, Glu, Meth, Gly) in human body fluids were tested to validate the selectivity and anti-interference ability of the new fluorescent probe. To highlight the selectivity of the new fluorescent probe, we set the competitive substances at a 10-fold concentration (300 µM). The changes in the FL intensity of the N-GQDs fluorescent probe with the presence of competitive substances can be ignored (Figure 6A), indicating the new probe is sensitive to 6-MP to some extent. Moreover, the anti-interfering ability of the probe in response to 6-MP with the presence of other competitive substances was also detected. Even when other interfering biomolecules existed in human body fluids, the luminescent response of the probe to 6-MP was basically unchanged (Figure 6B).

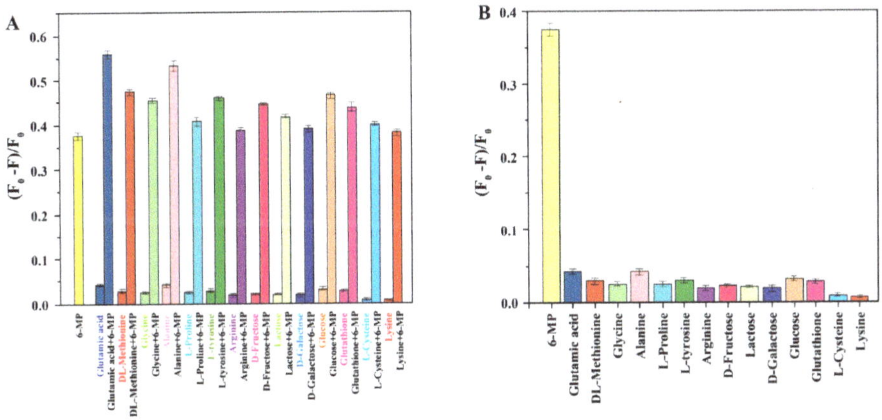

Figure 6. (**A**) Selective FL intensity response of N-GQDs to 6-MP (30 µM) and other potential interfering molecules (300 µM); (**B**) luminescent quenching efficiency of 6-MP (30 µM) with the presence of other potential interfering molecules (300 µM). All experiments were conducted at pH = 7.4 (λ_{ex} = 340 nm). The error bars show the standard deviations from three independent experiments.

3.5. Real Sample Detection

To validate the practicability of the new fluorescent probe, we applied it in 6-MP detection in human urine. First, urine was sampled from healthy volunteers without taking drugs. The urine samples were added with different concentrations of 6-MP and all diluted 100 times [34]. To validate the reliability and accuracy of the new method, we detected 6-MP concentrations in the real samples by using a standard addition method. The luminance intensity of the N-GQDs fluorescent probe in the spiked samples was recorded under excitation at 340 nm. The detected concentration was calculated through the established curve equation of the relationship between the sensor fluorescence intensity and the target substance concentration. The detected concentration was divided by the concentration of 6-MP added to the urine sample to obtain recovery (%). The recovery rate of urine samples was 98.05–102.68% (Table 1), indicating the new probe performs well in 6-MP detection from real samples. Compared with other potential 6-MP sensors, N-GQDs have advantages such as low production cost, nontoxicity, stable fluorescence performance, and the detection process does not require expensive detection instruments. Performance comparison of the analytical methods for the 6-MP assay shows this has a lower detection (Table S1). Hence, this new probe can be used into 6-MP detection in real human urine without any complex sample pretreatment.

Table 1. Recovery for the determination of 6-MP concentration detection in human urine.

Samples	Spiked (Mm)	Found (μM)	Recovery (%)	R.S.D (%, n = 3)
1	2	1.97	98.50	2.31
2	6	6.09	101.50	1.07
3	10	9.89	98.90	0.97
4	20	20.41	102.05	1.31
5	40	41.07	102.68	2.07
6	60	58.91	98.18	1.01

4. Conclusions

Graphene quantum dots (GQDs) were prepared from cellulose by a one-step hydrothermal method and were surface functionalized with amino groups. This method is outstanding, with a short synthesis time and the avoidance of a strong acid, strong base, or metal impurity, and is free from interference by inorganic salt residues or complex purification of products. This method presents a new clue for the recycled use of natural cellulose and offers a new green method for the synthesis of fluorescent quantum dots from biomass. The N-GQDs have high fluorescent stability and can be used to build a 6-MP detection fluorescent probe, which was used to sensitively detect 6-MP in human urine. Importantly, this method can be potentially expanded to the detection of other drugs with an ultraviolet absorption peak at 300–400 nm. This study can expand the application of N-GQDs into analysis of various drugs.

Supplementary Materials: The following supporting information can be downloaded at: https://www.mdpi.com/article/10.3390/ma17235852/S1, Figure S1: XRD patterns of N-GQDs; Figure S2: Fluorescence emission spectra of N-GQDs (λex= 340 nm) at 6-MP concentrations (0, 30, μM); Table S1: Performance comparison of the analytical methods for 6-MP assay; [5–7,11,13,18].

Author Contributions: Conceptualization, Q.X.; Methodology, Q.X. and J.D.; Software, Q.X.; Validation, G.Y.; Formal analysis, G.Y.; Data curation, Q.X. and J.D.; Writing—original draft, Q.X.; Writing—review & editing, X.Y. and R.Y.; Supervision, R.Y.; Project administration, X.Y.; Funding acquisition, R.Y. All authors have read and agreed to the published version of the manuscript.

Funding: This work was completed with the support of the Scientific and Technological Innovation Programs of Higher Education Institutions in Shanxi (No. 2022L623).

Institutional Review Board Statement: Research involving human participants or animals. The study did not involve any animals. Human urine was obtained from volunteers with full, informed consent and will not be used for other studies.

Informed Consent Statement: Not applicable.

Data Availability Statement: The original contributions presented in the study are included in the article and Supplementary Material, further inquiries can be directed to the corresponding authors.

Conflicts of Interest: The authors declare that they have no known competing financial interests or personal relationships that could have appeared to influence the work reported in this paper.

References

1. Qiu, J.; Cheng, R.; Zhang, J.; Sun, H.; Deng, C.; Meng, F.; Zhong, Z. Glutathione-Sensitive Hyaluronic Acid-Mercaptopurine Prodrug Linked via Carbonyl Vinyl Sulfide: A Robust and CD44-Targeted Nanomedicine for Leukemia. *Biomacromolecules* **2017**, *18*, 3207–3214. [CrossRef]
2. Dieck, C.L.; Tzoneva, G.; Forouhar, F.; Carpenter, Z.; Ambesi-Impiombato, A.; Sánchez-Martín, M.; Kirschner-Schwabe, R.; Lew, S.; Seetharaman, J.; Tong, L.; et al. Structure and Mechanisms of NT5C2 Mutations Driving Thiopurine Resistance in Relapsed Lymphoblastic Leukemia. *Cancer Cell* **2018**, *34*, 136–147.e6. [CrossRef] [PubMed]
3. Bouhnik, Y.; Scemama, G.; Taï, R.; Matuchansky, C.; Rambaud, J.C.; Lémann, M.; Modigliani, R.; Mary, J.-Y. Long-term follow-up of patients with Crohn's disease treated with azathioprine or 6-mercaptopurine. *Lancet* **1996**, *347*, 215–219. [CrossRef] [PubMed]
4. Yang, P.; Chen, Y.; Zhu, Q.; Wang, F.; Wang, L.; Li, Y. Sensitive chemiluminescence method for the determination of glutathione, l-cysteine and 6-mercaptopurine. *Microchim. Acta* **2008**, *163*, 263–269. [CrossRef]
5. Biparva, P.; Abedirad, S.M.; Kazemi, S.Y. Silver nanoparticles enhanced a novel TCPO–H_2O_2–safranin O chemiluminescence system for determination of 6-mercaptopurine. *Spectrochim. Acta Part A Mol. Biomol. Spectrosc.* **2015**, *145*, 454–460. [CrossRef]
6. Zhou, P.; He, L.; Gan, G.; Ni, S.; Li, H.; Li, W. Fabrication and evaluation of $[Co(phen)_2L]^{3+}$-modified DNA-MWCNT and SDS-MWCNT electrodes for electrochemical detection of 6-mercaptopurine. *J. Electroanal. Chem.* **2012**, *665*, 63–69. [CrossRef]
7. Shahrokhian, S.; Ghorbani-Bidkorbeh, F.; Mohammadi, A.; Dinarvand, R. Electrochemical determinations of 6-mercaptopurine on the surface of a carbon nanotube-paste electrode modified with a cobalt salophen complex. *J. Solid State Electrochem.* **2012**, *16*, 1643–1650. [CrossRef]
8. Cao, X.-N.; Lin, L.; Zhou, Y.-Y.; Shi, G.-Y.; Zhang, W.; Yamamoto, K.; Jin, L.-T. Amperometric determination of 6-mercaptopurine on functionalized multi-wall carbon nanotubes modified electrode by liquid chromatography coupled with microdialysis and its application to pharmacokinetics in rabbit. *Talanta* **2003**, *60*, 1063–1070. [CrossRef]
9. Hawwa, A.F.; Millership, J.S.; Collier, P.S.; McElnay, J.C. Development and validation of an HPLC method for the rapid and simultaneous determination of 6-mercaptopurine and four of its metabolites in plasma and red blood cells. *J. Pharm. Biomed. Anal.* **2009**, *49*, 401–409. [CrossRef]
10. Zakrzewski, R.; Borowczyk, K.; Łuczak, A.; Młynarski, W.; Trelińska, J. Determination of urinary 6-mercaptopurine and three of its metabolites by HPLC–UV coupled with the iodine–azide reaction. *Bioanalysis* **2013**, *5*, 869–877. [CrossRef]
11. Li, A.P.; Peng, J.D.; Zhou, M.; Zhang, J. Resonance light scattering determination of 6-mercaptopurine coupled with HPLC technique. *Spectrochim. Acta Part A Mol. Biomol. Spectrosc.* **2016**, *154*, 1–7. [CrossRef]
12. Rosenfeld, J.M.; Taguchi, V.Y.; Hillcoat, B.L.; Kawai, M. Determination of 6-mercaptopurine in plasma by mass spectrometry. *Anal. Chem.* **1977**, *49*, 725–727. [CrossRef]
13. Wang, L.; Zhang, Z. The study of oxidization fluorescence sensor with molecular imprinting polymer and its application for 6-mercaptopurine (6-MP) determination. *Talanta* **2008**, *76*, 768–771. [CrossRef]
14. Rabel, S.R.; Stobaugh, J.F.; Trueworthy, R. Determination of Intracellular Levels of 6-Mercaptopurine Metabolites in Erythrocytes Utilizing Capillary Electrophoresis with Laser-Induced Fluorescence Detection. *Anal. Biochem.* **1995**, *224*, 315–322. [CrossRef]
15. Han, G.; Liu, R.; Han, M.-Y.; Jiang, C.; Wang, J.; Du, S.; Liu, B.; Zhang, Z. Label-Free Surface-Enhanced Raman Scattering Imaging to Monitor the Metabolism of Antitumor Drug 6-Mercaptopurine in Living Cells. *Anal. Chem.* **2014**, *86*, 11503–11507. [CrossRef]
16. Chen, Z.; Zhang, G.; Chen, X.; Chen, J.; Liu, J.; Yuan, H. A fluorescence switch sensor for 6-mercaptopurine detection based on gold nanoparticles stabilized by biomacromolecule. *Biosens. Bioelectron.* **2013**, *41*, 844–847. [CrossRef]
17. Yuan, Y.; Wang, Y.; Liu, S.; Li, Y.; Duan, R.; Zhang, H.; Hu, X. Fluorescence quenching and spectrophotometric methods for the determination of 6-mercaptopurine based on carbon dots. *RSC Adv.* **2016**, *6*, 52255–52263. [CrossRef]
18. Gao, M.X.; Xu, J.L.; Li, Y.F.; Huang, C.Z. A rapid and sensitive spectrofluorometric method for 6-mercaptopurine using CdTe quantum dots. *Anal. Methods* **2013**, *5*, 673–677. [CrossRef]
19. Sun, L.; Li, S.; Ding, W.; Yao, Y.; Yang, X.; Yao, C. Fluorescence detection of cholesterol using a nitrogen-doped graphene quantum dot/chromium picolinate complex-based sensor. *J. Mater. Chem. B* **2017**, *5*, 9006–9014. [CrossRef]
20. Ganganboina, A.B.; Dutta Chowdhury, A.; Doong, R.A. N-Doped Graphene Quantum Dots-Decorated V2O5 Nanosheet for Fluorescence Turn Off–On Detection of Cysteine. *ACS Appl. Mater. Interfaces* **2018**, *10*, 614–624. [CrossRef] [PubMed]

21. Zhou, C.; He, X.; Ya, D.; Zhong, J.; Deng, B. One step hydrothermal synthesis of nitrogen-doped graphitic quantum dots as a fluorescent sensing strategy for highly sensitive detection of metacycline in mice plasma. *Sens. Actuators B Chem.* **2017**, *249*, 256–264. [CrossRef]
22. Xu, Q.; Gong, Y.; Zhang, Z.; Miao, Y.; Li, D.; Yan, G. Preparation of graphene oxide quantum dots from waste toner, and their application to a fluorometric DNA hybridization assay. *Microchim. Acta* **2019**, *186*, 483. [CrossRef] [PubMed]
23. Zhang, L.; Li, L.; Ma, C.; Ge, S.; Yan, M.; Bian, C. Detection of α-fetoprotein with an ultrasensitive electrochemiluminescence paper device based on green-luminescent nitrogen-doped graphene quantum dots. *Sens. Actuators B Chem.* **2015**, *221*, 799–806. [CrossRef]
24. Jia, X.; Li, J.; Wang, E. One-pot green synthesis of optically pH-sensitive carbon dots with upconversion luminescence. *Nanoscale* **2012**, *4*, 5572–5575. [CrossRef]
25. Kwon, W.; Rhee, S.-W. Facile synthesis of graphitic carbon quantum dots with size tunability and uniformity using reverse micelles. *Chem. Commun.* **2012**, *48*, 5256–5258. [CrossRef]
26. Edison, T.N.J.I.; Atchudan, R.; Shim, J.-J.; Kalimuthu, S.; Ahn, B.-C.; Lee, Y.R. Turn-off fluorescence sensor for the detection of ferric ion in water using green synthesized N-doped carbon dots and its bio-imaging. *J. Photochem. Photobiol. B Biol.* **2016**, *158*, 235–242. [CrossRef] [PubMed]
27. Yang, Z.; Xu, M.; Liu, Y.; He, F.; Gao, F.; Su, Y.; Wei, H.; Zhang, Y. Nitrogen-doped, carbon-rich, highly photoluminescent carbon dots from ammonium citrate. *Nanoscale* **2014**, *6*, 1890–1895. [CrossRef] [PubMed]
28. Chu, X.; Wang, S.; Cao, Y. A new fluorescence probe comprising nitrogen-doped graphene quantum dots for the selective and quantitative determination of cerium(iv). *N. J. Chem.* **2020**, *44*, 797–806. [CrossRef]
29. Deng, X.; Feng, Y.; Li, H.; Du, Z.; Teng, Q.; Wang, H. N-doped carbon quantum dots as fluorescent probes for highly selective and sensitive detection of Fe^{3+} ions. *Particuology* **2018**, *41*, 94–100. [CrossRef]
30. Zhang, Y.; Hu, W.; Li, B.; Peng, C.; Fan, C.; Huang, Q. Synthesis of polymer-protected graphene by solvent-assisted thermal reduction process. *Nanotechnology* **2011**, *22*, 345601. [CrossRef]
31. Lin, L.; Rong, M.; Lu, S.; Song, X.; Zhong, Y.; Yan, J.; Wang, Y.; Chen, X. A facile synthesis of highly luminescent nitrogen-doped graphene quantum dots for the detection of 2,4,6-trinitrophenol in aqueous solution. *Nanoscale* **2015**, *7*, 1872–1878. [CrossRef] [PubMed]
32. Zhu, S.; Yan, X.; Sun, J.; Zhao, X.-E.; Wang, X. A novel and sensitive fluorescent assay for artemisinin with graphene quantum dots based on inner filter effect. *Talanta* **2019**, *200*, 163–168. [CrossRef] [PubMed]
33. Su, D.; Wang, M.; Liu, Q.; Qu, Z.; Su, X. A novel fluorescence strategy for mercury ion and trypsin activity assay based on nitrogen-doped graphene quantum dots. *N. J. Chem.* **2018**, *42*, 17083–17090. [CrossRef]
34. Li, Y.; Huang, H.; Ma, Y.; Tong, J. Highly sensitive fluorescent detection of dihydroxybenzene based on graphene quantum dots. *Sens. Actuators B Chem.* **2014**, *205*, 227–233. [CrossRef]

materials

MDPI

Review

Hydrogel–Flexible Electronics Integrated Platforms for Diabetic Wound Management

Zhenjun Liu [1], Huanping Zhang [1], Yuqing Li [2], Shengxi Xu [3], Ning Fu [1,*], Fang Wang [1,*] and Wansong Chen [2,*]

1 Henan Province Engineering Research Center of Key Materials and Technologies for Low-Altitude Aircraft Battery Systems, College of Chemistry and Environmental Engineering, Anyang Institute of Technology, Anyang 455000, China; liuzhenjun@ayit.edu.cn (Z.L.); 20160463@ayit.edu.cn (H.Z.)

2 Hunan Provincial Key Laboratory of Micro & Nano Materials Interface Science, College of Chemistry and Chemical Engineering, Central South University, Changsha 410083, China; osh1105913482@163.com

3 School of Biomedical Engineering, The University of New South Wales, Sydney, NSW 2000, Australia; z5656600@ad.unsw.edu.au

* Correspondence: funing@ayit.edu.cn (N.F.); 20160461@ayit.edu.cn (F.W.); chenws@csu.edu.cn (W.C.)

Abstract

Diabetic wounds are a major clinical challenge, driven by hyperglycemia, oxidative stress, persistent inflammation, and bacterial infection. Conventional dressings offer limited benefit, creating demand for advanced therapeutic strategies. This review analyzes hydrogel-based wound dressings and flexible electronic devices. Hydrogels are categorized by angiogenesis promotion, antioxidant activity, anti-inflammatory regulation, antibacterial action, and electrical conductivity. Flexible electronics are examined for adaptability, sensitivity, and real-time monitoring potential. Hydrogels maintain moist environments, support tissue regeneration, and deliver multifunctional bioactivity. Growth factor-loaded and electroactive hydrogels promote angiogenesis. Reactive oxygen species (ROS)-responsive systems restore redox balance. Anti-inflammatory and antibacterial hydrogels regulate macrophages and reduce infection risk. Conductive hydrogels accelerate healing through electrical stimulation. Flexible electronics provide continuous monitoring, intelligent feedback, and remote management, enhancing treatment precision. Their integration with hydrogels represents a paradigm shift from passive dressings to active diagnostic and therapeutic systems. Challenges remain in material design, interfacial stability, and long-term biocompatibility. These issues guide future innovation and clinical translation, offering a foundation for smart diabetic wound management.

Keywords: diabetic wounds; hydrogels; flexible electronics

1. Introduction

Diabetes is defined as a chronic disease characterized by hyperglycemia. It represents an increasingly severe public health challenge with significant impacts on human health [1–5]. By the year 2045, it is estimated that the global population of individuals diagnosed with diabetes will reach 783 million. It is estimated that up to 25% of individuals will develop diabetic foot disease [6]. China currently has a substantial diabetic population, with tens of millions of patients diagnosed with diabetic foot disease (Figure 1) [7].

The term 'diabetic foot' encompasses pathologies including foot deformities, infections, and ulcerations, resulting from chronic diabetic neuropathy and vascular abnormalities in the lower limbs. Due to peripheral neuropathy with reduced pain sensation, minor injuries such as calluses and blisters often progress to ulcers without timely detection [8–10]. Persistent hyperglycemia severely disrupts wound healing. Compared to normal healing, diabetic wounds exhibit reduced endothelial cell proliferation and migration, impaired angiogenesis, and heightened inflammatory responses [11–13]. This healing impairment increases susceptibility to chronic ulcers, which may ultimately necessitate amputation ranging from toes to the lower leg. Traditional wound management relies on passive materials such as gauze and bandages, which primarily provide physical coverage and exudate absorption. These dressings lack dynamic responsiveness to the wound microenvironment and cannot intervene in the pathological processes that impede healing [14–16]. Consequently, there exists a demand for wound dressings that have the capability to absorb exudate while maintaining moisture levels within the wound, in addition to exhibiting multifunctional characteristics. These include antibacterial properties, which can contribute to the healing process for wounds afflicted with diabetes.

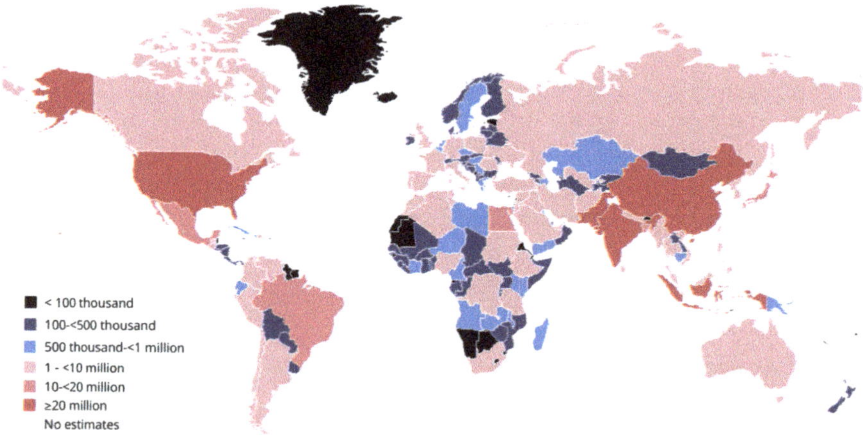

Figure 1. Global distribution of adults (20–79 years) with diabetes, 2024. Adapted with permission [7]. Copyright 2025, International Diabetes Federation.

Recent years have seen a paradigm shift in chronic wound management, driven by integrating functionalized hydrogels with flexible electronics. Hydrogels, defined by high water content, three-dimensional networks, biocompatibility, and tunable mechanical properties, have emerged as optimal wound dressing materials [17,18]. Through molecular modifications and responsive design, they can deliver multiple therapeutic functions including antibacterial [19], antioxidant [20,21], and pro-angiogenic effects [22,23]. Concurrently, flexible electronic devices, distinguished by mechanical flexibility and high sensitivity, enable real-time wound monitoring, intelligent feedback, and remote interaction. The synergistic integration of these technologies enhances treatment precision and timeliness, driving a transition from passive dressings to active diagnostic and therapeutic systems.

Nevertheless, the utilization of integrated hydrogel–flexible electronics systems continues to encounter substantial challenges with respect to their material composition, interfacial stability, and systemic integration. The following key research topics have been identified: the synergistic design of multifunctional materials, stable interfacial signal transmission, and long-term biocompatibility. This work provides a systematic review of the integration of functionalized hydrogels and flexible electronics from a materials science perspective, exploring their potential applications and the engineering constraints that exist in the management

https://doi.org/10.3390/ma19030509

of diabetic wounds. It also outlines prospective future directions to provide a theoretical foundation to support technological innovation and clinical translation in this field.

2. Challenges in Diabetic Wound Healing

2.1. Normal Wound Healing Process

Normal skin wound healing is a dynamic and highly coordinated biological process that primarily consists of four sequential yet overlapping stages: hemostasis, inflammation, proliferation, and tissue remodeling (as illustrated in Figure 2). These stages interact in a temporally regulated manner, making skin wound healing one of the most intricate regenerative processes in the human body [24,25].

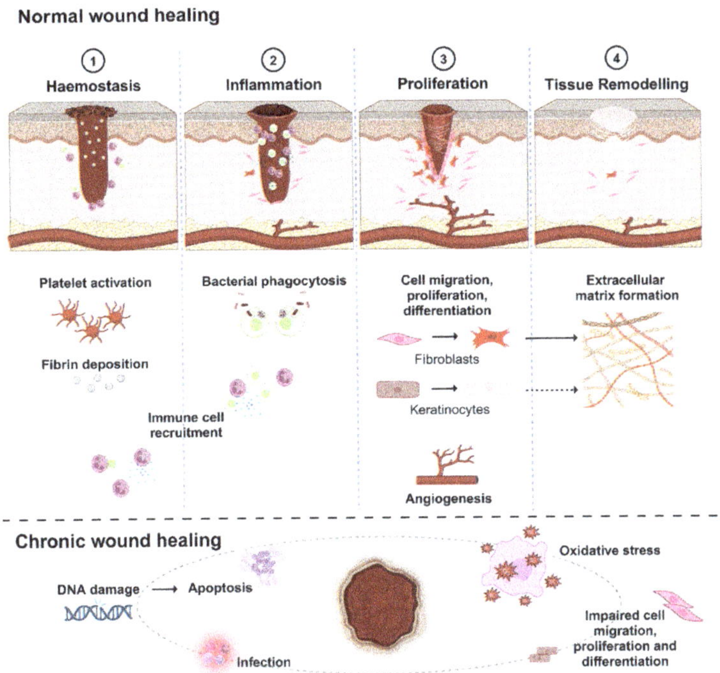

Figure 2. The normal wound healing process: ① bleeding and hemostasis, ② inflammation, ③ proliferation, and ④ remodeling [25].

Following integumentary injury, initial physiological responses involve vasoconstriction and platelet activation at the wound site to minimize hemorrhage and initiate fibrin clot formation. Inflammatory cells such as neutrophils and macrophages are subsequently recruited to the clot, serving functions including antimicrobial defense, debris clearance, and stimulation of cell proliferation and angiogenesis. Upon resolution of inflammation, endothelial cells near the wound proliferate and migrate, forming new blood vessels that deliver nutrients and oxygen essential for tissue regeneration. Fibroblasts then invade the clot to form granulation tissue while re-epithelialization proceeds; as epithelial cells proliferate and migrate across the wound bed, healing nears completion, typically restoring skin barrier function and tensile strength [26,27].

https://doi.org/10.3390/ma19030509

2.2. Wound Healing Process in Diabetes

Diabetic patients are particularly susceptible to chronic or non-healing wounds as a result of dysregulated blood glucose levels. Although the precise pathogenesis of diabetic wounds remains incompletely understood, current evidence suggests the involvement of multiple intrinsic and extrinsic factors. Existing research has identified several key contributors, including hyperglycemic toxicity [28], the accumulation of ROS induced by oxidative stress [29,30], sustained and excessive inflammatory cell infiltration [31], and recurrent bacterial infections [32]. It is evident that these elements contribute to an elongation of the four customary stages of normal wound healing, thereby impeding the healing process of diabetic wounds and leading to their progression into chronic wounds [12].

2.2.1. Hyperglycemia

Hyperglycemia, a common pathological state in diabetes, impairs wound healing through multiple mechanisms. Elevated glucose in the wound microenvironment reduces macrophage infiltration, compromising the inflammatory phase essential for healing initiation [33,34]. Under hyperglycemic conditions, glucose undergoes non-enzymatic glycation with proteins, forming advanced glycation end products (AGEs). AGEs induce collagen glycation, creating cross-linked structures that reduce extracellular matrix hydrophilicity, thereby impairing angiogenesis and granulation tissue formation and ultimately prolonging healing [33,34].

2.2.2. Accumulation of Reactive Oxygen Species

Normal wound healing involves redox signaling in which physiological ROS levels promote fibroblast activity, extracellular matrix synthesis, and angiogenesis via VEGF upregulation [35–37]. Maintaining optimal ROS concentrations is thus essential for healing and tissue homeostasis. However, diabetic wounds exhibit elevated ROS associated with hyperglycemia [38]. This accumulation stems from impaired glycolytic metabolism and AGE formation. AGEs induce glycation of mitochondrial respiratory chain proteins, resulting in excessive production of ROS (e.g., H_2O_2 and $\cdot O_2^-$). These elevated ROS species then activate NADPH oxidase, amplifying oxidative stress while simultaneously promoting glucose oxidation. This dual action accelerates further AGE formation, establishing a self-perpetuating deleterious cycle that impairs healing in diabetic wounds [39–41].

2.2.3. Inflammatory Infiltration

Persistent inflammatory infiltration characterizes diabetic wounds, where dysregulated cytokine balance impairs healing [42]. Chronic tissue damage and repair failure drive excessive accumulation of immune cells, particularly neutrophils and macrophages, perpetuating a pro-inflammatory cytokine cascade [34]. A central mechanism underlying this persistent inflammatory state is the disrupted phenotypic transition of macrophages from the pro-inflammatory M1 state to the anti-inflammatory M2 phenotype (as shown in Figure 3) [43,44]. During the normal process of wound healing, M1 macrophages predominate in the initial phase (approximately 3 days), subsequently transitioning to M2 macrophages. These M2 macrophages release angiogenesis-related factors and anti-inflammatory cytokines, which promote vascular regeneration. In the context of diabetic wounds, however, macrophages characteristically sustain an M1 phenotype, thereby perpetuating a state of inflammatory infiltration [45]. Furthermore, diabetes promotes excessive neutrophil extracellular trap production. Although these traps contribute to pathogen clearance and debridement, their overproduction damages epithelial and endothelial cells, delaying healing [46,47].

https://doi.org/10.3390/ma19030509

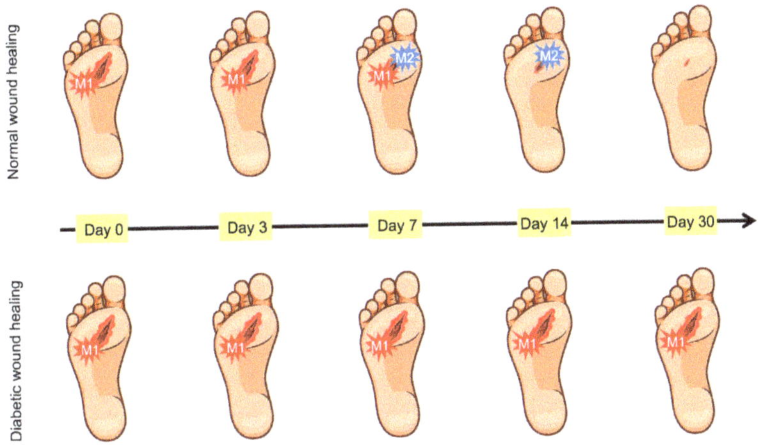

Figure 3. Macrophage phenotypes in normal and diabetic wounds.

2.2.4. Bacterial Infection

Bacterial infection significantly impairs chronic diabetic wound healing [48]. While intact epidermis efficiently resists contamination, disrupted healing in diabetic wounds creates portals for colonization [49]. Diabetic wounds, characterized by persistently elevated glucose levels in the wound microenvironment, provide a particularly favorable niche for bacterial proliferation. Adherent bacteria often form structured biofilms that facilitate quorum-sensing communication and enhance survival through physical protection and antibiotic resistance. Biofilms also stimulate localized inflammatory responses that further promote biofilm development, establishing a self-reinforcing cycle of infection and inflammation that contributes to wound chronicity [50,51].

3. Hydrogel-Based Diabetic Wound Therapy

Hydrogels are regarded as optimal materials for the treatment of diabetic wounds. Hydrogels are hydrophilic three-dimensional polymer networks. They exhibit excellent swelling properties, which provide a long-term stable moist environment during wound repair, thereby facilitating granulation tissue growth. The mechanical properties of these materials are comparable to those of soft tissue extracellular matrices, and their excellent biocompatibility supports wound tissue regeneration and re-epithelialization. As a consequence, hydrogels are regarded as the most competitive dressing material available at the present time. In recent years, there has been a notable increase in research activity focusing on the design and preparation of advanced hydrogel wound dressings [52–54].

3.1. Angiogenesis-Promoting Hydrogel Wound Dressings

In diabetic patients, hyperglycemia-induced endothelial dysfunction has been shown to impair angiogenesis and compromise vascular integrity, ultimately resulting in delayed or aberrant wound healing. Accordingly, the effective stimulation of angiogenesis holds substantial therapeutic value in the management of diabetic wounds [55,56]. Based on their angiogenic mechanisms and structural design strategies, current angiogenic hydrogels can be broadly categorized into two types: those loaded with growth factors and those incorporating pro-angiogenic materials.

Exogenous growth factor delivery has been extensively demonstrated to promote angiogenesis and accelerate wound repair. Among these, VEGF plays a pivotal role in revas-

https://doi.org/10.3390/ma19030509

cularization, a therapeutic approach aimed at enhancing oxygen supply within wound sites. This insight has spurred the investigation of VEGF-based therapies for diabetic wound treatment [57]. However, direct protein administration faces several limitations, including poor bioavailability of macromolecular proteins, rapid degradation, and high production costs, which collectively hinder clinical efficacy [58]. To overcome VEGF instability and deficiency under the oxidative and inflammatory conditions characteristic of diabetic wounds, hydrogels have been employed as sustained-release platforms. These hydrogel matrices act as protective reservoirs, enabling controlled and prolonged release of bioactive proteins, thereby maintaining therapeutic VEGF levels over time. For instance, Zhao et al. developed a supramolecular hydrogel patch with robust mechanical properties by combining a temperature-responsive poly(N-isopropylacrylamide) filler with a mixed supramolecular scaffold composed of N-acryloylglycine and 1-vinyl-1,2,4-triazole [59]. Under elevated temperatures associated with inflammatory responses, this hydrogel system exhibited temperature-triggered release of active agents. Such responsiveness was shown to facilitate re-epithelialization and enhance angiogenesis within the granulation tissue of diabetic wounds.

Emerging evidence suggests that, beyond their role in promoting cellular proliferation, certain biomaterials possess intrinsic capabilities to regulate the stability, bioactivity, and spatial distribution of cytokines and growth factors. The incorporation of such biomaterials into therapeutic platforms not only alleviates concerns regarding drug degradation and bioavailability but also enables direct modulation of local tissue repair and regeneration processes. For example, Li et al. developed a multifunctional hydrogel based on L-arginine-conjugated chitosan, harnessing the pro-angiogenic properties of L-arginine alongside the tissue-repairing functions of chitosan [34]. Their study demonstrated that this hydrogel effectively promoted re-epithelialization, collagen deposition, and angiogenesis in diabetic wound models. In addition to the intrinsic bioactivity of the constituent materials, specific physicochemical properties of hydrogels such as electrical conductivity and piezoelectricity can be strategically utilized to further promote angiogenesis [60,61]. Reports have shown that electric fields can stimulate endothelial cell proliferation and upregulate the expression of VEGF, thereby activating downstream signaling cascades essential for neovascularization during wound healing [62].

Building upon this principle, Wang et al. synthesized a self-powered electroactive composite hydrogel based on polyvinyl alcohol (PVA) [61]. Dipole interactions between PVA and polyvinylidene fluoride (PVDF) chains facilitated the formation of the electroactive β phase of PVDF. The piezoelectric properties of PVDF enabled the conversion of mechanical energy generated by rat movement into electrical energy, which in turn amplified the endogenous electric field within diabetic wound sites. This bioelectrical stimulation was shown to significantly enhance cellular proliferation, migration, and angiogenesis.

3.2. Antioxidant Hydrogel Wound Dressings

Diabetic wounds often show an imbalance between oxidative stress and antioxidant defense. This condition results in chronically elevated ROS, disrupted cellular metabolism, and accelerated senescence and apoptosis. Given the dynamic fluctuations in ROS concentrations at the wound site, modulating hydrogel degradation in response to ROS levels has emerged as a promising strategy for achieving intelligent and responsive therapeutic interventions in diabetic wound care [63]. The following section outlines current methodologies for the construction of antioxidant hydrogels. A critical step in this process involves grafting antioxidant moieties onto the hydrogel matrix. Additionally, antioxidant components may be incorporated into the hydrogel polymer chains through polymerization or grafting reactions [64,65]. For instance, Cao et al. introduced polydopamine (PDA) coated manganese dioxide (MnO_2), known for its antioxidant activity, into a hydrogel system to synthesize a bioactive glass (BG)-based nanocomposite

https://doi.org/10.3390/ma19030509

hydrogel (MnO_2@PDA-BGs), as illustrated in Figure 4A [66]. The superoxide dismutase-like and catalase-like enzymatic properties of MnO_2 facilitate the conversion of hydrogen peroxide (H_2O_2) and superoxide anions ($\cdot O_2^-$) into water and oxygen, thereby mitigating oxidative stress and alleviating hypoxic conditions at the wound site. In a complementary approach, Sun et al. developed an antioxidant hydrogel using a different design strategy, as shown in Figure 4B [67]. In this system, methacryloyl gelatin (GelMA) was graft-modified with 4-carboxyphenylboronic acid (CPBA) and subsequently photopolymerized with (-)-epigallocatechin gallate (EGCG) to form the GelMA-CPBA/EGCG hydrogel. The phenylboronic acid (PBA) groups of GelMA-CPBA interact with the ortho-dihydroxyl groups of EGCG to form glucose-responsive borate ester bonds. Under hyperglycemic conditions typical of diabetic wounds, these bonds undergo cleavage, triggering the release of EGCG. The released EGCG acts as a natural antioxidant, effectively scavenging excess ROS and contributing to the restoration of redox balance.

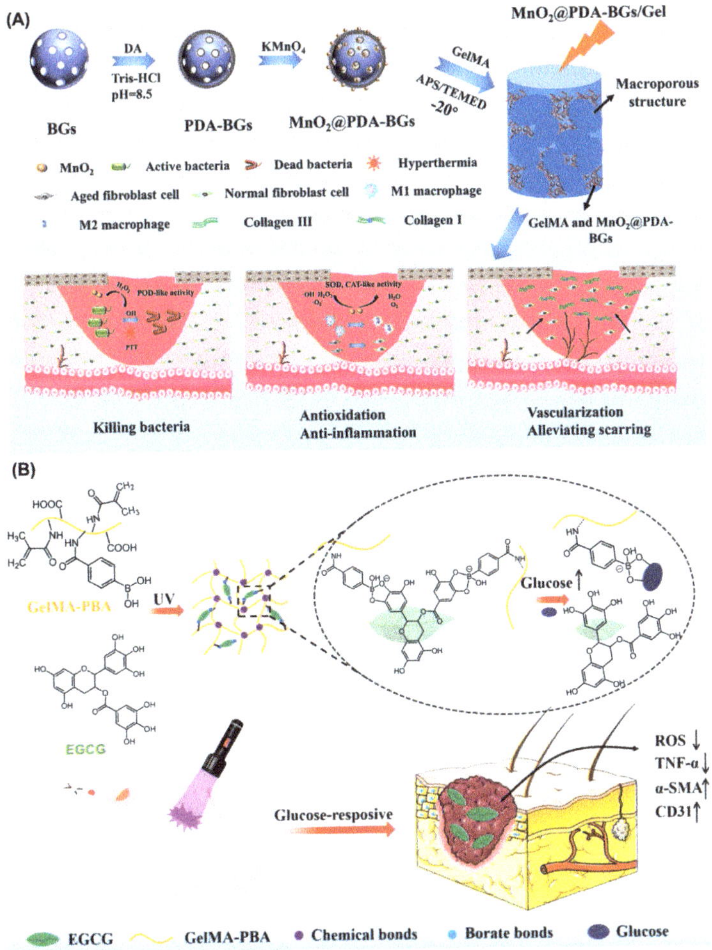

Figure 4. (**A**) Schematic illustration of MnO_2@PDA-BG hydrogel synthesis and diabetic wound treatment. Adapted with permission [66]. Copyright 2023, Wiley-VCH. (**B**) Schematic of the GelMA-CPBA/EGCG hydrogel synthesis and its application in diabetic wound treatment via ROS scavenging for anti-inflammation. Adapted with permission [67]. Copyright 2023, Wiley-VCH.

https://doi.org/10.3390/ma19030509

3.3. *Anti-Inflammatory Hydrogel Wound Dressings*

Macrophages play a pivotal bridging role in both the inflammatory and proliferative phases of wound healing. Accordingly, most anti-inflammatory hydrogel wound dressings are designed to regulate macrophage polarization, promoting the transition from the pro-inflammatory M1 phenotype to the anti-inflammatory M2 phenotype [68,69]. For instance, Jiang et al. developed a *B. striata* polysaccharide methacrylate (BSPMA)-GelMA hydrogel that effectively modulates macrophage phenotypes. This hydrogel was shown to suppress the expression of the pro-inflammatory cytokine tumor necrosis factor-alpha while upregulating the anti-inflammatory cytokine interleukin-10 in diabetic wound models, indicating its considerable therapeutic potential [70]. Fu et al. further advanced this concept by designing a three-dimensional porous polyurethane and hyaluronic acid hybrid hydrogel (PUHA-RGD), which was inspired by the structural features of the extracellular matrix. This hydrogel downregulates genes associated with cytokine and cytokine receptor interactions, thereby inducing reprogramming of the immune microenvironment. Both in vitro and in vivo experiments confirmed its ability to regulate macrophage polarization, suppress inflammation, and promote wound healing in diabetic models [71]. To overcome the limitations associated with conventional hydrogen therapy, Wu and colleagues engineered a hydrogel patch incorporating live algae and bacteria (Figure 5). Under hypoxic conditions, chlorella algae embedded within the patch utilize solar energy to synthesize biohydrogen. The generated hydrogen acts as an anti-inflammatory agent, reprogramming M1 macrophages at diabetic wound sites into M2 macrophages, thereby restoring a wound microenvironment conducive to healing [72].

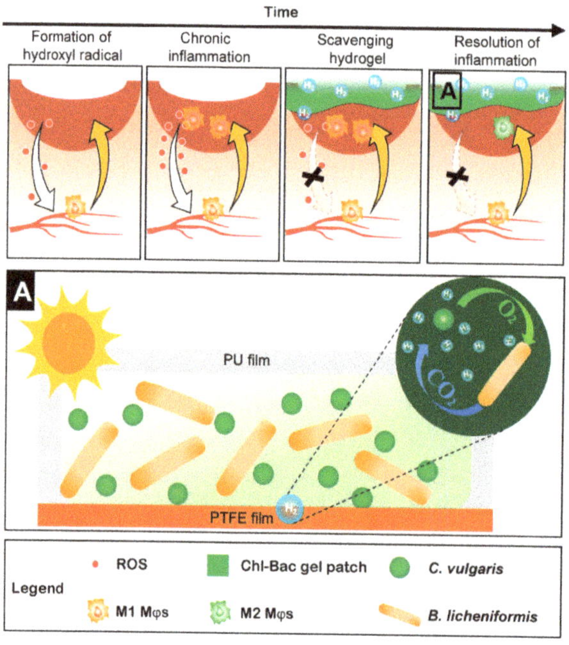

Figure 5. Schematic mechanism of hydrogen in diabetic wound therapy. Adapted with permission [72]. Copyright 2022, American Chemical Society.

https://doi.org/10.3390/ma19030509

3.4. Antibacterial Hydrogel Wound Dressings

Due to compromised immune function, diabetic patients exhibit a reduced capacity to resist bacterial infections and therefore require external antimicrobial interventions [73]. The classification of antimicrobial hydrogels is primarily based on their construction strategies and underlying antimicrobial mechanisms. These hydrogels can be broadly categorized into three types: (1) hydrogels loaded with antibiotics, (2) hydrogels incorporating inorganic nanoparticles, and (3) hydrogels possessing intrinsic antimicrobial properties [74–76].

Currently, antibiotics remain the most widely used antimicrobial agents in clinical practice, and antibiotic-loaded hydrogel wound dressings have attracted considerable research attention. For example, Malkoch et al. developed a dual-antibiotic delivery hydrogel dressing with tunable mechanical properties. This system incorporates two active pharmaceutical ingredients: hydrophilic neomycin sodium and hydrophobic ciprofloxacin. The hybrid design enables rapid release of neomycin sodium while sustaining the release of ciprofloxacin, thereby enhancing overall antimicrobial efficacy [77]. Despite their effectiveness, prolonged antibiotic use may lead to the development of bacterial resistance, which poses a significant challenge in the management of diabetic wounds [78,79]. To address this issue, increasing attention has been directed toward inorganic nanomaterial-based antimicrobial strategies. Incorporating inorganic nanoparticles into hydrogel matrices has been shown to improve antimicrobial performance and maintain long-term antibacterial activity, thereby mitigating the risk of resistance development. Among these, silver nanoparticles are the most extensively studied and applied [80]. For instance, Xiao et al. developed an injectable collagen–silver nanoparticle hydrogel, leveraging the broad-spectrum antibacterial activity of silver nanoparticles against multidrug-resistant bacteria. This hydrogel demonstrated potent antibacterial efficacy and exhibited promising potential for promoting wound regeneration [81]. In addition to externally loaded agents, the antimicrobial properties of hydrogels can also arise from the intrinsic characteristics of their precursor materials. Cationic polysaccharides such as chitosan, commonly used as hydrogel precursors, possess inherent antibacterial activity. This is primarily achieved through electrostatic interactions with negatively charged bacterial membranes or biomolecules, leading to membrane disruption and interference with bacterial metabolism [82,83].

3.5. Conductive Hydrogel Wound Dressings

As early as 1969, a series of reports indicated that electrical stimulation exerted a beneficial effect on wound healing processes [84]. Subsequent clinical studies have demonstrated that electrical stimulation can induce re-epithelialization in both skin and corneal wounds by promoting the migration and proliferation of fibroblasts, keratinocytes, and epithelial cells, while concurrently reducing edema formation. These findings underscore the therapeutic potential of electrical stimulation in the fields of skin tissue engineering and wound management [85,86]. Directed cell migration is a critical component of effective wound healing. Electrotaxis, defined as the directional movement of cells in response to electric fields, is a complex and cell type-specific phenomenon. At wound sites, both endogenous and externally applied electric fields can guide cellular migration and thereby facilitate tissue repair (Figure 6) [87,88].

https://doi.org/10.3390/ma19030509

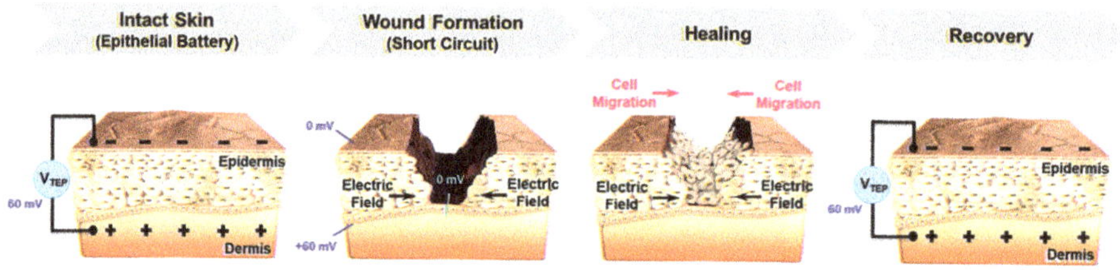

Figure 6. Trans-epithelial potential and electric field at the wound site before and after the healing process. Adapted with permission [88]. Copyright 2020, Wiley-VCH.

In recent years, conductive hydrogels with high electrical conductivity and excellent biocompatibility have been developed to address the limitations of conventional metallic electrodes, which often fail to achieve conformal contact with irregular wound surfaces due to mechanical mismatch. These hydrogels enable efficient delivery of electrical stimulation directly to the wound bed. For example, Li et al. designed an ion-conductive hydrogel that, under exogenous electric field stimulation, was shown to promote macrophage polarization, stimulate angiogenesis, and enhance collagen deposition in diabetic wound models [89]. Beyond serving as passive carriers for external stimulation, conductive hydrogels can also act as active conduits for endogenous electrical signals. Their intrinsic conductivity facilitates intercellular communication and supports the directional migration of epithelial cells, thereby promoting re-epithelialization and accelerating wound closure [90].

4. Flexible Electronic Devices

4.1. Overview of Flexible Electronic Devices

The integration of electronic devices into human life has been significantly accelerated by continuous advancements in high-tech innovation. The development of emerging electronic technologies, such as electronic skin, nanogenerators, brain–computer interfaces, and implantable medical devices, remains an active area of research and application. Compared with traditional rigid electronics based on silicon, which are increasingly constrained by the limitations of Moore's Law, flexible electronics offer a range of advantageous properties. These include reduced thickness, lightweight structure, low mechanical modulus, and the ability to stretch. Such features allow flexible electronics to conform seamlessly to various surfaces, rendering them effectively imperceptible in mechanical terms. These devices can be directly attached to the skin, integrated into garments, or implanted within the body, corresponding to epidermal electronics, wearable electronics, and implantable electronics, respectively. Owing to their mechanical adaptability, wearability, and compatibility with diverse application scenarios, flexible electronics have found widespread use in health monitoring systems, artificial electronic skin, and energy harvesting or storage platforms (Figure 7) [91–93].

https://doi.org/10.3390/ma19030509

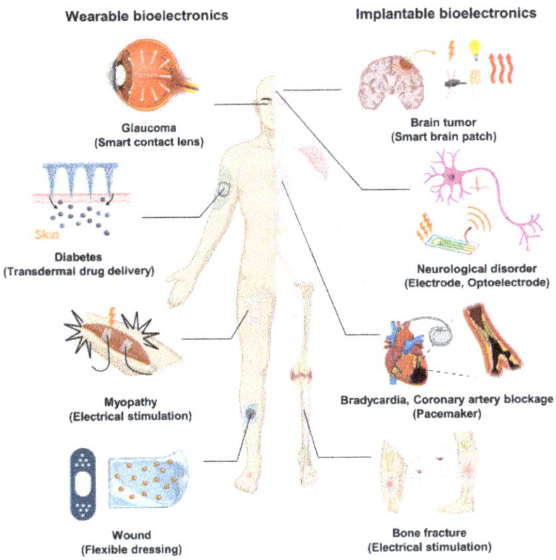

Figure 7. Applications of flexible electronics in medical fields. Adapted with permission [93]. Copyright 2023, American Chemical Society.

In the medical field, flexible electronics provide an optimal platform for patient self-management and personalized healthcare, as illustrated in Figure 7. This technology enables non-invasive, continuous monitoring of vital physiological signals in real time and in situ, thereby enhancing patient comfort and compliance. The data obtained through these systems are directly relevant to clinical decision-making, particularly in the context of preventive care and early disease diagnosis [94]. Moreover, flexible electronics have demonstrated significant advantages in the long-term tracking and management of chronic health conditions, including diabetes, cardiovascular diseases, and metabolic disorders. This capability is especially important for elderly individuals and patients suffering from persistent illnesses [95]. In the context of diabetic wound care, flexible electronics can be broadly categorized into two functional types based on their primary roles: devices designed for the detection of diabetic wounds and devices engineered for therapeutic intervention.

4.2. Flexible Electronic Devices for Diabetic Wound Monitoring

Real-time monitoring and assessment of wound status are essential for the effective diagnosis and management of diabetic wounds. However, current clinical practices for wound evaluation predominantly rely on the intermittent removal of dressings to perform visual inspections or wound culture tests [96]. This limitation underscores the urgent need for the development of advanced medical devices tailored for wound management. The adoption of intelligent monitoring systems by clinicians enables continuous evaluation of wound healing dynamics, thereby facilitating the timely adjustment and optimization of therapeutic strategies. In the domain of flexible electronics, researchers have dedicated considerable effort to the development of diagnostic flexible electronic devices capable of recording alterations in various wound biomarkers, including temperature [97], pH [98], glucose [99] and lactate, etc. [100]. For example, Amay J. Bandodkar and colleagues designed a miniature, wireless, battery-free wound monitoring device capable of measuring lactate concentrations at the wound site in real time [100]. Lactate plays a critical role in initiating the wound healing cascade, with physiological levels typically ranging from 5 to 15 millimoles per liter (mM). Deviations from this range have been

https://doi.org/10.3390/ma19030509

shown to impair healing: elevated lactate levels can induce severe hypoxia and delay tissue repair, whereas insufficient lactate levels may hinder the initiation of the healing process. By continuously monitoring lactate concentrations in diabetic mouse wound models and analyzing the resulting data, the researchers achieved an 83% accuracy rate in predicting wound closure outcomes.

As research has progressed, flexible electronic devices capable of simultaneously monitoring multiple wound parameters have also been developed [101]. For instance, Huang et al. introduced a diagnostic wound dressing (Figure 8) designed for long-term diabetic wound management [102]. The hydrogel substrates demonstrated strong adhesion, self-healing capacity, and antibacterial activity, arising from abundant hydrogen bonding, ionic interactions, and cationic chain segments (Figure 8). This system integrates multiplexed sensors for the concurrent detection of key wound biomarkers, such as glucose, pH, and temperature, across their clinically relevant ranges, thereby enabling comprehensive documentation of the healing progress. In addition to biochemical monitoring, the smart dressing also supports real-time acquisition of electrophysiological signals, such as electrocardiograms, electromyograms, and electroencephalograms. These data streams provide clinicians with valuable insights into the patient's overall health status and assist in clinical decision-making.

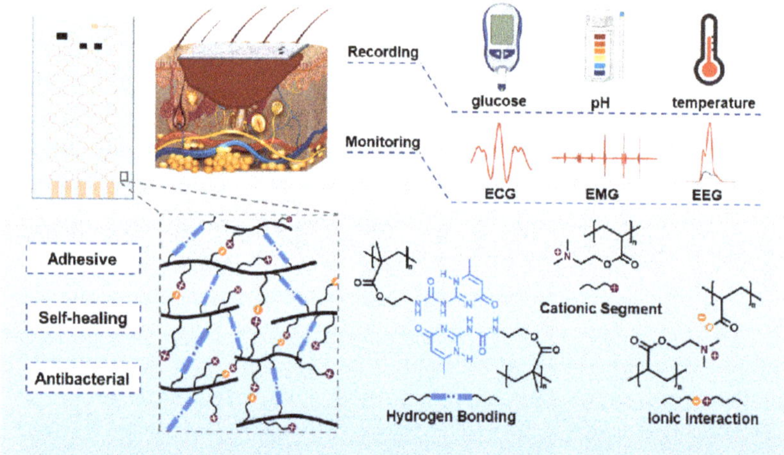

Figure 8. Schematic diagram of a monitored diabetic wound dressing. Adapted with permission [102]. Copyright 2024, Royal Society of Chemistry.

Recent advancements have pushed the boundaries of wound monitoring toward more integrated and intelligent systems. For instance, multimodal sensing platforms now combine various sensing mechanisms within a single device. A notable example is a battery-free, multifunctional microfluidic Janus wound dressing (MM-JWD) that integrates multiple colorimetric sensors for simultaneous monitoring of temperature, pH, and uric acid at the wound site [103]. This system employs smartphone-based image analysis for remote, real-time assessment of wound status, representing a significant step forward in point-of-care wound diagnostics. Another innovative approach involves microgel-based sensing dressings that utilize cholesterol liquid crystals, pH indicators, and glucose-sensitive enzymes encapsulated within hydrogel microspheres [104]. These "pixelated" sensors provide visual colorimetric feedback on wound temperature, pH, and glucose levels, which can be quantified through smartphone RGB analysis. This technology enables

https://doi.org/10.3390/ma19030509

continuous monitoring without frequent dressing changes, offering a practical solution for long-term wound management.

For specialized monitoring applications, photoacoustic sensing systems have emerged as a promising technology. Researchers have developed a wearable PVA/sucrose hydrogel patch incorporating nitrazine yellow, a pH-responsive dye with excellent photoacoustic properties [105]. When combined with a portable, miniaturized photoacoustic analysis device, this system can accurately detect wound pH across a wide and clinically relevant range, unaffected by blood color interference—a significant advantage over traditional colorimetric methods. Additionally, large-scale, high-resolution temperature mapping has been achieved through flexible amorphous silicon temperature sensor arrays with high spatial resolution and sensitivity [106]. These arrays enable dynamic localization of inflammation regions across large wound areas, providing detailed thermal maps that correlate with healing progression and infection status.

4.3. Flexible Electronic Devices for Diabetic Wound Treatment

Compared with wound monitoring alone, accelerating the healing of diabetic wounds through pharmaceutical intervention holds greater clinical significance. The incorporation of flexible electronic devices into wound monitoring systems offers a distinctive strategy for the active management of diabetic wounds. To achieve precise drug delivery and enable on-demand therapeutic responses, extensive research has focused on integrating flexible electronics into smart wound dressings [107]. Currently, thermoregulation represents the predominant approach in this domain, primarily utilizing thermoresponsive materials such as hydrogels as drug carriers. Temperature modulation via electronic components has been shown to enable controlled and timely release of therapeutic agents [108,109]. For example, Khademhosseini et al. developed a textile-based drug delivery system incorporating flexible thread microheaters (Figure 9). In this design, drug-loaded thermoresponsive hydrogels were coated onto specially engineered thermally conductive fibers, which were subsequently woven into fabric using textile manufacturing techniques. Upon electrical activation, the fabric facilitated targeted drug release tailored to the specific needs of diabetic wound environments [110]. Beyond temperature regulation, additional strategies such as electrical stimulation [111] and magnetoelectric stimulation [112] have also been explored.

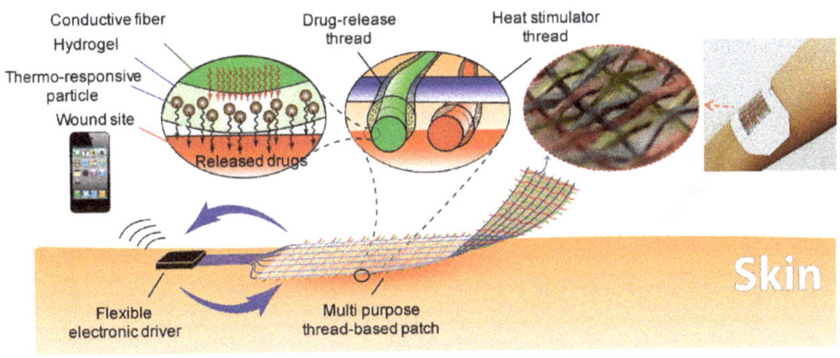

Figure 9. Schematic diagram of a drug delivery textile dressing based on a flexible threaded micro-heater. Adapted with permission [110]. Copyright 2017, Wiley-VCH.

In recent years, growing attention has been directed toward the development of fully integrated closed-loop systems that combine wound biomonitoring with therapeutic delivery. These systems aim to achieve active, minimally invasive treatment of diabetic wounds. Therapeutic flexible electronics have emerged as a promising alternative for managing

 https://doi.org/10.3390/ma19030509

chronic diseases, offering real-time, continuous, high-quality, and responsive treatment in a controllable and user-friendly format [93,113]. As a representative example, Jiang et al. designed a fully integrated wearable flexible electronic device capable of wireless monitoring and active therapy (Figure 10). This system comprises three core components: a customized mobile application, wearable electronic modules, and a therapeutic patch. The patch is fabricated from a multifunctional conductive polymer hydrogel, prepared by embedding polydopamine–polypyrrole (PDA-PPy) nanofibrils into a polyacrylamide (PAM) network through an in situ process. Benefiting from its unique architecture and the polycationic backbone of PDA-PPy nanofibrils, the hydrogel exhibits high drug-loading efficiency, optical transparency, strong skin adhesion, electrical conductivity, and broad-spectrum antimicrobial activity. In addition, the working electrode for glucose sensing was functionalized with dendrimers incorporating the electron mediator ferrocene-cored poly(ethylenimine) (Fc-PEI) and glucose oxidase (GOx), thereby enhancing the sensitivity and selectivity of biomarker detection. By integrating smart hydrogel materials with wearable bioelectronic platforms, the system enables real-time monitoring of wound biomarkers such as pH and glucose. Furthermore, the device supports electrical stimulation and iontophoresis-based insulin delivery, thereby achieving comprehensive and on-demand therapeutic intervention [114].

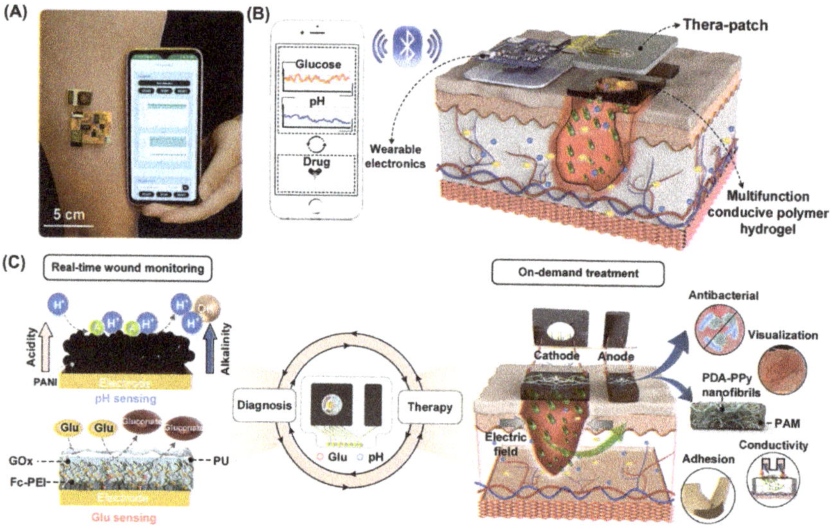

Figure 10. Schematic of a wearable device for wireless diagnostic and therapeutic integration for diabetic wound management. (**A**) Photograph of the integrated system. (**B**) Schematic showing the smartphone app, wearable device, and Thera-patch for monitoring and therapy. (**C**) Closed-loop strategy for real-time sensing and on-demand insulin delivery. Adapted with permission [114]. Copyright 2024, Wiley-VCH.

The field has witnessed remarkable innovations in self-powered therapeutic systems that eliminate the need for external batteries. A prominent example is the flexible self-powered bandage based on PVA/Ecoflex/MXene/borate/glycerol hydrogel, which demonstrates excellent power output and maintains performance under extreme conditions [87]. This bandage serves dual functions as a motion sensor for human activity monitoring and as an electrical stimulation device for accelerating diabetic wound repair through enhanced cell proliferation, collagen deposition, and angiogenesis. Similarly,

 https://doi.org/10.3390/ma19030509

triboelectric nanogenerators (TENGs) have been integrated into wound dressings to convert biomechanical energy into therapeutic electrical pulses. Research on bilateral flexible TENG-based hydrogel skin patches has shown that chitosan/PVA/ZnO nanoparticle hydrogels combined with TENG-generated electrical stimulation significantly promote wound healing, with slower electrical pulses proving more effective than rapid ones [115].

Advanced closed-loop therapeutic platforms represent the cutting edge of integrated wound management. One innovative system combines a wearable pulse piezoelectric nanogenerator (PENG) with a phosphatase and tensin homolog (PTEN) inhibitor-loaded electroresponsive hydrogel [116]. This setup converts animal movement into an electric field that simultaneously stimulates wound tissue and triggers on-demand drug release when the field intensity exceeds a defined threshold. This approach enhances cellular electrotaxis while avoiding the negative effects of sustained PTEN inhibition. Another breakthrough is a wireless, programmable, refillable hydrogel bioelectronic device that integrates sensing, electrical stimulation, and controlled drug release (e.g., metformin) within a single platform [117]. This system demonstrated accelerated wound healing in diabetic rat models over 21 days, showcasing the potential of programmable electroceutical interventions.

For more targeted interventions, microneedle-based bioelectronic systems have shown exceptional promise. Inspired by beetle antennae, researchers developed a wearable, battery-free, wireless intelligent dendritic rivet-like multilayer hydrogel electroactive microneedle bioelectronic device (GP-eMN) [118]. This system enables precise drug delivery, TENG-driven electrical stimulation, and visual assessment of wound conditions through simultaneous detection of uric acid, pH, and glucose. The bioinspired structure ensures mechanical interlocking with the dermis for effective, sustained drug delivery and rapid interstitial fluid uptake.

4.4. Integrated Hydrogel–Electronics Platforms for Theranostics

The most significant advancements in smart wound management emerge from deep material and functional integration, where hydrogels and flexible electronics are not merely combined but are engineered as synergistic components of unified therapeutic systems. This convergence creates platforms with capabilities exceeding the sum of their parts, enabling unprecedented levels of intelligent wound management.

4.4.1. Conductive and Responsive Hydrogel Architectures

At the foundation of integrated systems lies the development of hydrogels with intrinsic electronic functionality. Rather than attaching separate electrodes, researchers are engineering hydrogels that themselves serve as conductive elements. For instance, heparin-dopamine conjugated hydrogels incorporating reduced graphene oxide (Hep-PDA-rGO) achieve high conductivity while maintaining excellent biocompatibility [119]. These materials function simultaneously as epidermal strain sensors and active wound dressings, promoting angiogenesis through electrical conductivity while providing antibacterial and antioxidant effects. Similarly, Ag@polydopamine-functionalized boronate-linked chitosan hydrogels demonstrate multifunctional capabilities including strain sensing, rapid self-healing, tissue adhesion, and potent antibacterial activity against both *E. coli* and *S. aureus* [120]. These hydrogels accelerate wound healing while monitoring patient movements.

All-natural conductive supramolecular hydrogels represent another innovative direction. Systems like the GT5-DACD2-B hydrogel, formed through Schiff base reactions between gelatin and dialdehyde-β-cyclodextrin, incorporate fusidic acid (FA) via host–guest inclusion for inhibition of *S. aureus* proliferation [121]. Enhanced with tannic acid for

 https://doi.org/10.3390/ma19030509

antibacterial synergy and borax for conductivity, these fully natural materials offer excellent biocompatibility while functioning as flexible sensors for motion, facial expression, and speech recognition.

4.4.2. Bioresponsive Closed-Loop Systems

True innovation emerges when hydrogel–electronics integration enables autonomous, biomarker-responsive therapeutic actions. A groundbreaking example is the closed-loop patch (CLP) with biomimetic infection sensing, which detects bacterial hyaluronidase (HAase) secreted by pathogens like *S. aureus* [122]. The system uses an engineered hyaluronic acid hydrogel that degrades in response to HAase, changing the capacitance of interdigitated electrodes and triggering release of titanium hydride nanoparticles for sonodynamic antibacterial therapy. This "detect-and-treat" approach provides early infection warning before visible symptoms appear and significantly accelerates wound healing.

Multi-stimuli responsive hydrogels offer sophisticated control over therapeutic delivery. A four-in-one pH/glucose-responsive engineered hydrogel (PPy&L-arg@PVA-TSPBA) utilizes dynamic boronate ester bonds that cleave in response to the high glucose and low pH characteristic of infected diabetic wounds [123]. This triggers controlled release of antioxidant polypyrrole (PPy) and pro-angiogenic L-arginine, while the photothermal properties of PPy enable near-infrared activated bacterial elimination. The system synergistically addresses bacterial infection, oxidative stress, chronic inflammation, and impaired angiogenesis through a single integrated platform.

4.4.3. Integrated Intelligent Platforms

Advanced integration extends to 3D architectures that combine structural support with sensing and therapeutic functions. A notable example is the 3D hybrid scaffold with nanofiber yarns embedded in injectable hydrogel, created by weaving aligned polyacrylonitrile/reduced graphene oxide nanofibers and encapsulating them in a Schiff base-crosslinked hydrogel [124]. This construct provides directional guidance for cell alignment, antibacterial properties, and conductive pathways for strain sensing—enabling simultaneous wound monitoring and promotion of healing. In addition, machine learning-assisted visual monitoring hydrogels like the GelMA and chitosan methacrylate (CMCSMA) hydrogel incorporate phenol red as a pH indicator, enabling smartphone-based colorimetric analysis [125]. Machine learning algorithms process the spectral signals to reliably assess wound pH, optimizing treatment management based on wound status.

The convergence of hydrogel and flexible electronics technologies has led to diverse functional systems for diabetic wound management, each with distinct mechanisms and applications. To systematically compare these emerging technologies, Table 1 categorizes representative flexible electronic devices based on their primary functions, technical approaches, and operating mechanisms. This functional comparison provides a structured overview of current research trends and technological innovations in the field.

https://doi.org/10.3390/ma19030509

Table 1. Functional Comparison of Flexible Electronic Devices for Diabetic Wound Management.

Functions	Technical Subcategory	Mechanisms	References
Monitoring and Sensing	Single-parameter sensing	Temperature mapping	[106]
		pH monitoring	[105]
	Multi-parameter sensing	Microfluidic sensing of temperature, pH and uric acid	[103]
		Microgel sensing of temperature, pH and glucose	[104]
Active Therapy	Electrical stimulation	Self-powered electrical stimulation through biomechanical energy conversion	[87,126]
		Programmable electrical stimulation with parameter control	[117,127]
		TENG-based electrical stimulation from mechanical motion	[115]
	Controlled Drug delivery	Electrically triggered on-demand drug release using conductive hydrogels	[116]
		Minimally invasive transdermal delivery via bioinspired microneedles	[118]
Integrated Theranostics	Conductive/Responsive Hydrogel Architectures	High-conductivity hydrogel platform combining sensing and stimulation	[119–121,128]
	Bioresponsive Closed-Loop Systems	Closed-loop infection response with automated detection and treatment	[122]
		Glucose–pH dual-responsive release of therapeutic agents	[123]
	Integrated intelligent platforms	3D hybrid scaffold providing structural guidance and real-time monitoring	[124,129]
		Machine learning-assisted visual wound monitoring and assessment	[125]

5. Challenges and Future Prospects

The sophisticated functionalities envisioned for integrated hydrogel–flexible electronics systems present a stark contrast to the established standards of care in current clinical practice. The commercial market for diabetic wound management is dominated by passive, single-function hydrogel dressings. Products such as IntraSite™ Gel (Smith & Nephew, Watford, UK; modified carboxymethylcellulose for moisture retention and autolytic debridement), Purilon® Gel (Coloplast, Humlebaek, Denmark; calcium alginate and carboxymethylcellulose sodium for exudate absorption), and antimicrobial variants like Silvercel® (Cardinal Health, Dublin, OH, USA; alginate and carboxymethylcellulose hydrogel with ionic silver) are foundational in managing wound hydration and infection. However, these dressings operate on purely physicochemical principles without electronic components, lacking the capacity for real-time monitoring or feedback-controlled therapy. Few products that deeply and functionally integrates a therapeutic hydrogel matrix with flexible electronics for diagnostic–therapeutic synergy have yet reached the global medical market. Prototypes demonstrating such integration remain in preclinical or early clinical development. Bridging this gap and translating such integrated systems into clinical tools requires overcoming multifaceted challenges, which are analyzed in the following sections.

https://doi.org/10.3390/ma19030509

One of the foremost limitations lies in material stability and functional durability, which critically influence overall system performance. Many functional hydrogels are prone to degradation or loss of bioactivity under ambient conditions, with particular concern surrounding the stability of growth factors and natural antioxidants. These vulnerabilities require urgent resolution. Similarly, flexible electronic components are susceptible to signal drift, interface delamination, and mechanical fatigue during extended use, which can compromise continuous functionality and data fidelity. Future research should prioritize the development of advanced materials endowed with self-healing properties, dynamic responsiveness, and multifunctional capabilities. In parallel, optimization of interface architectures and micro-adhesion strategies will be essential to improve system robustness and operational longevity.

In addition to material-related constraints, the complexity associated with multifunctional integration imposes substantial demands on fabrication processes and system reliability. As the number of functional modules increases, system architectures become more intricate, leading to fabrication workflows that are both cumbersome and cost-intensive. These factors contribute to reduced stability and scalability. A key engineering challenge in this context is achieving a balance between multifunctionality and structural simplification, while promoting process standardization. Promising strategies to address this challenge include the adoption of modular design principles, the application of microfabrication technologies, and the utilization of programmable materials. These approaches may pave the way for future platforms characterized by high integration, enhanced reliability, and streamlined manufacturing.

Beyond material and engineering considerations, individual variability and the demand for treatment precision present substantial challenges to the integration of hydrogel–flexible electronics systems. Patients with diabetes exhibit considerable heterogeneity in wound types, healing dynamics, and physiological responses, making standardized treatment protocols insufficient to meet personalized clinical needs. To address this complexity, integrated systems must possess adaptive capabilities that allow dynamic adjustment of therapeutic strategies in response to real-time wound conditions, thereby enabling truly individualized care. Future directions may involve the incorporation of three-dimensional printing technologies in combination with biofeedback mechanisms to construct customizable treatment platforms, enhancing both clinical adaptability and therapeutic efficacy.

Another critical barrier to clinical translation lies in the absence of standardized validation protocols and regulatory frameworks. Most current research remains confined to in vitro models or animal studies, lacking systematic clinical validation and long-term follow-up data. Simultaneously, the regulatory classification of integrated systems remains ambiguous, as they do not conform to conventional definitions of pharmaceuticals or standalone medical devices. This ambiguity underscores the need for the development of novel evaluation criteria and approval pathways. Establishing robust regulatory frameworks and standardized clinical trial protocols will be essential for advancing industrial translation and ensuring patient safety.

Artificial intelligence-assisted diagnostic and therapeutic systems are anticipated to play a pivotal role in the evolution of next-generation convergent platforms. By integrating data acquired through flexible electronics with AI algorithms, it will be possible to achieve automated wound status recognition, personalized treatment planning, and predictive prognosis modeling. These capabilities have the potential to transform wound care into a domain of intelligent, data-driven healthcare. The development of personalized treatment platforms will become a central research focus, aiming to design customized systems

https://doi.org/10.3390/ma19030509

tailored to individual wound types, metabolic profiles, and immune characteristics, thereby achieving precision therapy and efficient tissue repair.

Continued progress in novel materials and interface engineering is anticipated to further enhance system performance. The development of hydrogel materials with multiresponsive behavior, self-healing capacity, and intrinsic biological activity, along with the optimization of interface adhesion and signal transmission between hydrogels and electronic components, will be critical for ensuring long-term operational stability. The advancement of this field will be driven by interdisciplinary collaboration across materials science, electronic engineering, clinical medicine, and data science. Such integration has the potential to establish comprehensive research platforms and clinical translation pathways, accelerating the transformation of experimental innovations into practical medical applications.

In conclusion, while substantial progress has been achieved in the integration of functional hydrogels with flexible electronics, widespread clinical adoption will require coordinated advancement across multiple dimensions. These include material innovation, system-level integration, intelligent algorithm development, and regulatory and clinical translation. Future research should prioritize the enhancement of system stability, the incorporation of intelligent decision-making frameworks, and the establishment of rigorous clinical validation protocols. The ultimate goal is to facilitate the seamless transition of these technologies from laboratory research to bedside application.

6. Conclusions

The integration of functionalized hydrogels with flexible electronics represents a convergent paradigm at the intersection of materials science and biomedical engineering, offering a novel technological pathway for the intelligent management of diabetic wounds. This paper presents a systematic review of the fundamental principles underlying hydrogel design, the material systems employed, and the strategies used to integrate flexible electronic devices. It further examines the engineering challenges associated with interfacial integration and material construction, and evaluates the practical performance and clinical potential of these hybrid systems through representative application examples.

Through the combination of functional molecular modification and the deliberate design of dynamic response mechanisms, hydrogel materials can be endowed with a wide range of therapeutic functions, including antibacterial, antioxidant, and pro-angiogenic properties. Flexible electronic devices, equipped with high-sensitivity sensing and remote interaction capabilities, enable real-time monitoring and intelligent feedback of wound microenvironments. The synergistic integration of these two components has been shown to improve treatment precision and responsiveness, thereby driving a shift in wound care from traditional passive dressings toward active diagnostic and therapeutic platforms.

Despite notable progress, several challenges remain, including issues related to material stability, the complexity of system integration, personalized adaptability, and the lack of established clinical translation pathways. Addressing these challenges will require sustained multidisciplinary collaboration to support the development of advanced materials, the incorporation of intelligent algorithms, and the creation of robust clinical validation frameworks. Such efforts are essential to facilitate the transition from laboratory prototypes to clinically deployable products.

The fusion of functional hydrogels with flexible electronics constitutes a rapidly emerging field, propelled by technological innovation and growing clinical demand. Its potential to serve as a foundational platform for diabetic wound management is considerable, with implications for advancing personalized, intelligent, and precision medicine. Continued research in this area is expected to yield dual benefits: improving patient quality of life and

https://doi.org/10.3390/ma19030509

establishing scalable technological paradigms for the treatment of chronic diseases. The emergence of these integrated systems signals the beginning of a new era defined by the deep convergence of smart biomaterials and flexible electronic technologies.

Author Contributions: Conceptualization, Z.L. and N.F.; resources, H.Z. and Y.L.; writing—original draft preparation, Z.L. and S.X.; writing—review and editing, Z.L., F.W. and W.C.; visualization, Y.L. and S.X.; supervision, N.F. and F.W.; project administration, Z.L. and W.C.; funding acquisition, H.Z. All authors have read and agreed to the published version of the manuscript.

Funding: This work was supported by the Postgraduate Education Reform and Quality Improvement Project of Henan Province (YJS2025GZZ54).

Institutional Review Board Statement: Not applicable.

Data Availability Statement: No new data were created or analyzed in this study. Data sharing is not applicable to this article.

Conflicts of Interest: The authors declare no conflicts of interest.

Abbreviations

The following abbreviations are used in this manuscript:

AGEs	Advanced glycation end products
CPBA	4-Carboxyphenylboronic acid
EGCG	(-)-Epigallocatechin gallate
GelMA	Methacryloyl gelatin
PBA	Phenylboronic acid
PVA	Polyvinyl alcohol
PVDF	Polyvinylidene fluoride
ROS	Reactive oxygen species
VEGF	Vascular endothelial growth factor

References

1. Joslin, E.P. The prevention of diabetes mellitus. *JAMA* **2021**, *325*, 190. [CrossRef]
2. Gromada, J.; Chabosseau, P.; Rutter, G.A. The α-cell in diabetes mellitus. *Nat. Rev. Endocrinol.* **2018**, *14*, 694–704. [CrossRef]
3. Tilg, H.; Moschen, A.R.; Roden, M. NAFLD and diabetes mellitus. *Nat. Rev. Gastroenterol. Hepatol.* **2016**, *14*, 32–42. [CrossRef] [PubMed]
4. Conte, C.; Cipponeri, E.; Roden, M. Diabetes mellitus, energy metabolism, and COVID-19. *Endocr. Rev.* **2024**, *45*, 281–308. [CrossRef]
5. Tomic, D.; Shaw, J.E.; Magliano, D.J. The burden and risks of emerging complications of diabetes mellitus. *Nat. Rev. Endocrinol.* **2022**, *18*, 525–539. [CrossRef] [PubMed]
6. Hart, T.; Milner, R.; Cifu, A. Management of a diabetic foot. *JAMA* **2017**, *318*, 1387–1388. [CrossRef] [PubMed]
7. IDF. *IDF Diabetes Atlas*, 11th ed.; IDF: Brussels, Belgium, 2025; Available online: https://diabetesatlas.org/resources/idf-diabetes-atlas-2025/ (accessed on 8 December 2025).
8. Grennan, D. Diabetic foot ulcers. *JAMA* **2019**, *321*, 114. [CrossRef]
9. Mishra, S.C.; Chhatbar, K.C.; Kashikar, A. Diabetic foot. *BMJ* **2017**, *359*, j5064. [CrossRef]
10. Tobalem, M.; Uçkay, I. Evolution of a diabetic foot infection. *N. Engl. J. Med.* **2013**, *369*, 2252. [CrossRef]
11. Xu, K.; Deng, S.; Zhu, Y.; Yang, W.; Chen, W.; Huang, L.; Zhang, C.; Li, M.; Ao, L.; Jiang, Y.; et al. Platelet rich plasma loaded multifunctional hydrogel accelerates diabetic wound healing via regulating the continuously abnormal microenvironments. *Adv. Healthc. Mater.* **2023**, *12*, 2301370. [CrossRef]
12. Xiong, Y.; Chu, X.; Yu, T.; Knoedler, S.; Schroeter, A.; Lu, L.; Zha, K.; Lin, Z.; Jiang, D.; Rinkevich, Y.; et al. Reactive oxygen species-scavenging nanosystems in the treatment of diabetic wounds. *Adv. Healthc. Mater.* **2023**, *12*, 2300779. [CrossRef]
13. Bokinni, Y. Barbados is in the grip of a diabetic foot amputation crisis. *BMJ* **2024**, *385*, q350. [CrossRef] [PubMed]
14. Dong, R.; Guo, B. Smart wound dressings for wound healing. *Nano Today* **2021**, *41*, 101290. [CrossRef]
15. Chen, Y.; Wang, X.; Tao, S.; Wang, Q.; Ma, P.-Q.; Li, Z.-B.; Wu, Y.-L.; Li, D.-W. Research advances in smart responsive-hydrogel dressings with potential clinical diabetic wound healing properties. *Mil. Med. Res.* **2023**, *10*, 37. [CrossRef] [PubMed]

 https://doi.org/10.3390/ma19030509

16. Shang, S.; Zhuang, K.; Chen, J.; Zhang, M.; Jiang, S.; Li, W. A bioactive composite hydrogel dressing that promotes healing of both acute and chronic diabetic skin wounds. *Bioact. Mater.* **2024**, *34*, 298–310. [CrossRef]
17. Yang, Y.; Zhong, S.; Meng, F.; Cui, X. Multi-Functional hydrogels to promote diabetic wound Healing: A review. *Chem. Eng. J.* **2024**, *497*, 154855. [CrossRef]
18. Duarte, J.; Mascarenhas-Melo, F.; Pires, P.C.; Veiga, F.; Paiva-Santos, A.C. Multifunctional hydrogels-based therapies for chronic diabetic wound healing. *Eur. Polym. J.* **2024**, *211*, 113026. [CrossRef]
19. Zhang, X.; Yan, Y.; Xu, H.; Liu, J.; Fan, J.; Gan, L.; Wan, H.; Zhou, G. Tannic acid-mediated hydrogel crosslinking boosts diabetic wound healing via antibacterial and pro-angiogenesis. *Eur. Polym. J.* **2025**, *236*, 114168. [CrossRef]
20. Zhao, H.; Liu, J.; Li, C.; Cheng, P.; Liu, P.; Liu, Y.; Guan, F.; Yao, M. Horseradish peroxidase/hydrogen peroxide-driven photothermally enhanced hydrogel loaded with black rice extract for treating infected skin wound. *Int. J. Biol. Macromol.* **2025**, *305*, 140827. [CrossRef]
21. Yang, Y.; Ma, Y.; Wang, H.; Li, C.; Li, C.; Zhang, R.; Zhong, S.; He, W.; Cui, X. Chitosan-based hydrogel dressings with antibacterial and antioxidant for wound healing. *Int. J. Biol. Macromol.* **2024**, *280*, 135939. [CrossRef]
22. Liu, L.; Zheng, J.; Li, S.; Deng, Y.; Zhao, S.; Tao, N.; Chen, W.; Li, J.; Liu, Y.-N. Nitric oxide-releasing multifunctional catechol-modified chitosan/oxidized dextran hydrogel with antibacterial, antioxidant, and pro-angiogenic properties for MRSA-infected diabetic wound healing. *Int. J. Biol. Macromol.* **2024**, *263*, 130225. [CrossRef]
23. Wang, Z.; Wang, L.; Wang, S.; Chen, H.; Wang, D.; Li, A.; Huang, Y.; Pu, Y.; Xiong, X.; Lui, X.; et al. The extracellular matrix promotes diabetic oral wound healing by modulating the microenvironment. *Biomater. Res.* **2025**, *29*, 0169. [CrossRef] [PubMed]
24. Ma, J.; Fang, Y.; Yu, H.; Yi, J.; Ma, Y.; Lei, P.; Yang, Q.; Jin, L.; Wu, W.; Li, H.; et al. Recent advances in living algae seeding wound dressing: Focusing on diabetic chronic wound healing. *Adv. Funct. Mater.* **2024**, *34*, 2308387. [CrossRef]
25. Hunt, M.; Torres, M.; Bachar-Wikstrom, E.; Wikstrom, J.D. Cellular and molecular roles of reactive oxygen species in wound healing. *Commun. Biol.* **2024**, *7*, 1534. [CrossRef]
26. Rodrigues, M.; Kosaric, N.; Bonham, C.A.; Gurtner, G.C. Wound healing: A cellular perspective. *Physiol. Rev.* **2019**, *99*, 665–706. [CrossRef]
27. Wang, G.; Lin, Z.; Li, Y.; Chen, L.; Reddy, S.K.; Hu, Z.; Garza, L.A. Colonizing microbiota is associated with clinical outcomes in diabetic wound healing. *Adv. Drug Deliv. Rev.* **2023**, *194*, 114727. [CrossRef]
28. Gan, Y.; Liang, B.; Gong, Y.; Wu, L.; Li, P.; Gong, C.; Wang, P.; Yu, Z.; Sheng, L.; Yang, D.-P.; et al. Mxene-based mild hyperthemia microneedle patch for diabetic wound healing. *Chem. Eng. J.* **2024**, *481*, 148592. [CrossRef]
29. Liu, Y.; Zhu, M.; Ou, J.; Li, K.; Ju, X.; Tian, Y.; Niu, Z. Multi-responsive sodium hyaluronate/tannic acid hydrogels with ROS scavenging ability promote the healing of diabetic wounds. *Int. J. Biol. Macromol.* **2024**, *278*, 134896. [CrossRef]
30. Black, H.S. A synopsis of the associations of oxidative stress, ROS, and antioxidants with diabetes mellitus. *Antioxidants* **2022**, *11*, 2003. [CrossRef]
31. Zhang, Y.; Wu, C.; Xu, Y.; Chen, Z.; Li, L.; Chen, J.; Ning, N.; Guo, Y.; Yang, Z.; Hu, X.; et al. Conductive hydrogels with hierarchical biofilm inhibition capability accelerate diabetic ulcer healing. *Chem. Eng. J.* **2023**, *463*, 142457. [CrossRef]
32. Buch Pranali, J.; Chai, Y.; Goluch Edgar, D. Treating polymicrobial infections in chronic diabetic wounds. *Clin. Microbiol. Rev.* **2019**, *32*, 10–128. [CrossRef]
33. Hu, Y.; Li, H.; Lv, X.; Xu, Y.; Xie, Y.; Yuwen, L.; Song, Y.; Li, S.; Shao, J.; Yang, D. Stimuli-responsive therapeutic systems for the treatment of diabetic infected wounds. *Nanoscale* **2022**, *14*, 12967–12983. [CrossRef]
34. Wang, H.; Xu, Z.; Zhao, M.; Liu, G.; Wu, J. Advances of hydrogel dressings in diabetic wounds. *Biomater. Sci.* **2021**, *9*, 1530–1546. [CrossRef] [PubMed]
35. Dunnill, C.; Patton, T.; Brennan, J.; Barrett, J.; Dryden, M.; Cooke, J.; Leaper, D.; Georgopoulos, N.T. Reactive oxygen species (ROS) and wound healing: The functional role of ROS and emerging ROS-modulating technologies for augmentation of the healing process. *Int. Wound J.* **2017**, *14*, 89–96. [CrossRef]
36. Wen, M.; Ouyang, J.; Wei, C.; Li, H.; Chen, W.; Liu, Y.-N. Artificial enzyme catalyzed cascade reactions: Antitumor immunotherapy reinforced by NIR-II light. *Angew. Chem. Int. Ed.* **2019**, *58*, 17425–17432. [CrossRef]
37. Dong, Y.; Wang, Z. ROS-scavenging materials for skin wound healing: Advancements and applications. *Front. Bioeng. Biotechnol.* **2023**, *11*, 1304835. [CrossRef] [PubMed]
38. Qu, M.; Xu, W.; Zhou, X.; Tang, F.; Chen, Q.; Zhang, X.; Bai, X.; Li, Z.; Jiang, X.; Chen, Q. An ROS-scavenging treg-recruiting hydrogel patch for diabetic wound healing. *Adv. Funct. Mater.* **2024**, *34*, 2314500. [CrossRef]
39. Jia, G.; Li, Z.; Le, H.; Jiang, Z.; Sun, Y.; Liu, H.; Chang, F. Green tea derivative-based hydrogel with ROS-scavenging property for accelerating diabetic wound healing. *Mater. Des.* **2023**, *225*, 111452. [CrossRef]
40. He, Y.; Liu, K.; Guo, S.; Chang, R.; Zhang, C.; Guan, F.; Yao, M. Multifunctional hydrogel with reactive oxygen species scavenging and photothermal antibacterial activity accelerates infected diabetic wound healing. *Acta Biomater.* **2023**, *155*, 199–217. [CrossRef]

 https://doi.org/10.3390/ma19030509

41. Shu, X.; Shu, J.; Wang, Y.; Deng, H.; Zhang, J.; He, J.; Wu, F. ROS-scavenging hydrogel to accelerate wound healing and reduce scar formation. *Chem. Eng. J.* **2023**, *474*, 145941. [CrossRef]
42. Zhang, Y.; Luo, J.; Zhang, Q.; Deng, T. Growth factors, as biological macromolecules in bioactivity enhancing of electrospun wound dressings for diabetic wound healing: A review. *Int. J. Biol. Macromol.* **2021**, *193*, 205–218. [CrossRef]
43. Yuan, Y.; Fan, D.; Shen, S.; Ma, X. An M2 macrophage-polarized anti-inflammatory hydrogel combined with mild heat stimulation for regulating chronic inflammation and impaired angiogenesis of diabetic wounds. *Chem. Eng. J.* **2022**, *433*, 133859. [CrossRef]
44. Cai, Y.; Chen, K.; Liu, C.; Qu, X. Harnessing strategies for enhancing diabetic wound healing from the perspective of spatial inflammation patterns. *Bioact. Mater.* **2023**, *28*, 243–254. [CrossRef]
45. Louiselle, A.E.; Niemiec, S.M.; Zgheib, C.; Liechty, K.W. Macrophage polarization and diabetic wound healing. *Transl. Res.* **2021**, *236*, 109–116. [CrossRef]
46. Yang, S.; Gu, Z.; Lu, C.; Zhang, T.; Guo, X.; Xue, G.; Zhang, L. Neutrophil extracellular traps are markers of wound healing impairment in patients with diabetic foot ulcers treated in a multidisciplinary setting. *Adv. Wound Care* **2020**, *9*, 16–27. [CrossRef] [PubMed]
47. Phillipson, M.; Kubes, P. The healing power of neutrophils. *Trends Immunol.* **2019**, *40*, 635–647. [CrossRef] [PubMed]
48. Chai, C.; Zhang, P.; Ma, L.; Fan, Q.; Liu, Z.; Cheng, X.; Zhao, Y.; Li, W.; Hao, J. Regenerative antibacterial hydrogels from medicinal molecule for diabetic wound repair. *Bioact. Mater.* **2023**, *25*, 541–554. [CrossRef]
49. Simões, D.; Miguel, S.P.; Ribeiro, M.P.; Coutinho, P.; Mendonça, A.G.; Correia, I.J. Recent advances on antimicrobial wound dressing: A review. *Eur. J. Pharm. Biopharm.* **2018**, *127*, 130–141. [CrossRef] [PubMed]
50. Siddiqui, A.R.; Bernstein, J.M. Chronic wound infection: Facts and controversies. *Clin. Dermatol.* **2010**, *28*, 519–526. [CrossRef]
51. James, G.A.; Swogger, E.; Wolcott, R.; Pulcini, E.d.; Secor, P.; Sestrich, J.; Costerton, J.W.; Stewart, P.S. Biofilms in chronic wounds. *Wound Repair Regen.* **2007**, *16*, 37–44. [CrossRef]
52. Jia, B.; Li, G.; Cao, E.; Luo, J.; Zhao, X.; Huang, H. Recent progress of antibacterial hydrogels in wound dressings. *Mater. Today Bio* **2023**, *19*, 100582. [CrossRef] [PubMed]
53. Zhang, X.; Qin, M.; Xu, M.; Miao, F.; Merzougui, C.; Zhang, X.; Wei, Y.; Chen, W.; Huang, D. The fabrication of antibacterial hydrogels for wound healing. *Eur. Polym. J.* **2021**, *146*, 110268. [CrossRef]
54. Zhuo, S.; Liang, Y.; Wu, Z.; Zhao, X.; Han, Y.; Guo, B. Supramolecular hydrogels for wound repair and hemostasis. *Mater. Horiz.* **2024**, *11*, 37–101. [CrossRef]
55. Wei, Q.; Wang, Y.; Ma, K.; Li, Q.; Li, B.; Hu, W.; Fu, X.; Zhang, C. Extracellular vesicles from human umbilical cord mesenchymal stem cells facilitate diabetic wound healing through mir-17-5P-mediated enhancement of angiogenesis. *Stem Cell Rev. Rep.* **2021**, *18*, 1025–1040. [CrossRef] [PubMed]
56. Galkowska, H.; Wojewodzka, U.; Olszewski, W.L. Chemokines, cytokines, and growth factors in keratinocytes and dermal endothelial cells in the margin of chronic diabetic foot ulcers. *Wound Repair Regen.* **2006**, *14*, 558–565. [CrossRef]
57. Tejedor, S.; Wågberg, M.; Correia, C.; Åvall, K.; Hölttä, M.; Hultin, L.; Lerche, M.; Davies, N.; Bergenhem, N.; Snijder, A.; et al. The combination of vascular endothelial growth factor A (VEGF-A) and fibroblast growth factor 1 (FGF1) modified mRNA improves wound healing in diabetic mice: An ex vivo and in vivo investigation. *Cells* **2024**, *13*, 414. [CrossRef] [PubMed]
58. Zha, W.; Wang, J.; Guo, Z.; Zhang, Y.; Wang, Y.; Dong, S.; Liu, C.; Xing, H.; Li, X. Efficient delivery of VEGF-A mRNA for promoting diabetic wound healing via ionizable lipid nanoparticles. *Int. J. Pharm.* **2023**, *632*, 122565. [CrossRef]
59. Chen, C.; Wang, Y.; Zhang, H.; Zhang, H.; Dong, W.; Sun, W.; Zhao, Y. Responsive and self-healing structural color supramolecular hydrogel patch for diabetic wound treatment. *Bioact. Mater.* **2022**, *15*, 194–202. [CrossRef]
60. Guan, L.; Ou, X.; Wang, Z.; Li, X.; Feng, Y.; Yang, X.; Qu, W.; Yang, B.; Lin, Q. Electrical stimulation-based conductive hydrogel for immunoregulation, neuroregeneration and rapid angiogenesis in diabetic wound repair. *Sci. China Mater.* **2023**, *66*, 1237–1248. [CrossRef]
61. Wang, L.; Yu, Y.; Zhao, X.; Zhang, Z.; Yuan, X.; Cao, J.; Meng, W.; Ye, L.; Lin, W.; Wang, G. A biocompatible self-powered piezoelectric poly(Vinyl alcohol)-based hydrogel for diabetic wound repair. *ACS Appl. Mater. Interfaces* **2022**, *14*, 46273–46289. [CrossRef]
62. Wu, C.; Long, L.; Zhang, Y.; Xu, Y.; Lu, Y.; Yang, Z.; Guo, Y.; Zhang, J.; Hu, X.; Wang, Y. Injectable conductive and angiogenic hydrogels for chronic diabetic wound treatment. *J. Control. Release* **2022**, *344*, 249–260. [CrossRef] [PubMed]
63. Shi, C.; Zhang, Y.; Wu, G.; Zhu, Z.; Zheng, H.; Sun, X.; Heng, Y.; Pan, S.; Xiu, H.; Zhang, J.; et al. Hyaluronic acid-based reactive oxygen species-responsive multifunctional injectable hydrogel platform accelerating diabetic wound healing. *Adv. Healthc. Mater.* **2024**, *13*, 2302626. [CrossRef]
64. Xu, Z.; Liu, G.; Huang, J.; Wu, J. Novel glucose-responsive antioxidant hybrid hydrogel for enhanced diabetic wound repair. *ACS Appl. Mater. Interfaces* **2022**, *14*, 7680–7689. [CrossRef]

 https://doi.org/10.3390/ma19030509

65. Li, J.; Zhao, M.; Liang, J.; Geng, Z.; Fan, Y.; Sun, Y.; Zhang, X. Hollow copper sulfide photothermal nanodelivery platform boosts angiogenesis of diabetic wound by scavenging reactive oxygen species. *ACS Appl. Mater. Interfaces* **2024**, *16*, 4395–4407. [CrossRef]
66. Zhu, S.; Li, M.; Wang, Z.; Feng, Q.; Gao, H.; Li, Q.; Chen, X.; Cao, X. Bioactive glasses-based nanozymes composite macroporous cryogel with antioxidative, antibacterial, and pro-healing properties for diabetic infected wound repair. *Adv. Healthc. Mater.* **2023**, *12*, 2302073. [CrossRef]
67. Chen, F.; Qin, J.; Wu, P.; Gao, W.; Sun, G. Glucose-responsive antioxidant hydrogel accelerates diabetic wound healing. *Adv. Healthc. Mater.* **2023**, *12*, 2300074. [CrossRef] [PubMed]
68. Shi, T.; Lu, H.; Zhu, J.; Zhou, X.; He, C.; Li, F.; Yang, G. Naturally derived dual dynamic crosslinked multifunctional hydrogel for diabetic wound healing. *Compos. Part B* **2023**, *257*, 110687. [CrossRef]
69. Li, F.; Liu, T.; Liu, X.; Han, C.; Li, L.; Zhang, Q.; Sui, X. Ganoderma lucidum polysaccharide hydrogel accelerates diabetic wound healing by regulating macrophage polarization. *Int. J. Biol. Macromol.* **2024**, *260*, 129682. [CrossRef]
70. Liu, J.; Qu, M.; Wang, C.; Xue, Y.; Huang, H.; Chen, Q.; Sun, W.; Zhou, X.; Xu, G.; Jiang, X. A dual-cross-linked hydrogel patch for promoting diabetic wound healing. *Small* **2022**, *18*, 2106172. [CrossRef]
71. Feng, Y.; Xiao, K.; Chen, J.; Lin, J.; He, Y.; He, X.; Cheng, F.; Li, Z.; Li, J.; Luo, F.; et al. Immune-microenvironment modulatory polyurethane-hyaluronic acid hybrid hydrogel scaffolds for diabetic wound treatment. *Carbohydr. Polym.* **2023**, *320*, 121238. [CrossRef] [PubMed]
72. Chen, H.; Guo, Y.; Zhang, Z.; Mao, W.; Shen, C.; Xiong, W.; Yao, Y.; Zhao, X.; Hu, Y.; Zou, Z.; et al. Symbiotic algae–bacteria dressing for producing hydrogen to accelerate diabetic wound healing. *Nano Lett.* **2022**, *22*, 229–237. [CrossRef]
73. Xu, A.; Zhang, N.; Su, S.; Shi, H.; Lu, D.; Li, X.; Zhang, X.; Feng, X.; Wen, Z.; Ma, G.; et al. A highly stretchable, adhesive, and antibacterial hydrogel with chitosan and tobramycin as dynamic cross-linkers for treating the infected diabetic wound. *Carbohydr. Polym.* **2024**, *324*, 121543. [CrossRef]
74. Khunmanee, S.; Choi, A.; Ahn, I.Y.; Kim, W.J.; Bae, T.H.; Kang, S.H.; Park, H. Effective wound healing on diabetic mice by adhesive antibacterial GNPs-lysine composited hydrogel. *iScience* **2024**, *27*, 108860. [CrossRef] [PubMed]
75. Fang, Y.; Nie, T.; Li, G.; Wang, L.; Du, J.; Wu, J. Multifunctional antibiotic hydrogel doped with antioxidative lycopene-based liposome for accelerative diabetic wound healing. *Chem. Eng. J.* **2024**, *480*, 147930. [CrossRef]
76. Li, S.; Dong, S.; Xu, W.; Tu, S.; Yan, L.; Zhao, C.; Ding, J.; Chen, X. Antibacterial hydrogels. *Adv. Sci.* **2018**, *5*, 1700527. [CrossRef] [PubMed]
77. Fan, Y.; Lüchow, M.; Zhang, Y.; Lin, J.; Fortuin, L.; Mohanty, S.; Brauner, A.; Malkoch, M. Nanogel encapsulated hydrogels as advanced wound dressings for the controlled delivery of antibiotics. *Adv. Funct. Mater.* **2021**, *31*, 2006453. [CrossRef]
78. Sun, B.; Wu, F.; Wang, X.; Song, Q.; Ye, Z.; Mohammadniaei, M.; Zhang, M.; Chu, X.; Xi, S.; Zhou, N.; et al. An optimally designed engineering exosome–reductive cof integrated nanoagent for synergistically enhanced diabetic fester wound healing. *Small* **2022**, *18*, 2200895. [CrossRef]
79. Zhou, X.; Zhao, B.; Wang, L.; Yang, L.; Chen, H.; Chen, W.; Qiao, H.; Qian, H. A glucose-responsive nitric oxide release hydrogel for infected diabetic wounds treatment. *J. Control Release* **2023**, *359*, 147–160. [CrossRef]
80. Jiang, H.; Xu, Q.; Wang, X.; Shi, L.; Yang, X.; Sun, J.; Mei, X. Preparation of antibacterial, arginine-modified Ag nanoclusters in the hydrogel used for promoting diabetic, infected wound healing. *ACS Omega* **2023**, *8*, 12653–12663. [CrossRef] [PubMed]
81. Fu, C.; Fan, Y.; Liu, G.; Li, W.; Ma, J.; Xiao, J. One-step fabrication of an injectable antibacterial collagen hydrogel with in situ synthesized silver nanoparticles for accelerated diabetic wound healing. *Chem. Eng. J.* **2024**, *480*, 148288. [CrossRef]
82. Li, X.; Jing, X.; Yu, Z.; Huang, Y. Diverse antibacterial treatments beyond antibiotics for diabetic foot ulcer therapy. *Adv. Healthc. Mater.* **2023**, *12*, 2300375. [CrossRef]
83. Lin, Y.; Xu, J.; Dong, Y.; Wang, Y.; Yu, C.; Li, Y.; Zhang, C.; Chen, Q.; Chen, S.; Peng, Q. Drug-free and non-crosslinked chitosan/hyaluronic acid hybrid hydrogel for synergistic healing of infected diabetic wounds. *Carbohydr. Polym.* **2023**, *314*, 120962. [CrossRef]
84. Wolcott, L.E.; Wheeler, P.C.; Hardwicke, H.M.; Rowley, B.A. Accelerated healing of skin ulcer by electrotherapy: Preliminary clinical results. *South. Med. J.* **1969**, *62*, 795–801. [CrossRef] [PubMed]
85. Amirabdollahian, F.; Ash, R. The effect of supplemental zinc on the zinc intake of British adults. *Public Health Nutr.* **2008**, *11*, 650. [CrossRef]
86. Lin, F.; Baldessari, F.; Gyenge, C.C.; Sato, T.; Chambers, R.D.; Santiago, J.G.; Butcher, E.C. Lymphocyte electrotaxis in vitro and in vivo. *J. Immunol.* **2008**, *181*, 2465–2471. [CrossRef]
87. Sun, L.; Wang, Z.; Kang, H.; Luo, P.; Su, J.; Wei, W.; Zhou, P.; Yu, A.; Dai, H. A flexibility self-powered band-aid for diabetes wound healing and skin bioelectronics. *Chem. Eng. J.* **2024**, *481*, 148096. [CrossRef]

 https://doi.org/10.3390/ma19030509

88. Korupalli, C.; Li, H.; Nguyen, N.; Mi, F.-L.; Chang, Y.; Lin, Y.-J.; Sung, H.-W. Conductive materials for healing wounds: Their incorporation in electroactive wound dressings, characterization, and perspectives. *Adv. Healthc. Mater.* **2021**, *10*, 2001384. [CrossRef]
89. Zhang, H.; Hu, H.; Dai, Y.; Xin, L.; Pang, Q.; Zhang, S.; Ma, L. A conductive multifunctional hydrogel dressing with the synergistic effect of ROS-scavenging and electroactivity for the treatment and sensing of chronic diabetic wounds. *Acta Biomater.* **2023**, *167*, 348–360. [CrossRef] [PubMed]
90. Jin, X.; Shang, Y.; Zou, Y.; Xiao, M.; Huang, H.; Zhu, S.; Liu, N.; Li, J.; Wang, W.; Zhu, P. Injectable hypoxia-induced conductive hydrogel to promote diabetic wound healing. *ACS Appl. Mater. Interfaces* **2020**, *12*, 56681–56691. [CrossRef] [PubMed]
91. Zhang, Y.; Zhang, T.; Huang, Z.; Yang, J. A new class of electronic devices based on flexible porous substrates. *Adv. Sci.* **2022**, *9*, 2105084. [CrossRef]
92. Yang, H.; Zheng, H.; Duan, Y.; Xu, T.; Xie, H.; Du, H.; Si, C. Nanocellulose-graphene composites: Preparation and applications in flexible electronics. *Int. J. Biol. Macromol.* **2023**, *253*, 126903. [CrossRef]
93. Zhang, Z.; Zhu, Z.; Zhou, P.; Zou, Y.; Yang, J.; Haick, H.; Wang, Y. Soft bioelectronics for therapeutics. *ACS Nano* **2023**, *17*, 17634–17667. [CrossRef]
94. Lin, R.; Lei, M.; Ding, S.; Cheng, Q.; Ma, Z.; Wang, L.; Tang, Z.; Zhou, B.; Zhou, Y. Applications of flexible electronics related to cardiocerebral vascular system. *Mater. Today Bio* **2023**, *23*, 100787. [CrossRef]
95. Zhang, T.; Liu, N.; Xu, J.; Liu, Z.; Zhou, Y.; Yang, Y.; Li, S.; Huang, Y.; Jiang, S. Flexible electronics for cardiovascular healthcare monitoring. *Innovation* **2023**, *4*, 100485. [CrossRef]
96. Gould, L.; Li, W.W. Defining complete wound closure: Closing the gap in clinical trials and practice. *Wound Repair Regen.* **2019**, *27*, 201–224. [CrossRef]
97. Guo, H.; Bai, M.; Zhu, Y.; Liu, X.; Tian, S.; Long, Y.; Ma, Y.; Wen, C.; Li, Q.; Yang, J.; et al. Pro-healing zwitterionic skin sensor enables multi-indicator distinction and continuous real-time monitoring. *Adv. Funct. Mater.* **2021**, *31*, 2106406. [CrossRef]
98. Zhang, S.; Wang, L.; Xu, T.; Zhang, X. Luminescent MOF-based nanofibers with visual monitoring and antibacterial properties for diabetic wound healing. *ACS Appl. Mater. Interfaces* **2023**, *15*, 9110–9119. [CrossRef] [PubMed]
99. Khor, S.M.; Choi, J.; Won, P.; Ko, S.H. Challenges and strategies in developing an enzymatic wearable sweat glucose biosensor as a practical point-of-care monitoring tool for type II diabetes. *Nanomaterials* **2022**, *12*, 221. [CrossRef] [PubMed]
100. Garland, N.T.; Song, J.W.; Ma, T.; Kim, Y.J.; Vázquez-Guardado, A.; Hashkavayi, A.B.; Ganeshan, S.K.; Sharma, N.; Ryu, H.; Lee, M.-K.; et al. A miniaturized, battery-free, wireless wound monitor that predicts wound closure rate early. *Adv. Healthc. Mater.* **2023**, *12*, 2301280. [CrossRef]
101. Wang, X.; Yang, Y.; Zhao, W.; Zhu, Z.; Pei, X. Recent advances of hydrogels as smart dressings for diabetic wounds. *J. Mater. Chem. B* **2024**, *12*, 1126–1148. [CrossRef]
102. Hou, Z.; Wang, T.; Wang, L.; Wang, J.; Zhang, Y.; Zhou, Q.; Zhang, Z.; Li, P.; Huang, W. Skin-adhesive and self-healing diagnostic wound dressings for diabetic wound healing recording and electrophysiological signal monitoring. *Mater. Horiz.* **2024**, *11*, 1997–2009. [CrossRef]
103. Xiang, Y.; Zhao, Y.; Jiang, Y.; Feng, J.; Zhang, X.; Su, W.; Zhou, Y.; Wei, P.; Low, S.S.; Li, H.N. Battery-free and multifunctional microfluidic janus wound dressing with biofluid management, multi-indicator monitoring, and antibacterial properties. *ACS Appl. Mater. Interfaces* **2024**, *16*, 50321–50334. [CrossRef]
104. Xiao, Y.; Xu, K.; Zhao, P.; Ji, L.; Hua, C.; Jia, X.; Wu, X.; Diao, L.; Zhong, W.; Lyu, G.; et al. Microgels sense wounds' temperature, pH and glucose. *Biomaterials* **2025**, *314*, 122813. [CrossRef]
105. Guo, L.; Zhang, X.; Zhao, D.-M.; Chen, S.; Zhang, W.-X.; Yu, Y.-L.; Wang, J.-H. Portable photoacoustic analytical system combined with wearable hydrogel patch for PH monitoring in chronic wounds. *Anal. Chem.* **2024**, *96*, 11595–11602. [CrossRef]
106. Liu, J.; Li, Z.; Sun, M.; Zhou, L.; Wu, X.; Lu, Y.; Shao, Y.; Liu, C.; Huang, N.; Hu, B.; et al. Flexible bioelectronic systems with large-scale temperature sensor arrays for monitoring and treatments of localized wound inflammation. *Proc. Natl. Acad. Sci. USA* **2024**, *121*, e2412423121. [CrossRef] [PubMed]
107. Hu, Y.W.; Wang, Y.H.; Yang, F.; Liu, D.X.; Lu, G.H.; Li, S.T.; Wei, Z.X.; Shen, X.; Jiang, Z.D.; Zhao, Y.F.; et al. Flexible organic photovoltaic-powered hydrogel bioelectronic dressing with biomimetic electrical stimulation for healing infected diabetic wounds. *Adv. Sci.* **2023**, *11*, 2307746. [CrossRef] [PubMed]
108. Mostafalu, P.; Tamayol, A.; Rahimi, R.; Ochoa, M.; Khalilpour, A.; Kiaee, G.; Yazdi, I.K.; Bagherifard, S.; Dokmeci, M.R.; Ziaie, B.; et al. Smart bandage for monitoring and treatment of chronic wounds. *Small* **2018**, *14*, 1703509. [CrossRef]
109. Gong, M.; Wan, P.; Ma, D.; Zhong, M.; Liao, M.; Ye, J.; Shi, R.; Zhang, L. Flexible breathable nanomesh electronic devices for on-demand therapy. *Adv. Funct. Mater.* **2019**, *29*, 1902127. [CrossRef]

https://doi.org/10.3390/ma19030509

110. Mostafalu, P.; Kiaee, G.; Giatsidis, G.; Khalilpour, A.; Nabavinia, M.; Dokmeci, M.R.; Sonkusale, S.; Orgill, D.P.; Tamayol, A.; Khademhosseini, A. A textile dressing for temporal and dosage controlled drug delivery. *Adv. Funct. Mater.* **2017**, *27*, 1702399. [CrossRef]
111. Wang, X.-F.; Li, M.-L.; Fang, Q.-Q.; Zhao, W.-Y.; Lou, D.; Hu, Y.-Y.; Chen, J.; Wang, X.-Z.; Tan, W.-Q. Flexible electrical stimulation device with Chitosan-Vaseline® dressing accelerates wound healing in diabetes. *Bioact. Mater.* **2021**, *6*, 230–243. [CrossRef]
112. Ke, Q.; Zhang, X.; Yang, Y.; Chen, Q.; Su, J.; Tang, Y.; Fang, L. Wearable magnetoelectric stimulation for chronic wound healing by electrospun $CoFe_2O_4$@CTAB/PVDF dressings. *ACS Appl. Mater. Interfaces* **2024**, *16*, 9839–9853. [CrossRef]
113. Jiang, Y.; Trotsyuk, A.A.; Niu, S.; Henn, D.; Chen, K.; Shih, C.-C.; Larson, M.R.; Mermin-Bunnell, A.M.; Mittal, S.; Lai, J.-C.; et al. Wireless, closed-loop, smart bandage with integrated sensors and stimulators for advanced wound care and accelerated healing. *Nat. Biotechnol.* **2022**, *41*, 652–662. [CrossRef]
114. Gong, X.; Yang, J.; Zheng, Y.; Chen, S.; Duan, H.; Gao, J.; Haick, H.; Yi, C.; Jiang, L. Polymer hydrogel-based multifunctional theranostics for managing diabetic wounds. *Adv. Funct. Mater.* **2024**, *34*, 2315564. [CrossRef]
115. Ziyazadeh, M.; Vafaiee, M.; Mohammadpour, R.; Ehtesabi, H. Dual-sided and flexible triboelectric nanogenerator-based hydrogel skin patch for promoting wound healing. *Nano Energy* **2025**, *134*, 110558. [CrossRef]
116. Fu, S.; Yi, S.; Ke, Q.; Liu, K.; Xu, H. A self-powered hydrogel/Nanogenerator system accelerates wound healing by electricity-triggered on-demand phosphatase and tensin homologue (PTEN) Inhibition. *ACS Nano* **2023**, *17*, 19652–19666. [CrossRef] [PubMed]
117. Du, N.; Fan, Y.; Zhang, Y.; Huang, H.; Lyu, Y.; Cai, R.; Zhang, Y.; Zhang, T.; Guan, Y.; Nan, K. Wireless, Programmable, and Refillable Hydrogel Bioelectronics for Enhanced Diabetic Wound Healing. *Adv. Sci.* **2024**, *11*, 2407820. [CrossRef]
118. Liu, Y.; Luo, X.; Li, L.; Chen, L.; Qiao, Z.; Si, C.; Haiyan, J.; Liu, X. Wearable, battery-free, and wireless microneedle-based bioelectronics for robustly-integrated chronic wound management and therapeutic diagnosis. *Nano Energy* **2025**, *138*, 110909. [CrossRef]
119. Dou, Y.; Zhang, Y.; Zhang, S.; Ma, S.; Zhang, H. Multi-functional conductive hydrogels based on heparin–polydopamine complex reduced graphene oxide for epidermal sensing and chronic wound healing. *J. Nanobiotechnol.* **2023**, *21*, 343. [CrossRef]
120. Shi, W.; Li, H.; Xu, C.; Wu, G.; Chen, J.; Zhang, J.; Liang, L.; Wu, Q.; Liang, Y.; Li, G.; et al. Ag@polydopamine-functionalized borate ester-linked chitosan hydrogel integrates monitoring with wound healing for epidermal sensor. *npj Flex. Electron.* **2024**, *8*, 82. [CrossRef]
121. Li, M.; Wang, S.; Li, Y.; Meng, X.; Wei, Y.; Wang, Y.; Chen, Y.; Xiao, Y.; Cheng, Y. An integrated all-natural conductive supramolecular hydrogel wearable biosensor with enhanced biocompatibility and antibacterial properties. *ACS Appl. Mater. Interfaces* **2024**, *16*, 51618–51629. [CrossRef]
122. Huang, X.; Cheng, S.; Gong, F.; Yang, X.; Pei, Z.; Cui, X.; Hou, G.; Yang, N.; Han, Z.; Chen, Y.; et al. A closed-loop patch based on bioinspired infection sensor for wound management. *Nano Today* **2024**, *57*, 102400. [CrossRef]
123. Du, B.; Ren, X.; Wang, X.; Wen, Y.; Yang, J.; Li, L.; Bai, P.; Lang, F.; Li, L.; Zhang, R. Four-in-one pH/glucose-responsive engineered hydrogel for diabetes wound healing. *Nano Today* **2025**, *62*, 102725. [CrossRef]
124. Qiu, W.; Wang, Q.; Li, M.; Li, N.; Wang, X.; Yu, J.; Li, F.; Wu, D. 3D hybrid scaffold with aligned nanofiber yarns embedded in injectable hydrogels for monitoring and repairing chronic wounds. *Compos. Part B* **2022**, *234*, 109688. [CrossRef]
125. Deng, D.; Liang, L.; Su, K.; Gu, H.; Wang, X.; Wang, Y.; Shang, X.; Huang, W.; Chen, H.; Wu, X.; et al. Smart hydrogel dressing for machine learning-enabled visual monitoring and promote diabetic wound healing. *Nano Today* **2025**, *60*, 102559. [CrossRef]
126. Peng, L.; Tan, M.; Wu, C. Thermoelectric Hydrogel Composite Dressing for Promoting Diabetic Ulcer Repair and Loading Targeted Modified Exosome and Preparation Method Thereof. CN119925682, 6 May 2025.
127. Hui, J.; Liu, W.; Fan, D.; Zheng, X.; Yao, B.; Gao, Q. Preparation Method of Hydrogel Dressing Capable of Treating Diabetic Wounds in Combination with Electrical Stimulation. CN117653775, 8 March 2024.
128. Zhang, Y.; Tang, L.; Sun, X.; Wang, P.; Li, N. Preparation Method and Application of Conductive Hydrogel Flexible Electrode. CN118718028, 1 October 2024.
129. Wu, H.; Yang, G.; Wen, J.; Gong, H. 3D printing Hydrogel Electrode for Diagnosis and Treatment of Diabetic Neuropathy. CN118000742, 10 May 2024.

https://doi.org/10.3390/ma19030509

MDPI AG
Grosspeteranlage 5
4052 Basel
Switzerland
Tel.: +41 61 683 77 34

Materials Editorial Office
E-mail: materials@mdpi.com
www.mdpi.com/journal/materials

www.ingramcontent.com/pod-product-compliance
Lightning Source LLC
Chambersburg PA
CBHW061604210926
51850CB00009B/2208
* 9 7 8 3 7 2 5 8 8 1 2 9 1 *